Narrowband Land-Mobile Radio Networks

The Artech House Mobile Communications Library

John Walker, *Series Editor*

For a complete listing of *The Artech House Telecommunications Library*, turn to the back of this book

Narrowband Land-Mobile Radio Networks

Jean-Paul Linnartz

Artech House
Boston • London

Library of Congress Cataloging-in-Publication Data

Linnartz, Jean Paul
  Narrowband Land-Mobile Radio Networks
  Includes bibliographical references and index.
  ISBN 0-89006-645-0
  1. Mobile communication systems. 2. Mobile radio stations. I. Title.

TK6570.M6L56   1993   92-32244
621.3845'6—dc20   CIP

British Library Cataloguing in Publication Data

Linnartz, Jean Paul
  Narrowband Land-Mobile Radio Networks
  I. Title.
  621.3845

ISBN 0-89006-645-0

© 1993 ARTECH HOUSE, INC.
685 Canton Street
Norwood, MA 02062

International Standard Book Number: 0-89006-
Library of Congress Catalog Card Number:

10 9 8 7 6 5 4 3 2

Contents

Preface

Throughout history, the human race has shown a particular interest in communication. The demand for communication is closely related to mobility. Currently, a rapid growth of "mobile communication" has been seen, made feasible by recent technological developments. Mobile communication involves radio communication, in which at least one of the terminals can be in motion. Congestion of the radio frequencies allocated for mobile services requires highly efficient use of the electromagnetic spectrum.

Coverage

This book addresses the performance of mobile radio systems with narrowband radio channels. The term narrowband is used indicate that the transmission bandwidth is assumed to be relatively small compared to the coherence bandwidth of the radio channel. In particular, the effects of multipath fading and shadowing and of mutual interference between mobile users is investigated. To this end, the relevant propagation mechanisms are described initially, focusing on statistical channel models. For the planning of real systems, the suitability of some propagation models using terrain data is empirically evaluated. The statistical channel description is used to compute the performance of mobile radio links. Expressions are given for the outage probability and for the average duration of fades, taking into account mutual interference between users. A simplified model for the bit error rate in digital transmission is proposed, and numerical results are given for mobile telephone networks.

To study the performance of packet-switched mobile networks, the statistics of randomly arriving data packets are described for a number of access protocols. Using the assessed properties of physical radio links, the capacity of mobile data networks is computed, focusing on the spatial distribution of users over the service area. The network performance is calculated as a function of the distance between the mobile user and the fixed base station.

Methods are developed to find the effects of mutual interference between mobile terminals in a network of known dimensions. This allows selection of the most spectrum-efficient design that satisfies a prescribed network performance. Hence, the fundamental techniques presented in this book apply to spectrum-efficient spatial design and planning of infrastructures for mobile communication.

Acknowledgments

My interest in mobile radio communication was raised during my Ir. (M.Sc. E.E.) research project under the guidance of Dr. Jens C. Arnbak at Eindhoven University of Technology. This project has been the motivation for continuing my research at the Telecommunications and Traffic-Control Systems (T.V.S.) Group at Delft University of Technology, where I again enjoyed Dr. Arnbak's inspiring advice and support. This led to the completion my Ph.D. thesis, entitled "Effects of Fading and Interference in Narrowband Land-Mobile Networks," which was used as the basis for this book. I also greatly appreciated the numerous discussions with Dr. Ramjee Prasad, and I wish to thank him for the stimulating research environment he created. Further, I would like to thank Ir. Geert Awater, Ir. Adri Kegel, and the numerous students who in some way or another contributed to the contents of this book, particularly Hugo Goossen, Ramin Hekmat, Aart J.'t Jong, Pascal P.C. de Klerk, Kees van der Plas, Robert-Jan Venema, and John J.P. Werry.

The propagation experiments presented in Chapter 2 were obtained at the Physics and Electronics Laboratory F.E.L.-T.N.O., The Hague, in 1987 and 1988, where I benefited from the advice of Ir. Frank J.M. van Aken.

Modifications and additions to the original Ph.D thesis have been carried out at the University of California at Berkeley. I wish to acknowledge the discussions with my colleagues and graduate students, which significantly contributed to my insight and helped to improve the manuscript.

The cover and Figures 1.1 to 1.3, 1.6, 2.3 to 2.4, and 3.1 were drawn by my father, Martin Linnartz. It is with great pleasure that I include these pencil drawings, though not as a direct justification or visualization of the scientific results reported in this dissertation: an artist's interpretation, visualized in these drawings, fits the abstract and delimited characteristics of this introduction to theoretical analyses. By describing real-life systems with idealized, mathematical models and interpretations, scientific research appears to have clear analogies with the work of an artist. Both the artist and the researcher observe and analyze systems (natural or artificial) and formulate their own, modeled world. Despite all limitations, such models yield valuable insight into the system being investigated. To me, this is the most appealing aspect of scientific research.

It would have been impossible for me to complete this book without the support and friendship of all the others in The Netherlands and in California who contributed to making the preparation of this book an enjoyable time.

Jean-Paul Linnartz
Berkeley, California
June 1992

Chapter 1
Introduction

1.1 HISTORICAL SETTING

Communication systems using electrical and electronic technolgy have a significant impact on modern society. In early history, as the courier speeding from Marathon to Athens in 490 B.C. illustrates, information could be exchanged only by the physical transport of messages. Only a few examples exist of nonelectrical communication techniques for transfer of information via infrastructures other than those for physical transport: smoke signals, signal flags in maritime operations, and the semaphore, among others [1]. Early attempts to communicate visual signals by means of the semaphore, a pole with movable arms, were made in the 1830s in France. A similar experimental system was used by the Dutch during the "ten days campaign" against the Belgian revolt in 1831–1832. In 1837, the House of Representatives passed a resolution requesting the Secretary of the Treasury to investigate the feasibility of setting up such a system in the United States [2]. The market interest in enhanced communication systems was also clearly illustrated by the fact that in 1860 the Pony Express started regular physical message services over land in the U.S. At the same time, however, electronic systems for communication had started to develop.

*Tele*communication is defined by the *International Telecommunication Union* (ITU) as the transmission, emission, or reception of any signs, signals, or messages by electromagnetic systems [3]. The demonstration of (electrical) telegraphy by Joseph Henry and Samuel F.B. Morse in 1832 followed the discovery of electromagnetism by Hans Christian Ørsted and André-Marie Ampère early in the 1820s. In the 1840s, telegraph networks were built on the East Coast of the U.S. and in California. Rapid extension of their use followed; the first transatlantic cable was laid in 1858. In 1864, James Clerk Maxwell postulated wireless propagation, which was verified and demonstrated by Heinrich Hertz in 1880 and 1887, respectively. Marconi and Popov started experiments with the radio-telegraph shortly thereafter, and Marconi patented a complete wireless system in 1897. In 1876, Alexander Graham Bell patented the telephone. The invention of the diode by Fleming in 1904 and the

---

triode by Lee de Forest in 1906 made possible rapid development of long-distance (radio) telephony. The invention of the transistor by Bardeen, Braittain, and Shockley, which later led to the development of integrated circuits, paved the way for miniaturization of electronic systems. The continued advances in microelectronic circuits have recently made the rapid development of mobile and personal communication systems feasible. Such systems offer person-to-person communication, giving freedom of movement for the users and, if desired, eliminating the ineffective calls experienced with the fixed telephony service when the user is away from his or her terminal. Moreover, new services, particularly those employing mobile *data* communication, are becoming feasible, such as *automatic vehicle location* (AVL) for fleet management, electronic mail, remote access to databases, vehicle printers, or automatic repetition of the messages if the driver has been away from the vehicle. Further, data communication makes encryption and data processing possible.

The effect of telecommunications on society at large [4, 5] may turn out to be as significant as the invention of the printing press, which initiated the changes in society in the sixteenth century and contributed to the end of feudalism [6, 7]. The introduction of new services, with due anticipation of their direct and indirect consequences, often requires conscious policies [8]. Nonetheless, user acceptance of new opportunities appears at least as important as political consensus. One discipline relevant to the understanding of the impact of the new developments, and thus also to formulating an information policy, whether liberal or more regulated, is the dynamic market theory [9]: each innovation has a critical period during its introduction. Case studies [10] indicate that the simultaneous introduction of a new service *and* a new technology has often failed in the past, notwithstanding any impressive advantages the new service would offer [8]. For instance, in many countries, except possibly in France, the introduction of videotex—a *new* service using a *new* technology—was not very successful. On the other hand, facsimile is a clear example of the successful introduction of a new service based on existing technology and an existing infrastructure, namely, the telephone network. Similarly, teletext offered a new service using existing transmitters for television broadcasting.

The successful introduction of the compact disk (CD) is an example of a new technology for an existing service previously offered by long-play records. Even the introduction of the electrical telegraph can be interpreted as an improved technological solution to a service pioneered by the semaphore and the Pony Express service. As the concurrent experiments with semaphore communication show, the electrical telegraph was thus not a purely technology-push development: a clear interest (market pull) in communication to bridge physical distances was present. The acceptance of these innovations came about as a novel and better technical implementation of a service previously offered with inferior techniques.

In the 1980s, we have seen a very successful innovation of the existing telephone service: cellular radio. Further technological advances in mobile communication are developed apace, while experiments with new services and applications are also performed with currently known, mature, though sometimes inadequate, technology. Examples of the latter range from the not very successful experiments with mobile data services using wireline telephone modems over cellular voice channels, to exchanging routine information (e.g., about traffic jams) in spoken form. However, such pilot systems and experiments appear to be indispensable. Introduction of new technologies and infrastructures such as packet-switching (store-and-forward) mobile data networks may only be profitable and justifiable once experimental systems prove their usefulness to potential users and are accepted by them.

The main body of this book addresses some of the technical problems experienced in designing new systems for mobile (voice and data) communication. The aspect of mobility related to communication will be discussed in the next section.

1.2 COMMUNICATION AND MOBILITY

Evident in early history is the economic relation between mobility, trade, and transport on the one hand, and exchange of information on the other. Shortly after the first coins were minted in Lydia in the seventh century B.C., professional currency exchange was made by the many monetary sovereignties at the important commercial crossroads [11]. Doing business at crossroads ensured not only a continuous stream of clients, but also information from all relevant parts of the ancient trading world.

The first news-gathering services were established by the Medici, Fugger, and Welser families, who set up their own messenger services to assist cross-border commercial operations. Ships and gazettes reported on wars, crop failures, gold finds or natural catastrophes in foreign territories. In the seventeenth century, Louis Elsevier located the Copenhagen offices of his book printers and sellers company in the local Stock Exchange, which ensured rapid communication with international partners. Here, the novel newspaper *Amsterdam Corantos* was available, with the most up-to-date news items, exchange rates, and commodity values collected worldwide by crews of the Dutch merchant fleet. The fleet of ships transported not only goods, including printed matter and maps, but also business information and newly discovered topographical data. Thus, shipping activities not only *distributed* information (e.g., in printed form), they also *contributed* information [12]. The very combination of these two information services allowed countries like Holland to develop into important international trading nodes during the first part of the seventeenth century. This historical example illustrates that information was originally communicated via the infrastructure for physical transport (see Figure 1.1), and that information was merely transferred between locations of human activities, such as cities.

With the advent of telecommunications, the spatial relationship between information and physical transport changed dramatically. In the early telegraph systems, the origination and destination of information traffic continued to be mainly determined by the existing locations of human activities. However, although messages were still mostly sent from city to city, they no longer used media primarily

Figure 1.1 Historically, the infrastructure for communication coincided with the infrastructure for physical transport.

intended for physical transport (Figure 1.2). Introduction of the most basic telecommunication facilities appeared to greatly improve the efficiency of logistics. At present, the improved techniques for the exchange of data allowed by computer communications (e.g., *electronic data interchange* (EDI)) are believed to significantly enhance the efficiency of logistics operations or to remove market imperfections [13]. The growth of facsimile technology is explained by the classical need for exchanging messages (between fixed locations) in written or printed form and the ubiquity of an electronic infrastructure, namely, the public circuit-switched telephone network. Still, facsimile messages are rarely sent to mobile users; origin and destination of the messages often coincide with origin, destination, or major nodes in the logistics infrastructure.

Mobile communication (i.e., communication in which the originator of the message or its recipient (or both) can be in motion) is one of the earliest applications of radio communication techniques. Radio further modified the spatial relationship between the infrastructures for telecommunication and physical transport. Not only are the media spatially separated, the origination and destination of messages generally no longer coincide with the origin or destination of physical transport. For instance, Marconi's wireless telegraph was used in ship-to-shore communications. The possibility of exchanging information with vessels during their motion from one

5

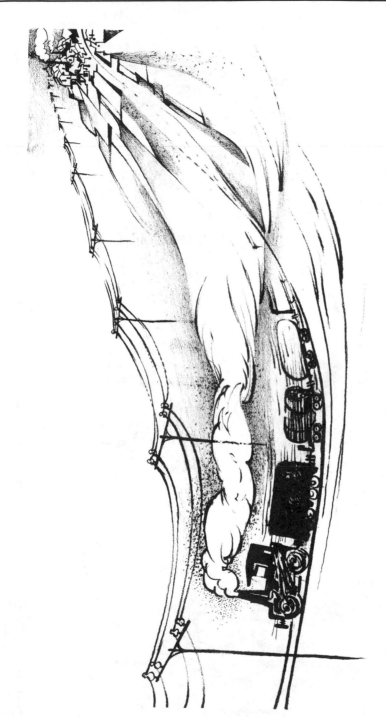

Figure 1.2 The invention of the telegraph caused separation of the infrastructures for communication and physical transport.

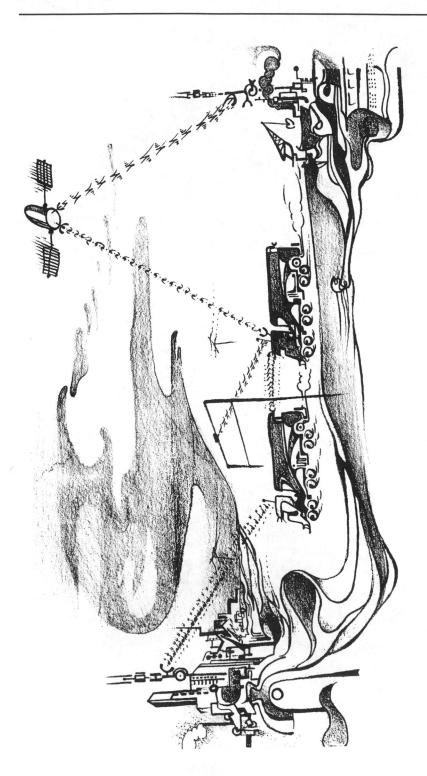

Figure 1.3 Telematic services in traffic and transport require a dense infrastructure for communication, which is closely related to the road infrastructure.

point to another generalized the spatial relationship between the flow of information traffic and the infrastructure for physical transport into a more dynamic and area-oriented mode of operation.

As illustrated in Figure 1.3, some recently proposed applications of telematics in traffic and transportation [13–19] would also require support of intensive information streams along roads and highways. The growth of the road traffic and the increasing inconvenience and environmental damage caused by congestion require better use of the infrastructure for physical transport. Improved management or pricing techniques will require communication between vehicles and roadside base stations or between vehicles participating in road traffic [13]. For instance, traveller information and optimum route selection require broadcasting of traffic data [14, 18]. AVL and fleet management can offered using satellite or land-base mobile radio networks [19]. According to the U.S. Congress, an *Intelligent Vehicle Highway System* (IVHS) [16, 17], with fully automatic vehicle control, is to be demonstrated by 1997.

Clearly, these applications will require extensive use of mobile communications in addition to extending conventional services (such as telephony) to mobile subscribers. Efficient use of the available radio spectrum and effective spatial management of tele-traffic appear to be essential. The above observations illustrate the interaction of mobility and telecommunications throughout history. In dealing with mobile radio communication, we will particularly address issues of effective spatial use of the radio spectrum in relation to the specific properties of the mobile radio channel.

1.3 TRAFFIC MANAGEMENT IN TELECOMMUNICATION NETS

Where the intensity of the transport of information, people, or goods grows beyond some extent, "traffic" arises, and mutual interaction between various individual contributions to this traffic becomes an issue. Traffic management is concerned with weighing the individual interests of the traffic participants and the settlement of potential mutual conflicts.

In some recent policy notes on *road* traffic management (e.g., [15]), the word "management" is used for a wide variety of measures taken to optimize the use of resources, including prior planning and design, *static management*, and real-time control, *dynamic management*. In static management, systems and infrastructures are designed and planned to offer optimum performance within certain cost or environmental constraints. On the other hand, in dynamic management, the existing resources are used in such a way that the fluctuating traffic intensities are optimally supported.

In telecommunication networks, the issue of management (in the above wide sense) of intense collective streams of information has been studied for many decades and is still a relevant topic of research. Before discussing the management of tele-communication traffic in more detail, the layered organization of the exchange of messages in telecommunication networks is described.

1.3.1 Layered Organization of Telecommunication Nets

The *Open Systems Interconnection* (OSI) reference model plays an important role in standardizing protocols for exchange of data between complex information systems [20]. While originally intended to design computer protocols in a structured manner, the model recently proved helpful in formulating managerial and political issues related to telecommunications and mass media (e.g., [8, 21]). In the following sections, the model is used to sketch the role of traffic management tools.

The OSI model consists of seven layers (1 to 7 in Table 1.1), each of which can be specified to support reliable interconnection between two systems, I and II, irrespective of the technical implementation of lower layers. Each layer adds specific functions to lower layers to enrich the value of the offered services. Table 1.1 shows a division between the lower three and the upper four layers. The lower layers (1,

Table 1.1
(Extended) OSI Reference Model for Exchange of Information

System		Layer	Tasks/Aspects
I	II		
(9)	(9)	Information delivery	Legitimate delivery to parties in economic and social processes
(8)	(8)	Information structure	Structuring and organization of information
7	7	Application layer	Interface to application software and user services
6	6	Presentation layer	Handling and translation of structured data
5	5	Session layer	Coordination and synchronization of the session
4	4	Transport layer	"End-to-end" transfer of data
3	3	Network layer	Routing and switching
2	2	Data link layer	Error control (coding)
1	1	Physical layer	Control of physical circuits

Physical media for interconnection (cables, radio links, optical fibre, etc.)

2, and 3) address *transport* or *bearer* services. The higher layers (4 to 7) add user-oriented functions necessary for services, such as electronic mail, EDI, or videotex. The combination of layers 1 to 7 offer *value added* services.

During the 1980s, the distinction between the three lower and four upper layers was clearly drawn in legal and regulatory issues (e.g., in considering the extent of the PTT monopoly for transport services [21]). At present, the applications of information and computer systems become increasingly interwoven with telecommunication networks [22]; this interaction is known as *telematics*. For a more complete view of telematic services, it may therefore also be necessary to consider the specific user applications which are located above layer 7. A corresponding extension of the OSI layers has been proposed by de Jong [23]: social and juridical aspects of information transactions are addressed by the information structure (8) and the information delivery and access (9) layers.

1.3.2 Structured Traffic Management

Linnartz and Arnbak [24] discuss a layered structure similar to the model in Table 1.1 to study traffic and transport in the wide sense (i.e., including both telecommunications and physical transport). It was noted that in communication networks, and possibly also in infrastructures for physical traffic and transport, management of traffic can be organized in accordance with a layered structure.

1.3.2.1 Management Above OSI Layer 7

In medieval Europe, the widespread illiteracy and the lack of appropriate means of communication gave a monopoly of information exchange (at the highest layer in Table 1.1) to the nobility and clergy. The effect of such a control on information flow can be illustrated by the refusal of the Greek Orthodox Church to place clocks in the church towers. This deprived the population of the enhanced time efficiency that professional activities enjoyed in Roman Catholic Western Europe [25]. The effects of a liberal exchange of information became far more evident with the invention of the printing press in the fifteenth century and the ensuing developments in the sixteenth century [6, 7]. Printing provided a means of communicating huge amounts of information to a much larger readership. During the Reformation, the new possibilities offered by the novel technology were effectively exploited. Freedom of expression of thoughts, news gathering, and the exchange of information are now accepted as fundamental human rights in a modern democratic society. Nevertheless, the widespread use of telecommunications and its impact on society have led to continuous political discussions throughout the past decades. The strict definition of aims of public broadcasting in The Netherlands and many other countries is still based on national cultural considerations, though a more liberal international attitude is being enforced, partly because the traditional protection of national broadcasting organizations is undermined by technical developments such as satellite broadcasting [8, 26]. New technologies also inspire reconsideration of the longstanding exclusive monopolies granted to the national *postal, telephone, and telegraph administrations* (PTT) on telecommunication services [21]. Liberalization of the market for terminal equipment and deregulation in certain transport services, such as mobile radio, have allowed faster changes in the variety of telematic services offered to private and business users.

The tendency towards more liberal regulations for electronic exchange of information does not imply a relief from the need for management of information streams at the uppermost layer. On the contrary, the more information becomes abundantly available, the greater the need for management and structuring of authorized information traffic. For instance, requirements for information security, control of access to databases, reduction of distracting information given to car drivers, avoidance of information overload in control rooms during calamities, and selection of relevant documents in database searches are currently receiving attention from researchers.

According to a recent consultant survey [27], 230,000 labor hours are lost daily in The Netherlands (with 15 million inhabitants) because of ineffective answering or routing of telephone calls. This is expected to result in a yearly loss of about 1.8 billion U.S. dollars. It may be asserted that, at the highest layers, the adverse effects of imperfect management of traffic is usually not a consequence of any imperfect operation of telecommunication networks or services; these effects arise from inefficient organization of activities at the layers of information organization and delivery.

1.3.2.2 Management at the Upper OSI Layers

Traffic management at layers 1 through 7 of the OSI reference model is more within the traditional domain of the telecommunication engineer. At the upper layers, traffic management is generally of a static nature (e.g., attempting to avoid the exchange of excessively redundant data or spreading the peaks of the traffic loads in time).

During the development of digital mobile radio systems, with tight restrictions on spectrum use, efficient voice coding received ample attention. The standard wireline analog telephone circuit is recommended internationally to have a bandpass characteristic in the frequency range of 300 to 3400 Hz, with a signal-to-noise ratio of at least 34 dB [28]. To achieve this quality with standard digital pulse-code modulation (PCM) transmission, each voice channel requires a bit rate of 32 to 64 kb/s. To conserve spectrum in digital mobile radio, source coding is applied to reduce the bit rate below 16 kb/s [18]. The successes in voice coding during the last decade have provided the possibility of encoding voice into, say, 10 kb/s and still retain acceptable voice quality, including easy recognition of the speaker. In the pan-European

GSM network specifications, 9.6 kb/s (or half this rate for special applications) is being considered.

Another technique that improves the efficiency of spectrum usage is to transmit digital coded messages rather than use voice communication. To many groups of professional users of mobile communication, standardized data messages are of interest and may even replace routine voice communication. Measurements of the characteristics of traffic in typical mobile radio channels show that many messages are short and stereotyped [29]. Most user categories require spoken message lengths of about 15 seconds, except in taxi dispatch service (with shorter messages of about 8 seconds) and some services with longer messages. Short, stereotyped messages are well suited for being coded in a textual form. From a spectrum-efficiency point of view, exchange of coded messages is many orders of magnitude more efficient than transmission of voice, particularly if names of streets and other such information are to be transferred.

Tariff structures can be employed to manage telecommunication traffic at the transport layer. Higher tariffs during busy (business) hours and lower tariffs in the evening and at nighttime have proved to be a successful measure for spreading the peak load of the telephone net with substantial impact on the total volume of the busy-hour traffic, reducing the infrastructure investment required to meet peak traffic demands.

Despite their impact on the spectral efficiency of the network, these aspects of management at the higher OSI layers will not not be considered further within this book. However, they do have a number of implications for the design of the system at lower layers: since highly efficient voice coding techniques became available, digital communication has become highly competitive with analog voice communication in the attempts to conserve radio spectrum. Therefore, the discussion in the following chapters is not restricted to analog voice communication. Digital voice communication and some specific possibilities such as site diversity are addressed in a number of sections. Moreover, the advantages of computerized *data* messages compared to voice communication were the motivation for devoting Chapters 6 to 9 to packet-switching (store-and-forward) networks.

1.3.2.3 Management at the Network Layer

In telecommunication networks, static traffic management was first studied at the network layer, particularly by the Danish mathematician A.K. Erlang. In 1917, Erlang addressed the statistical behavior of a large ensemble of telephone subscribers [30]. His studies offered the potential to effectively dimension the capacity of switches and trunks to the collective demand and to ensure delivery of the required grade of service to the individual subscribers. The laws of Erlang played a key role in design and planning at the network layer.

During recent decades, studies similar to the telephony work by Erlang have been performed to investigate data traffic (e.g. [31–33]). Prior to 1970, wire connections were the principal means of data communication. Dial-up and leased telephone lines provided cost-effective and reasonably reliable means for interchange of data. It was recognized that this circuit-switched technology was inadequate for the needs of any computer communication network required to handle bursty traffic (i.e., traffic with a large peak-to-average ratio in the required data rate). Due to developments in computer hardware, around 1970 the cost of switching communication resources had dropped below the cost of the transmission resources (bandwidth) being switched [34]. Accordingly, the technology of packet-switched computer nets emerged as a cost-effective means of data exchange over shared wired channels.

The following chapters mainly address the steady-state performance of generic (voice and data) communication systems given a certain (fixed) traffic intensity. The effect of fluctuating traffic intensities and dynamic management of the communication resources *at the network layer* is outside the scope of the analyses. It offers an understanding of the static management of mobile communication networks with multiple users, particularly the dimensioning and spatial layout. Two widely different tele-traffic models are considered: Chapters 3 to 5 address *continuous wave* (CW) voice communication, while Chapters 6 to 9 address randomly arriving short data messages.

1.3.2.4 Management at the Physical Layer

For many decades, multiplexed communication channels (trunks) between switches with bundled traffic from many subscribers were expensive and, thus, limited in capacity. From the 1950s on, national PTTs established microwave links to overcome the bandwidth and repeater-spacing limitations of coaxial trunk links. Nowadays, optical fibers offer capacity well in excess of what can be reached by any reasonable expansion of the microwave radio relay network, and this technology is likely to dominate in all new trunk systems, though backup radio systems are maintained to ensure reliability.

Radio communication is expected to have a much greater impact on future communication networks by providing local access for remote or mobile subscribers to the network. In the former Federal Republic of Germany, about 50% to 70% of all investments in the telephone network, accumulated over several decades to a total of 60 million DM in 1980, was made in the subscriber access loops [8]. The density of telephone traffic covers many orders of magnitude [36]: from less than 150 erlangs/km^2 in villages and rural areas to perhaps 100,000 erlangs/km^2 in office buildings. Optimum infrastructures for telecommunication networks may depend on the spatial distribution of the subscribers and the characteristics of the environment [35, 37]. Figure 1.4 [36] relates the distribution of the telephone traffic intensity to an

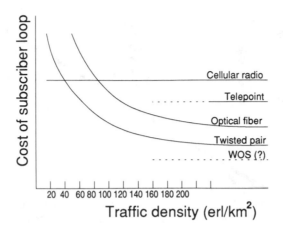

Figure 1.4 Impression of the relative cost of the subscriber loop versus the spatial traffic density [35].

impression of the cost of the subscriber loop. The cost of a wired subscriber loop is more or less proportional to the distance between the local switch and the subscriber terminal, so it tends to descrease with the subscriber density. This contrasts with the case for radio communication, where the cost of radio access does not significantly depend on the propagation distance or subscriber density. In *rural, thinly populated areas*, radio communication therefore offers more economic subscriber loops than wired lines (probably including repeater/amplifiers) to every remote subscriber. On the other hand, in *residential areas*, existing copper wires may remain more cost-effective than new radio access. However, new optical loops may eventually become competitive with installing or refurbishing the copper wire loop [38], particularly if the telephone network is integrated with networks for other services.

However, the decrease of costs per wired subscriber loop may not be extrapolated for *business areas*: cabled networks are often impractical because of frequent changes in the organizational structure and spatial distribution of activities. This may lead to prohibitively frequent changes in the telecommunication infrastructure. Radio communication, particularly *wireless office systems* (WOS) and tetherless access to *private branch exchanges* (PBX), may offer much more flexible and economic solutions. The increasing mobility of staff and the desire for more effective exchange of information thus stimulates the development of wireless systems.

The ensuing increasing demand for radio spectrum is likely to lead to severe shortages of frequency assignments in the near future [39]. Figure 1.5 gives an estimate of the required frequency spectrum as a function of the user density [36].

According to various sources (summarized in [39]), personal communication services may require 40 to 230 MHz of spectrum, depending on the number of subscribers, the technology employed, and the quantity of services. The *International*

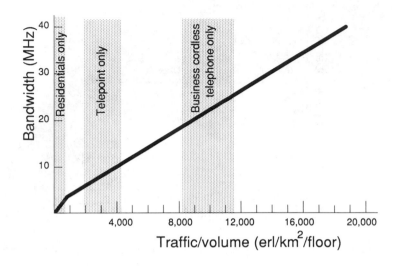

Figure 1.5 Required radio spectrum in megahertz versus traffic density in erlangs/square kilometers/ floor [36].

Radio Consultative Committee (CCIR) study group 8 recommends 60 MHz for low-power, personal tetherless access, plus 170 MHz for vehicular radio.

Spectrum efficiency can be improved by using techniques for dynamic management at the lower OSI layers. In trunking systems for voice communication with multiple closed user groups [40], voice channel frequencies are assigned only after a request for a call is received. This is in contrast to the frequency assignment methods used hitherto, where a frequency assigned to one user group may not be used by other user groups, even if the channel is idle. In packet-switched mobile data communications, use of frequencies is also commonplace [31, 32]. Allowing a large number of mobile terminals to have random access to a commonly used part of the radio spectrum requires management of the tele-traffic streams at the physical layer.

It can be concluded that, contrary to the common belief held some years ago, radio communication has not been killed by optical communication, since radio nets allow user mobility, provide ubiquitous access, and have a more attractive area-cost structure than fixed optical nets with their heavy installation costs.

From the above discussion of static and dynamic traffic management issues, it appears that many challenging technical problems are found in the management at the lower layers of digital radio networks, taking into account the physical characteristics of the shared radio channel. Some of the aspects of static management will be addressed in the following chapters. Dynamic spectrum management is implicit in the study of cellular telephone or random-access networks, where many users share common radio facilities according to actual needs at a given moment. In the case of a packet-switched (ALOHA) network, we explicitly study the dynamic behavior of (queuing of) packet traffic by addressing the stability of the net. This is included to verify the validity and accuracy of the stationary models and techniques employed here.

The next sections review the effects of mobility on communication systems and compare a number of radio techniques proposed and/or developed for mobile communications.

1.4 EFFECT OF MOBILITY ON COMMUNICATION SYSTEMS

User movements affect the communication system at many layers:

- At the *physical layer*, channel characteristics vary with the location of the user, and, because of mobility, vary in time. Because antennas mounted on vehicles have rather unfavorable positions and little clearance (Figure 1.6), the propagation characteristics in the mobile radio channel are notoriously poor. A mobile radio link is hindered by a number of propagation mechanisms, namely, multipath scattering from objects near the mobile antenna, shadowing by dominant obstacles, and attenuation mechanisms on the propagation path between transmitter and receiver [41]. Multipath reception causes rapid fluctuations of the received signal power, whereas shadowing is experienced as a slow amplitude effect (see Figure 1.7). Adaptive transmitting power control can mitigate the adverse effects of individual shadowing to some extent. Since most

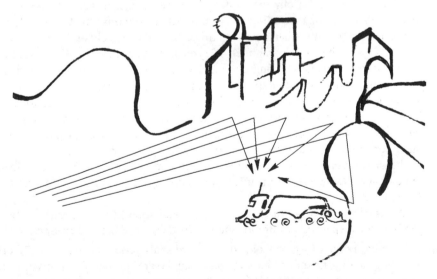

Figure 1.6 Multipath reception, caused by obstacles and terrain features in the vicinity of the mobile antenna.

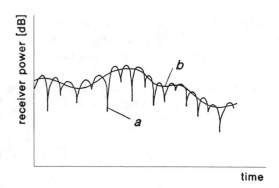

Figure 1.7 (a) Instantaneous and (b) mean received signal power (in decibels) versus time in a typical mobile radio channel with multipath reception and shadowing.

mobile networks are interference-limited rather than noise-limited, interfering signals are likely to produce substantially greater adverse collective impact if adaptive power control is used to overcome shadowing on their own transmission path. Mobile systems must live with signal fluctuations during transmission. Because of high demands on spectrum efficiency, extensive frequency reuse and mutual interference must be accepted. This means that the classical system analysis inspired by the stationary *additive white Gaussian noise* (AWGN) channel of conventional information theory may not be applicable.

- At the *data link layer*, channel coding is to be chosen in accordance with the specific character of the fading radio channel. In contrast to AWGN channels with errors randomly distributed in time, bursts (i.e., clusters) of errors are experienced in the mobile channel. A typical mobile channel allows relatively reliable communication during certain periods, interrupted by other periods of particularly poor communication. The latter are called *fades*.
- At the *network layer*, the routing of signals may change from time to time as the user moves through the service area of the network. This may require localization of the terminal (or localization of a specific user in more sophisticated personal communication systems) to find optimum routing.
- At the *presentation layer*, improved source coding techniques have helped to achieve acceptable spectrum efficiency with digital systems, as discussed in Section 1.3.2.2.
- At the *application layer* (and above), personal mobility and transportation of goods require specific communication facilities, as discussed in Section 1.2.

Despite this wide range of aspects of mobility, we shall generally focus on the effect that mobility has on channel behavior (physical layer). It will be shown in later chapters that this also indirectly influences the system performance at higher layers. Above all, it is the effect of mobility on the channel that influences the spectrum efficiency of mobile networks: ample design margins to maintain reliable operation despite fluctuating signal strengths often lead to less efficient use of the spectrum than would be possible with nonfading channels.

1.5 SURVEY OF MOBILE RADIO NETWORKS

A number of different techniques are used to implement land-mobile radio systems. The channel characteristics studied here address narrowband VHF or UHF propagation over ranges larger than a few tens of meters, but less than, say, several tens of kilometers. Most systems using this type of channel are cellular systems [41, 42]: frequency reuse is applied in a regular, systematic manner. The next subsection describes the cellular system concept and also briefly addresses the differences in two other common techniques for mobile communication with vehicles, namely, short-range communication and land-mobile satellite communication.

1.5.1 Cellular Networks

1.5.1.1 Principles of Cellular Frequency Reuse

In the cellular concept, frequencies allocated to the service are reused in a regular pattern of areas, called *cells*, each covered by one base station. In mobile-telephone nets, these cells are usually hexagonal [41, 42]. In radio broadcasting, a similar concept has been developed based on rhombic cells [43, 44]. Figure 1.8 depicts a typical example of a theoretical cellular structure for radio telephone networks.

To ensure that the mutual interference between users remains below a harmful level, adjacent cells use different frequencies. In fact, a set of C different frequencies $\{f_1, \ldots, f_C\}$ are used for each cluster of C adjacent cells. Cluster patterns and the corresponding frequencies are reused in a regular pattern over the entire service area. Thus, the total bandwidth for the system is C times the bandwidth occupied by a single cell.

The closest distance between the centers of two cells using the same frequency (in different clusters) is determined by the choice of the cluster size C and the layout of the cell cluster. This distance is called the *frequency re-use distance*. The reuse distance r_u, normalized to the size of each hexagon, is

$$r_u = \sqrt{3C} \tag{1.1}$$

For hexagonal cells (i.e., with "honeycomb" cell layouts commonly used in mobile radio), possible cluster sizes are $C = i^2 + ij + j^2$, with integer i and j ($C = 1, 3, 4, 7, 9, \ldots$). In practical FM or digital cellular networks for public radio telephony,

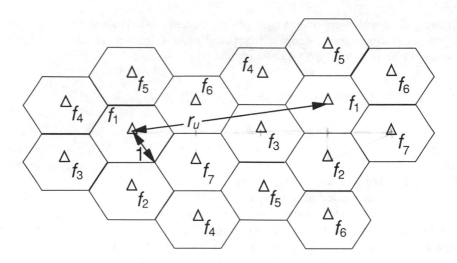

Figure 1.8 Idealized cell structure according to a hexagonal frequency repetition pattern with clusters of $C = 7$ cells.

the cluster size is on the order of $C = 7$ or 9, though with special techniques, such as diversity reception, smaller reuse distances can be used.

In most cellular systems, each base station can carry more than one telephone call in its cell. If the number of parallel channels per base station is denoted by M, the total bandwidth for the cellular net B_s is the product of the occupied bandwidth per channel B_T, the number of channels per cell M, and the cluster size C (thus, $B_s = MCB_T$). The spectrum efficiency SE of a cellular net can be defined as the carried traffic per cell A_c, expressed in erlangs, divided by the bandwidth of the total system B_s and divided by the area of a cell S_U. Mostly, the spectrum efficiency is expressed in erlangs/megahertz/square kilometer, rather than in the SI dimension erlangs/hertz/square meter. From

$$SE = \frac{A_c}{B_s S_u} = \frac{A_c}{B_T C M S_u} \tag{1.2}$$

we observe that the spectrum efficiency decreases with the cluster size C. Chapters 3 and 5 show that the system performance (e.g., expressed in terms of the outage probability of bit error rate experienced by the user) improves with increasing reuse distance, so, because of (1.1), it improves with the cluster size. Hence, achieving high system performance and efficient use of the radio spectrum are conflicting objectives for a network designer.

When a mobile terminal moves outside the coverage area of a base station, the network management is assumed to take appropriate measures. A handover to another base station is required to ensure sufficient quality of reception, including acceptable interference power levels. A mobile user experiences the worst average performance if the terminal is located at the boundary of two cells where the distances to base stations are maximum. The performance of the network is therefore often specified for this worst case [46].

As can be seen from the above discussion, cellular engineering combines traffic assignment, interference management, and spectrum conservation [47, 48]. Cellular mobile radio differs from previous design strategies for mobile radio in two aspects: frequency reuse and cell splitting. Conventional (noise-limited) radio systems were often designed with the objective of having each base station cover the largest possible area by using antennas mounted in high towers and using the maximum affordable power. Different frequencies were assigned to each base station. The layout of the base stations and their service areas did not change for the lifetime of the system. In contrast to this, with interference-limited cellular systems, the service area is divided into a large number of cells (as in Figure 1.8), each with its own base station. The power radiated by the base stations is kept to a minimum, and the antennas are located just high enough above the ground to achieve the desired coverage. This enables nonadjacent cells to reuse the same set of frequencies.

As the demand for services increases, the number of cells may at some time become insufficient to provide the required grade of service. Cell splitting (i.e., subdividing congested cells into smaller cells and reusing the frequencies in some efficient new pattern) can then be used to increase the traffic handled in a given area without the need for increasing the total bandwidth occupied by the entire system. Figure 1.9 gives an example of a technique used for sharing frequencies among cells of different sizes.

In a practical environment with nonuniform distributions of the traffic and more significant propagation impairments in certain parts of the service area, optimal solutions lead to tailor-made structures with cells of different size. Computer aids are often invoked, using topographical databases [49].

The fact that the user capacity of cellular networks can be increased by reducing cell sizes stimulates the development currently of micro and picocellular networks. In macrocellular networks, cell sizes usually range from 1 to 20 km. Typically, microcellular networks have cell sizes of 400 m to 2 km, and picocellular nets have cell sizes of 4 m to 200 m. For pico and microcellular networks, the term *personal communication* is often used: most macrocellular telephone nets are designed for use within cars and may be too bulky and power-consuming to be conveniently portable. With microcellular and picocellular networks, low transmitting power, say, less than 20 mW, can be used, allowing miniaturization of the handheld terminal [50, 51].

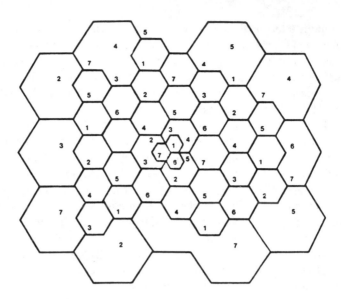

Figure 1.9 Two levels of cell splitting: each smaller cell is allocated the same number of channels (*M*) as the larger cell and is therefore able to support the same number of subscribers [18].

1.5.1.2 Cellular Radio Service

During the 1980s, the more advanced national PTTs focused their marketing attention on providing a telephone service to mobile subscribers. The first public radio telephone nets had been introduced during the 1970s based on the cellular concept for frequency assignment [18]. Some initial systems used the 150-MHz band and only supported operator-assisted calls. To provide enhanced user capacity with improved and automatic services, Sweden, Norway, Denmark, Belgium, The Netherlands, Switzerland, and Austria implemented slightly different versions of the *Nordic Mobile Telephone* (NMT) system [52]. These analog systems all use *frequency division multiple access* (FDMA). The system was set up in the 450-MHz band and was later also deployed in the 900-MHz band. At the end of the 1980s, the various versions of the NMT system had a total of nearly a million subscribers. By 1989, the analog *Total Access Communication System* (TACS) had approximately 750,000 subscribers in Great Britain. In France, the Radiocom network had about 165,000 subscribers, and in Germany, the C450-network (NETZ-C) had another 165,000 subscribers. In the U.S., the *Advanced Mobile Phone System* (AMPS) and *American Radio Telephone System* (ARTS) had a total of 2 million subscribers in 1988.

European introduction of the GSM digital cellular telephone net in the 900-MHz band is planned for the early the 1990s. The acronym GSM originally stood for *Groupe Special Mobile*, but later the name *Global System for Mobile Communication* was adopted. The system employs *time division multiple access* (TDMA) of eight subscriber signals per channel. The channel bit rate is 270.8 kb/s and *Gaussian minimum shift keying* (MSK) is used. The system is suitable for voice communication as well as for circuit-switched data communication. In the U.S., one of the digital cellular concepts proposes a channel bit rate of 48.6 kb/s, using $\pi/4$ *differential quadrature phase-shift keying* (DQPSK) modulation on a 830-MHz carrier [53]. A TDMA access scheme with three subscribers on each channel is proposed to ensure compatibility with the 30-kHz frequency spacing used for existing analog networks. Voice is coded into 8 kb/s, and, including error control coding and signaling, the bit rate per subscriber is 16.2 kb/s.

Cox [50] concluded from observing the developments in the late 1980s that the technology of personal communication is progressing along two completely separate evolutionary paths: high-power *vehicular* cellular mobile radio technology, as described in the above in this section, and low-power pedestrian and stationary *cordless* telephone technology. The two techniques are essentially different. The vehicular cellular technology has been developed during the last 45 years, and new systems are still designed to meet the standards of the wireline telephony service. In contrast to this, the history of the handheld cordless telephone is, at most, 15 years, and the system is designed with a focus on miniaturization at the expense of some services. Particularly, restrictions imposed by power consumption are relevant to handheld sets. Power-demanding digital processing operations, such as forward error-correction coding, speech coding, and multipath channel equalization, are extensively employed in the proposed system for future vehicle telephone nets, but still appear impractical in low-powered handheld sets. Reduction of power consumption during standby mode is one of the reasons why the user of a cordless set can initiate a telephone call, but cannot receive calls unless a paging system is included.

A number of systems for personal communications have been introduced or proposed. In the U.K., subscribers can make a telephone connection using a handset within a few hundred meters from a "telepoint" base station. The *Digital European Cordless Telephone* (DECT) network is a further development of such a personal system, aimed at pan-European standardization of wireless PBXs. The 1880- to 1900-MHz band has been designated for the DECT service. The development and design of such systems not only carefully addresses combining spectrum efficiency and performance objectives, but also has to meet constraints on power consumption.

Cellular voice networks are not limited to *public* telephone services. Closed user-group trunking networks are installed in a number of countries. The advantages to be gained from trunking mobile voice communication for closed user groups onto a set of common frequency channels are evident from the large quantity of traffic still handled on separate frequencies [52]. In the Federal Republic of Germany in 1989, when trunking was not widely used, more than 100,000 mobile radio networks for closed user groups supported traffic from more than 700,000 voice terminals in the business sector, plus 200,000 terminals in the public-services sector [52]. Nine frequency bands, with a total of 550 channels, were allocated. Companies in a city like Munich operated a total of 40,000 terminals in 4500 nets. This corresponds to an average of 100 to 150 users, usually belonging to four or five companies, per frequency. Combining this traffic (trunking) is seen as a new development in mobile radio, although the concept is a simple extension of Erlang's classical results for wired telephone service to the management of radio channel usage.

Despite its novelty and usefulness, it may be argued that the cellular telephone nets are not a full-blown telematic service, but just an extension of the *plain old telephone service* (POTS/PSTN) to mobile subscribers. The circuit-switching protocols of the networks have been designed for voice communication lasting a few minutes. Most analog networks are not very suitable for data transfer unless sophisticated modems are used. The GSM digital channels provide a gross data rate of 16 kb/s, which can be used for data communication with a bit rate of up to 9.6 kb/s. A few public packet-switched (store-and-forward) mobile networks for data communication are in operation (e.g., in the Nordic countries, Canada, and the U.S). Its communication protocol [54] has been optimized for the exchange of data messages with a relatively short length (i.e., on the order of a few hundreds of bits or even less), which corresponds to the contents of a written version of a typical voice message in mobile radio [29]. The initial version of the system operates in the 80-MHz band and uses FM modulation of a 1200-b/s MSK subcarrier. A cellular frequency reuse pattern has been adopted.

1.5.2 Short-Range Networks

Generally, the interest in short-range communication (i.e., communication over distances ranging from a few to a few hundred meters and mostly within line of sight) is growing rapidly [51]. One reason is that the user capacity of cellular networks increases if cell sizes are reduced. Furthermore, short-range communication is of interest to special services that can be offered by using mobile communication in traffic and transport [13, 16] (e.g. control of road traffic flows to avoid congestion by dynamic route guidance or by road pricing). Microwave line-of-sight or infrared links, leaky feeders, ultrasonic acoustical communication, and communication via inductive loops may be considered for these links. In some applications, communication sessions occur at specific sites, but extended coverage areas (e.g., by cellular repetition) are not always required. Other services, such as automatic vehicle control on intelligent highways, require exchange of information between vehicles, with distributed algorithms for management of the radio traffic.

In wireless office systems, communication also takes place at a short range, mostly via indoor radio or infrared lightwave channels. While the problem of radio access in mobile radio is related to that in WOSs [55, 56], in some respects they differ from each other. The traffic pattern in office automation is often very different from the occasional short standardized messages encountered in mobile radio, since file transfer or routine access to databases, or data delivery to a high-resolution video terminal may be supported. Hence, the required data rates may be much higher in WOS than in mobile radio. Power consumption required for digital signal processing may not be a limiting factor.

The design of systems with short-range communication links requires specific investigation of these propagation paths. The description of the short-range radio channel (including the microcellular channel) differs from the radio channel with ranges longer than a few hundred meters: propagation is characterized by short delay spreads and possibly even a dominant line-of-sight component (see Chapter 2). Moreover, in the case of WOS, the channel is almost stationary, whereas the mobile radio channel exhibits dynamic fading that can be fast compared to the duration of a data message.

1.5.3 Satellite Networks

Recently, mobile satellite communications has received increasing attention. The potential independence of a local earth-based infrastructure stimulates use of mobile satellite links. Particularly, international fleet management systems employ satellite links for the exchange of messages because satellites can also be used for continentwide paging and navigation on board the vehicle. The absence of a pan-European earth-based infrastructure for packet data communications has been an important stimulus for satellite communication links between vehicles and their home bases. If fleet management systems became more widely used in more local applications, such as taxi dispatch or packet-delivery services, satellite communication systems may experience the disadvantage of limited local frequency reuse. In due course, communication may thus need to be transferred to ground-based cellular mobile networks. Once again, satellite technology would thus have proven its important ability to offer an early service which only later becomes available in the standard fixed infrastructure. Earlier examples of this have been international satellite television program distribution and video conferencing, now also possible by broadband optical cables, intercontinental digital leased lines, and "hot lines" for journalists and statesmen traveling in less developed regions. In this book, we consider the principles of a mature, fully fledged ground-based local radio infrastructure. therefore, innovative mobile satellite networks are outside our scope here, however interesting they may be as a trigger for new global services.

1.6 SCOPE OF THIS BOOK

In conclusion, this book is confined to mobile communication in the VHF and UHF bands, over ranges on the order of a few tens (or hundreds) of meters to tens of kilometers. It is in this range that most cellular systems operate and that the congestion of frequencies is most apparent. Since spectrum efficiency is of major concern, mutual interference is considered a more important limitation to system performance than noise.

Only narrowband (i.e., frequency-nonselective) radio channels are studied. Recently, wideband transmission techniques, including spread-spectrum transmission of signals with a low user data rate, also started to receive considerable attention from researchers throughout the world [57, 58]. Wideband communication is made possible by developments in fast microelectronic circuits (e.g., for adaptive equalizers to combat selective fading). By the time the work reported in this book was completed, definite results for the superiority of either narrowband or wideband concepts were not yet available. It is the author's impression that no general judgment can yet be made on this issue and that the optimum choice will depend on many conditions. This book is not intended to advocate the use of narrowband transmission techniques one-sidedly, but the results may assist in weighing narrowband and wideband design options.

The OSI layered organization of communication networks has been the motivation for the following overall structure of this book. The physical medium for communication is described in Chapter 2 in terms of propagation mechanisms. Based on these models for the physical channel, Chapters 3 to 5 describe the performance of the radio link (physical layer). Chapters 6 to 9 are devoted to random multiple-access methods, taking into account the performance impairments at the physical layer. Most symbols and acronyms used in these chapters are listed in appendixes.

The first part of Chapter 2 presents a review of some models for narrowband propagation, used in studies of UHF/VHF mobile radio networks. Rayleigh fading, shadowing, and UHF groundwave propagation are discussed. These statistical models do not require specific terrain data, but are supplemented with results from rural propagation measurements in the VHF and UHF bands. The accuracy and validity of a number of relatively simple (both statistical and deterministic) propagation models are discussed.

Probabilities of signal outage are derived in Chapter 3, based on the fluctuations of received signal power levels discussed in Chapter 2. An outage of the wanted signal is assumed to occur if the *signal-to-interference* (C/I) ratio drops below a certain threshold value. A Rician- or Rayleigh-fading and shadowed wanted signal is considered in the presence of multiple fading interference signals and additive white Gaussian noise. The signal outage results are extended to site diversity (i.e., to the case of two cooperating receivers located at geographically distinct sites).

The time constants of fading in the mobile radio channel are considered in Chapter 4. The threshold crossing rate (i.e., the number of times per second a certain C/I ratio (called the *threshold*) is crossed) and the average nonfade durations are derived for a wanted Rayleigh-fading signal in the presence of multiple Rayleigh-fading interferers.

In Chapter 5, an approximate model is proposed for the study of bit errors in a mobile radio channel. The effects of fading of the wanted and interfering signals are taken into account. This chapter is intended to offer a mathematically tractable model for investigation of random-access networks in later chapters.

Throughout Chapters 2 through 5, the mobile radio channel and its effects on the reception of a wanted signal are discussed. This situation particularly concerns CW communications, such as telephony, where each receiver is expected to support wanted signals from one particular transmitter. In contrast to this, random-access protocols are reviewed in Chapter 6, particularly slotted ALOHA and *inhibit sense multiple access* (ISMA). At this stage, the distinction between wanted and interfering signals becomes less obvious than in Chapters 3 to 5: in random-access networks, signals from multiple users compete to capture a common receiver; this receiver is successful if it locks onto any one of the signals and correctly decodes the data message [59]. In the case of bursty data communication, the behavior of the communication protocols influences the amount of interference experienced. The performance of some medium-access schemes is evaluated, based on a generic description of the physical radio channel and of the data link layer. The well-known expression for steady-state throughput of data packets from an infinite population of terminals (e.g., see Kleinrock and Togabi [31]) are reformulated for the probability of successful reception of an a priori selected test packet. Moreover, generalized probabilities of successful reception are considered: the specific effects of the mobile radio channel are added later, in Chapters 7 to 9.

In Chapter 7, various models for receiver capture, based on the previous discussion in Chapters 3 to 5, are considered for a number of random-access protocols discussed in Chapter 6. Models considered include the probability of signal outage of the packet signal (compare Chapter 3), the influence of the packet duration (compare Chapter 4), and packet error rates for *binary phase shift keying* (BPSK) modulation (compare Chapter 5).

Chapter 8 focuses further on the capture model proposed in Chapter 3: correct reception is assumed to occur if the C/I ratio is above a certain threshold. The performance of the ALOHA protocol is assessed for a number of propagation models in terms of the total throughput, network stability, and delay. It is observed that most of the retransmissions occur in areas with poor propagation to the central receiver. This has a disadvantageous effect on the traffic capacity and stability of the network.

Packet-switched networks with cellular frequency reuse are addressed in Chapter 9. The spatial distribution of packet traffic generation in the service area is considered. The traffic to be offered to achieve a certain required throughput is studied, taking into account collisions with packets from the same cell, as well as interference between cochannel cells and noise. Since the results suggest that the frequency reuse distance can be very small, compared to cellular telephone networks, the case of contiguous frequency reuse (i.e., with the same frequency used everywhere in the entire service area) is also addressed. The common throughput of two receivers with data packets arriving from overlapping areas is computed.

REFERENCES

[1]   "From Semaphore to Satellite," *International Telecommunications Union*, Geneva, 1965.

[2]   Finn, B. "A Continent Bound by Wire," *IEEE Spectrum*, August 1990, pp. 58–59.

[3]   "Radio Regulations," Art. 1, I.T.U., Geneva, 1986.

[4]   Wright, K., "Trends in Communication: The Road to the Global Village," *Scientific American*, Vol. 262, No. 3, March 1990, pp. 57–66.

[5]   U.S. Congress, Office of Technology Assessment, "Critical Connections, Communication for the Future," OTA-CIT-407, Washington, D.C.: U.S. Government Printing Office, Jan. 1990.

[6]   Bulckens, B., and M. Termont, "Chrono 4, Middle Ages and New Time" (in Dutch), 2nd ed., Kapellen, Belgium: DNB—Uitgeverij Pelckmans, 1987, "Thanks to paper and the printing press, new ideas reached all of Europe," pp. 128.

[7]   Arnbak, J.C., "Gutenberg's Lesson" (in Dutch), *Information and Information Policy (i&i)*, Vol. 5, No. 2, Summer 1987, pp. 79–85.

[8]   Arnbak, J.C., J.J. van Cuilenburg, and E.J. Dommering, *Convergence and Divergence in Communications* (in Dutch: *Verbinding en Ontvlechting in de Communicatie (V.O.C.)*), Amsterdam: Otto Cramwinckel Uitgever, 1990, ISBN 90 71894 15 0.

[9]   de Jong, H.W., *Dynamic Market Theory* (in Dutch), Leiden, 1989.

[10]   Kaam, B.v., "Wasn't Teletekst Just Something Like Television?" (in Dutch), *Information and Information Policy (i&i)*, Vol. 8, No. 3, Authum, 1990, pp. 33–45.

[11]   "Telecommunications and International Finance," *British Telecommunications Engineering*, Vol. 6, Jan. 1988, pp. 265–267.

[12]   Arnbak, J.C., "Cable and Satellite in the Near Future" (in Dutch), *Information and Information Policy (i&i)*, No. 4, Winter 1983, pp. 26–32.

[13]   Policy Note, "Telematics Traffic and Transport" (in Dutch), Ministry of Transport and Public Works, The Hague, The Netherlands, Tweede Kamer (1989–1990) 21 449, Nos. 1 and 2, 26 Feb. 1990, ISSN 0921–7371.

[14]   Gleissner, E., "From Traffic Announcement to Measurement Station" (in German), *Funkschau*, No. 7, 22 March 1991, pp. 73–75,78.

[15]   *Masterplan for Dynamic Traffic Management* (in Dutch), Dept. of Traffic Studies, Rijkswaterstaat, Rotterdam, 1989.

[16]   Ristenblatt, M.P., "Communication Architectures for the Intelligent Vehicle Highway Systems Are Available," *IEEE Veh. Soc. Newsletter*, Vol. 39, No. 1, Feb. 1992, pp. 8–19.

[17]   Shladover, S.E., C.A. Desoer, J.K. Hedrick, M. Tomizuka, J. Walrand, W.B. Zhang, D.H. McMahon, H. Peng, S. Sheikholeslam, and N. McKeown, "Automatic Vehicle Control Developments in the PATH Program," *IEEE Trans. on Veh. Tech.*, Vol. 40, No. 1, Feb. 1991, pp. 114–130.

[18]   Walker J., ed., *Mobile Information Systems*, London: Artech House, 1990.

[19]   Skomal, E.N., *Automatic Vehicle Locating Systems*, New York: Van Nostrand Reinhold Company, 1981.

[20]   Recommendations X.200 in "Data Communication Networks Open Systems Interconnection (OSI), Model and Notation, Service Definition," *Blue Book*, Rec. X.200, X.219, Vol. VIII, Fascicle VIII.4, IXth Plenary Assembly CCITT, Melbourne, 14–25 Nov. 1988.

[21]   Committee Steenbergen, *Signals for Later, a New Direction for the PTT* (in Dutch), The Hague, 1985.

[22] Arnbak, J.C., "Many Voices, One Structure," Inaugural Lecture, Delft University of Technology, Delft University Press, 8 Oct. 1986. Also in *The Information Society*, Vol. 5, 1987, pp. 101–118.

[23] de Jong, C., "How Can I Help You? Demand and Supply in Telecommunication" (in Dutch), Inaugural Address, Delft University of Technology, Delft University Press, 6 June 1990.

[24] Linnartz, J.P.M.G., and J.C. Arnbak, "Traffic, With Speed Limits of the Velocity of Light; a Comparison Between Road Traffic and Telecommunication," *Proc. Symp. Telematics in Transportation and Traffic: Quality by Integration*," Fac. of Civil Eng., T.U. Delft, 20 March 1991, pp. 3.1–3.24, ISSN: LVV rapport 0920–0592. (Also see *Information and Information Policy*, Vol. 9, No. 4, Winter 1991, pp. 25–35.

[25] Van Houten, B.C., *Technology, Science and Society in Historical Perspective* (in Dutch), D.W. Vaags, J. Wemelsfelder, eds., Utrecht: Aula Paperback 87, Het Spectrum B.V., 1983, 1987.

[26] Donner, J.P.H., J.M. de Meij, and K.J.M. Mortelmans (Committee Donner), *Advice by the Committee on Ether Frequencies and Commercial Broadcasting* (In Dutch), The Hague, 27 Jan. 1992.

[27] "Three Billion Guilders Loss Caused by Non-optimum Use of Telephone Switches" (In Dutch), *Telecomletter (Telecombrief)*, No. 11, 15 June 1990.

[28] "General Characteristics of International Telephone Connections and Circuits," Recommendation G.101-G.181, *CCITT Blue Book*, Vol. III, Fascicle III.1, IXth Plenary Assembly, Melbourne, 14–25 Nov. 1988.

[29] Cohen, P., H.-H. Hoang, and D. Haccoun, "Traffic Characterization and Classification of Users of Land-Mobile Communications Channels," *IEEE Trans. on Veh. Tech.*, Vol. VT-33, 1984, pp. 276–284.

[30] Erlang, A.K., "Solution of Some Problems in the Theory of Probabilities of Significance in Automatic Telephone Exchanges," in *Elektroteknikeren*, Vol. 13, 1917; and in *Post Office Electrical. Engineers J.*, Vol 10, 1917.

[31] Kleinrock, L., and F.A. Togabi, "Packet Switching in Radio Channels: Part 1—Carrier Sense Multiple Acces Modes and Their Throughput-Delay Characteristics," *IEEE Trans. on Commun.*, Vol. COM-23, No.12, Dec. 1975, pp. 1400–1416.

[32] Kleinrock, L., "Computer Applications," *Queuing Systems*, Vol. 2, New York: Wiley, 1976.

[33] Van der Rhee, J., and F.C. Schoute, "ATM Traffic Capacity Modelling," *Philips Telecomm. and Data Syst. Rev. PTR*, Vol. 48, No. 2, June 1990, pp. 24–32.

[34] Roberts, L.G., "Data by the Packet," *IEEE Spectrum*, Vol. 11, No. 2, Feb. 74, pp. 46–51.

[35] Arnbak, J.C., "Cross-Sectoral Service Provision in a Densely Populated Country," Paper prepared for discussion at the OECD Committee for Information, Computer and Communication Policy (ICCP), Third Special Session on Telecom. Policy, Paris, 28–30 Nov. 1990.

[36] DECT Reference Document DTR-RES 301, Version 2.1, European Telecommunications Standards Institute (ETSI), 1990.

[37] Jansson, H., J. Swerup, and S. Wallinder, "Cellular Telephony: An Exploration of the Future" (in Dutch), *Electrical Engineering and Electronics*, No. 1, 1991, pp. 16–21.

[38] "The 21st Century Subscriber Loop," special issue of *IEEE Comm. Mag.*, Vol. 29, No. 3, March 1991.

[39] "Personal Communications Regulatory Issue: A Corporate View," special issue of *IEEE Comm. Mag.*, Vol. 29, No. 2, Feb. 1991.

[40] "Trunking for Private Mobile Radio Use in The Netherlands," *HDTP-Newsletter*, Ministry of Transport and Public Works, Groningen, The Netherlands, Feb. 1990.

[41] Jakes, W.C., Jr., ed., "Microwave Mobile Communications," New York: Wiley, 1974.

[42] MacDonald, V.H., "The Cellular Concept," *Bell Sys. Tech. J.*, Vol. 58, No. 1, Jan. 1979, pp. 15–41.

[43] Report 944, "Theoretical Network Planning," Recommendations and Reports of the CCIR, Broadcasting Service (Sound), Vol. X-1, XVIth Plenary Assembly CCIR Dubrovnik, 1986.

[44]   Prosch, T.A., "A Possible Frequency Planning Method and Related Model Calculations for the Sharing of VHF Band II (87.5–108 MHz) Between FM and DAB (Digital Audio Broadcast) Systems," *IEEE Trans. on Broadcasting*, Vol. BC-37, No. 2, June 1991, pp. 55–63.

[45]   Hatfield, D.N., "Measures of Spectral Efficiency in Land-Mobile Radio," *IEEE Trans. on Electromagn. Compat.*, Vol. EMC-19, No. 3, Aug. 1977, pp. 266–268.

[46]   Gosling, W., "Protection Ratio and Economy of Spectrum Use in Land Mobile Radio," *IEE Proc. F*, Vol. 127, 1980, pp. 174–178.

[47]   Oetting, J., "Cellular Mobile Radio: An Emerging Technology," *IEEE Comm. Mag.*, Vol. 21, No. 8, Nov. 1983, pp. 10–15.

[48]   Whitehead, J.F., "Cellular System Design: An Emerging Engineering Discipline," *IEEE Comm. Mag.*, Vol. 24, No. 2, Feb. 1986, pp. 8–15.

[49]   Lorenz, R.W., "Frequency Planning of Cellular Radio by the Use of a Topographical Data Base," *Proc. 35th IEEE Veh. Tech. Conf.*, Boulder, CO, 21–23 May 1985, pp. 1–5.

[50]   Cox, D.C., "Personal Communications—A Viewpoint," *IEEE Spectrum*, Nov. 1990, pp. 8, 11, 12, 14, 16, 18, 19, 20, 92.

[51]   Goodman, D.J., "Evolution of Wireless Information Networks," *European Transactions on Telecommunications ETT*, Vol. 2, No. 1, Jan./Feb. 1991, pp. 105–113.

[52]   Becker, K.-F., "Mobile Communications on the Road to Integration," *Telecom Report International*, Vol. 13, No. 5–6, 1990, pp. 166–169.

[53]   Raith, K., and J. Uddenfeldt, "Capacity of Digital Cellular TDMA Systems," *IEEE Trans. on Veh. Tech.*, Vol. 40, No. 2, May 1991, pp. 323–332.

[54]   Brentson, G., "Mobitex—A New Network for Mobile Data Communications," *Ericsson Review*, No. 1, 1989, pp. 33–39.

[55]   Kavehrad, M., and P.J. Mclane, "Spread Spectrum for Indoor Digital Radio," *IEEE Comm. Mag.*, Vol. 25, No. 6, June 1987, pp. 32–40.

[56]   Pahlavan, K., "Wireless Communications for Office Information Networks," *IEEE Comm. Mag.*, Vol. 23, No. 6, June 1985, pp. 19–26.

[57]   Viterby, A.J., "Wireless Digital Communication: A View Based on Three Lessons Learned," *IEEE Comm. Mag.*, Vol. 29, No. 9, Sept. 1991, pp. 33–36.

[58]   Gilhousen, K.S., I.M. Jacobs, R. Padovani, A.J. Viterbi, L.A. Weaver, Jr., and C.E. Wheathley III, "On the Capacity of a Cellular CDMA System," *IEEE Trans. on Veh. Tech.*, Vol. 40, No. 2, May 1991, pp. 303–312.

[59]   Linnartz, J.P. M.G., and J.J.P. Werry, "Error Correction and Error Detection Coding in a Fast Fading Narrowband Slotted ALOHA Network With BPSK Modulation," *Proc. International Symposium on Comm. Theory & Applications*, Crieff, U.K., Paper 37, 9–13 Sept. 1991.

[60]   Walrand, J., *Communication Networks: A First Course*, Aksen Associates, Series in Electrical and Computer Engineering, Homewood, IL, 1991.

Chapter 2
VHF and UHF Propagation in
Land-Mobile Communication

Land-mobile communication is burdened with particular propagation complications, compared to the channel characteristics in radio systems with fixed and carefully positioned antennas. The antenna height at a mobile terminal is usually very small, typically less than a few meters. Hence, the antenna is expected to have very little clearance, resulting in obstacles and reflecting surfaces in the vicinity of the antenna having a substantial influence on the characteristics of the propagation path. Moreover, the propagation characteristics change from place to place and, if the terminal moves, from time to time.

In generic system studies, the mobile radio channel is usually evaluated from statistical propagation models: no specific terrain data is considered, and channel parameters are modeled as stochastic variables. The generally accepted statistical model for narrowband mobile propagation is summarized and reviewed in the first part of this chapter (Section 2.1). The second part of this chapter (Sections 2.2 and 2.3) investigates deterministic propagation models (i.e., those requiring specific terrain data). More specifically, the assessment of a VHF/UHF propagation model appropriate for frequency assignment in rural areas is reported. Existing theoretical and empirical models for propagation phenomena, particularly diffraction and groundwave propagation, are verified with field measurements. Empirical models to account for the combined occurrence of multiple large-scale mechanisms are proposed. These results further verify the validity of the statistical models described in Section 2.1.

2.1 REVIEW OF STATISTICAL PROPAGATION MODELS

It is usually possible to distinguish between three mutually independent, multiplicative propagation phenomena: multipath fading, shadowing, and *large-scale* path loss [1-3]. Multipath propagation leads to rapid fluctuations of the phase and amplitude of the signal if the vehicle moves over a distance on the order of a wavelength or more. Multipath fading thus has a *small-scale* effect. Shadowing is a *medium-scale* effect: field strength variations occur if the antenna is displaced over distances larger than a few tens or hundreds of meters. The large-scale effects cause the received power to vary gradually due to signal attenuation determined by the geometry of the path profile in its entirety. This is in contrast to the local propagation mechanisms, which are determined by terrain features in the immediate vicinity of the antennas. The large-scale effects determine a power level averaged over an area of tens or hundreds of meters and are therefore denoted as the *area-mean* power $\bar{\bar{p}}_i$. The index i is introduced to denote a signal from transmitter i. The index $i = 0$ or j will be used to indicate a wanted (or reference) signal, whereas i ($i = 1, 2, \ldots,$ n) is mostly used to denote interference. In our studies of systems employing multiple receivers, the index A_i is used to denote signal power levels at receiver A from transmitter i. Shadowing introduces additional fluctuations, so the received local-mean power $\bar{p}_i$ varies around the area-mean. The term *local-mean* is used to denote the signal level averaged over a few tens of wavelengths, typically 40λ [2]. This ensures that the rapid fluctuations of the instantaneous received power due to multipath effects are largely removed. These propagation mechanisms will be described in the following sections.

2.1.1 Large-Scale Attenuation

The characteristics of the path profile largely influence the behavior of the radio link. In statistical models, the path loss is often expressed as a function of the distance between transmitter and receiver d, the corresponding antenna heights h_t and h_r, and the carrier frequency f_c. Although specific terrain data is not considered in statistical analyses, the average behavior of the attenuation as a function of range is influenced by terrain parameters. In VHF/UHF land-mobile radio, the principal contributing large-scale mechanisms are free-space loss, groundwave propagation, and diffraction. For the statistical models employed in generic system analysis, the effects of these large-scale mechanisms are usually simplified by assuming an average attenuation, increasing with distance according to the *attenuation power law*. The latter model will be used extensively in the system analyses in the next chapters.

Free Space Loss

In free space, the energy of radio waves diverges over an area proportional to the square of the propagation distance d. For isotropic antennas, the free space loss A_{fs}, expressed in decibels, is

$$A_{fs} = 20 \log \left(\frac{4\pi d}{\lambda} \right) \tag{2.1}$$

with λ the wave length.

Groundwave Loss

Waves traveling over land interact with the earth's surface. For propagation over a plane earth, a general, analytical expression of the received signal has been derived by Norton [4] and simplified by Bullington [5, 6]. For isotropic antennas, the field strength is written as the complex sum of a direct line-of-sight wave, a wave reflected from the earth's surface and a surface wave [1], namely,

$$E_i = E_{0_i}(1 + R_c e^{j\Delta} + (1 - R_c) F(\cdot) e^{j\Delta} + \cdots) \tag{2.2}$$

where R_c is the reflection coefficient, E_{0i} the theoretical field strength for propagation in free space, $F(\cdot)$ the complex surface wave attenuation, Δ the phase difference between the direct and reflected waves, and j the imaginary number $\sqrt{-1}$. In UHF land-mobile communication, the propagation loss can be approximated by taking account of only the first two terms in (2.2), since $|F(\cdot)| << 1$ for the usual values of h_t and h_r. The phasor sum of the direct wave and the ground-reflected wave is known as the *space wave* [7]. Its path loss, expressed in decibels, is

$$A_G = -20 \log\left(\frac{\lambda}{4\pi d}\left|1 + R_c e^{j\Delta}\right|\right) \tag{2.3}$$

If $d >> 5h_t h_r$, the phase difference Δ can be approximated by [1]

$$\Delta \approx \frac{4\pi h_t h_r}{\lambda d} \tag{2.4}$$

and for most types of terrain the reflection coefficient approaches $R_c \rightarrow -1$ [1, 8]. The corresponding excess loss A_R, relative to free-space attenuation, is

$$A_R \simeq -20 \log\left[2 \sin\left(\frac{2\pi h_t h_t}{\lambda d}\right)\right] \tag{2.5}$$

Except for communication at short range, one may assume $d\lambda >> 4h_r h_t$. The total path loss can then be approximated by

$$A_K = A_{fs} + A_R \simeq -20 \log\left(\frac{h_r h_t}{d^2}\right) \tag{2.6}$$

In this case, the field strength thus diminishes proportionately with the inverse of the fourth power of the distance. Experiments confirmed the fourth power law also for less perfect surface conditions [9]. At shorter range, the effect of ground-wave attenuation is less pronounced, so the decrease with the signal strength is less rapid [10] and is often well described by free-space loss.

The excess loss A_R is called *ground-reflection loss*. However, the attenuation is not caused by superposition of the direct wave with a *specular* reflection in any clearly defined point on the path: it is the interaction of the radio wave with the earth's surface over the entire path that contributes to the second term. In land-mobile radio, a substantial part of the path profile may lie within the first Fresnel zone, so almost the entire path profile contributes to the existence of the ground-reflected wave. Nevertheless, (2.5) and (2.6) can be derived from a *two-ray model* [1] with a direct wave and a specularly reflected wave with reflection coefficient $R_c = -1$.

Diffraction Loss

If the direct line of sight between the terminals is obstructed, radio waves experience attenuation caused by diffraction. An exact solution is known for the diffraction loss caused by an idealized obstacle with the shape of a knife edge located perpendicularly to the line of sight in free space. The geometry of the path is illustrated in Figure 2.1. The diffraction parameter v is defined as

$$v \triangleq h_m \sqrt{\frac{2}{\lambda}\left(\frac{1}{d_t} + \frac{1}{d_r}\right)} \qquad (2.7)$$

where h_m is the height of the obstacle, and d_t and d_r are the terminal distances from the obstacle (Figure 2.1).

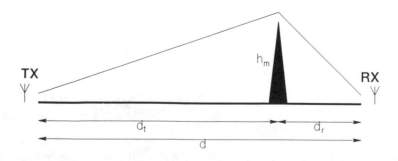

Figure 2.1 Path geometry for a single knife edge in free space.

This definition implies that the obstacle h_m obstructs $(1/2)v^2$ Fresnel zones of the radio path. The diffraction loss A_d, expressed in decibels, is closely approximated by [11]

$$A_d = \begin{cases} 6 + 9\,v - 1.27\,v^2 & 0 < v < 2.4 \\ 13 + 20 \log v & v > 2.4 \end{cases} \tag{2.8}$$

The total path loss is of the form $A_K = A_{fs} + A_d$. However, this *single knife-edge* (SKE) approximation of the path profile is not very realistic in VHF or UHF mobile radio communications; usually, the rest of the ground profile also influences the electromagnetic field at the receiver antenna. Approximation of a path profile by *multiple knife edges* (MKE) is a frequently applied refinement. Techniques to estimate the loss over such a sequence of knife edges have been proposed by Bullington [12], Deygout [13], and Epstein and Peterson [14].

In 1969, Okumura [15] presented a highly accurate propagation model for macrocellular mobile communication in urban areas. The method takes detailed terrain data into account. Empirical curves for diffraction loss in [15] are claimed to be more accurate than the theoretical models, especially closely behind a large obstacle ($d_r \ll d$). The latter observation is also made by Preller and Koch [16]: closely behind an obstacle, losses are difficult to predict since reflected radio waves may enhance the received signal strength to higher levels than would occur due to diffraction only.

For typical path profiles encountered in practical mobile communications, the large-scale attenuation is caused by a combination of the above three mechanisms interacting in a complicated manner [17, 18]. Theoretical discussion of the combined occurrence of reflection and diffraction loss is scarce. Bullington proposed an approximate technique to express the path loss over a knife edge located on a flat plane earth [12], but results are approximate. It is stated that the influence of groundwave attenuation is reduced if substantial diffraction loss occurs over the path [18]. In some rare events, the path loss over a path with a sharp (knife-edge type) obstacle is less than that in the absence of the obstacle [19]. Blomquist [18] suggested the empirical estimate of the path loss:

$$A_K = A_{fs} + \sqrt{A_d^2 + A_R^2} \tag{2.9}$$

But most models (e.g., [1, 13–15, 20]) assume a simpler addition of losses:

$$A_K = A_{fs} + A_d + A_R \tag{2.10}$$

where A_R is an empirically modified version of (2.5). Combined large-scale mechanisms will be discussed further in Section 2.2, where detailed terrain profiles are considered and empirical findings are reported.

If the detailed terrain profile is unknown, the total area-mean path attenuation A_K is usually written in the form of

$$A_K = 10\beta \log d + A_0 \qquad (2.11)$$

with A_0 and β empirical constants. As can be seen from (2.1) and (2.6), theoretical values are $\beta = 2$ in free space and $\beta = 4$ in the case of propagation over a highly conducting plane earth. The statistical relation between the diffraction loss and the path length d depends strongly on the type of terrain and the type of antenna positions considered. Egli [9] proposed the semiempirical model

$$A_K = 40 \log d + 20 \log (f_c/40\text{ MHz}) - 20 \log h_r h_t \qquad (2.12)$$

Evidently, this model is based on groundwave propagation, but an empirically determined frequency-dependent term, $20 \log (f_c/40\text{ MHz})$, is included. Other empirical values reported for the propagation exponent β in the UHF bands are 3 to 4 for distances in the range of 3 km $< d <$ 50 km [1, 2]. Anticipating field measurements that will be reported in Sections 2.2 and 2.3.3, Figure 2.2 presents measured average path loss for propagation at $f_c = 450$ MHz in the North German Plain. These curves would suggest that UHF propagation in forested areas is principally a diffraction

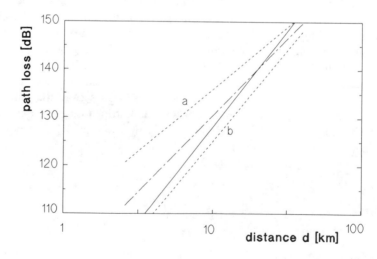

Figure 2.2 Area-mean path loss measured in the North German Plain at $f_c = 450$ MHz and antenna heights h_r, $h_t \approx 21$ m (—·—) (see Section 2.3.3). Forestal paths (– – –a), paths over open area (– – –b), and model by Egli [8] (——).

effect in free space; the exponent was found to be $\beta \approx 2.6$, only somewhat higher than the theoretical value of $\beta = 2$ for unobstructed propagation in free space. On the other hand, for relatively open areas, the plane-earth model is more appropriate ($\beta \approx 3.8$). The entire set of measurements showed a 32 log d trend ($\beta = 3.2$), which is in agreement with a number of other experimental results (e.g., [1]).

2.1.1.1 Statistical Model for Path Loss in Macrocells

In our analyses of multiuser networks from Chapter 3 onwards, all transmitters will be assumed to have identical transmitting power. For ease of notation, the normalized distance r_i and the normalized power $\bar{p}_i$ are defined such that

$$\bar{p}_i = r_i^{-\beta} \tag{2.13a}$$

where the distances r_i are usually normalized so that the wanted signal $0 < r_0 < 1$.

We will mostly consider the path loss law $\beta = 4$ in numerical examples of macrocellular networks, because this value is supported by the theoretical model for UHF-groundwave propagation, because it appeared realistic in certain environments [9], and because a number of scientific papers also focus on the case $\beta = 4$. Moreover, $\beta = 4$ produces some closed-form solutions for analyses of the random-access networks addressed in Chapter 8 and 9. It is to be noted that in interference-limited networks, a fast decay of signal power (large β) gives a more favorable performance than a slow decay (small β). As a result, our curves for $\beta = 4$ may be somewhat optimistic for areas with relatively irregular terrain where $\beta \approx 3$.

2.1.1.2 Statistical Model for Path Loss in Microcells

For communication in microcellular networks, the path loss for short-range communication (if $d\lambda < 4h_t h_t$) is also relevant. This requires an extension of the path loss law (2.13a) to include the transition from free-space propagation to groundwave propagation. One solution is a stepwise transition from $\beta \approx 2$ to $\beta \approx 4$ at a certain (turnover) distance d_g. If antennas are mounted below the tops of buildings, the street geometry may cause guided propagation of radio waves. So in some cases the attenuation exponent β may even be significantly less than 2. Measurements by Green [10] at $f_c = 900$ MHz in London showed a significant increase of the path loss exponent for distances larger than a few hundred meters ($h_t = 5$ m, $h_r = 1.5$ m). Harley [21], however, suggested a smooth transition, with

$$\bar{p}_i = r_i^{-\beta_1}\left(1 + \frac{r_i}{r_g}\right)^{-\beta_2} \tag{2.13b}$$

where r_g is the turnover distance normalized to the cell size. Empirical values for β_1 and β_2 have been reported in a number of papers (e.g., [10, 21–27]) (see also Table 2.1).

To reduce the interference between cochannel cells in a microcellular network, transmitting antennas should be kept low (e.g., at limited height on a lamp post), which ensures rapid decrease of signal power outside the cell, particularly if the turnover distance d_g falls within the cell size. In microcellular nets, low transmitting antennas are located in a scattering environment, and guided propagation through streets may occur. This may limit the accuracy of the groundwave model, even though turnover distances are measured on the order of a few hundreds of meters and appear to agree reasonably with $d_g = 4h_r h_t/\lambda$.

Table 2.1
Path Loss Models for Mobile VHF/UHF Propagation Using Low Transmit Antenna Heights
(h_t = 5 to 10 m)

Ref	Model	Area	β_1	β_2	d_g
(2.1)	FSL	Free space	2	0	∞
[8]	Egli	Average terrain	0	4	0
(2.1), (2.6)	Two-ray	Plane earth	2	2	$\dfrac{4h_r h_t}{\lambda}$
[9]	Green	London	1.7–2.1	2–7	200–300 m
[20]	Harley	Melbourne	1.5–2.5	−0.3–0.5	150 m
[22]	Pickhlotz et al.	Orlando, FL	1.3	3.5	90 m

2.1.2 Shadowing

Experiments reported by Egli [9] showed that for paths longer than a few hundred meters, the received (local-mean) power fluctuates with a *log-normal* distribution about the area-mean power, with the area-mean power in the form of (2.12). By log-normal it is meant that the local-mean power expressed in logarithmic values, such as decibels or nepers, has a normal (i.e., Gaussian) distribution. The *probability density function* (pdf) of the local-mean power is thus in the form of

$$f_{\bar{p}_j}(\bar{p}_j|\bar{\bar{p}}_j) = \frac{1}{\sqrt{2\pi}\,\sigma_s \bar{p}_j} \exp\left[-\frac{1}{2\sigma_s^2} \ln^2\left(\frac{\bar{p}_j}{\bar{\bar{p}}_j}\right) \right] \tag{2.14}$$

where σ_s is the logarithmic standard deviation of the shadowing, expressed in natural units. The standard deviation s_s expressed in decibels is found from $s_s = 4.34\sigma_s$ [28, 29]. In (2.14), the area-mean power $\bar{\bar{p}}_i$ corresponds to the *logarithmic* average μ_1 of the received local mean power, with μ_k defined as

$$\mu_k \triangleq \int_0^\infty [\ln(\bar{p}_j)]^k f_{\bar{p}_j}(\bar{p}_j | \bar{\bar{p}}_j) d\bar{p}_j \qquad (2.15)$$

so $\mu_1 = \ln(\bar{\bar{p}}_i)$. The logarithmic variance is found from $\sigma_s^2 = \mu_2 - \mu_1^2$.

The *linear* average and higher-order moments of the local-mean power are of the form [28, 29]

$$E[\bar{p}_j^m] \triangleq \int_0^\infty \bar{p}_j^m f_{\bar{p}_j}(\bar{p}_j | \bar{\bar{p}}_j) \, d\bar{p}_j = \bar{\bar{p}}_j^m \exp\left(m^2 \frac{\sigma_s^2}{2} \right) \qquad (2.16)$$

The linear average is thus not identical to the area-mean power, except in the limiting case of $\sigma_s = 0$.

For average terrain, Egli reported a logarithmic standard deviation of about $s_s \approx 8.3$ dB and $s_s \approx 12$ dB for VHF and UHF frequencies, respectively [9]. Such large fluctuations are caused not only by local shadow attenuation by obstacles in the vicinity of the antenna, but also by large-scale effects leading to a coarse estimate of the area-mean power. This log-normal fluctuation was called *large-area shadowing* by Marsan, Hess, and Gilbert [30]; over semicircular routes in Chicago, with fixed distance to the base station, s_s was found to range from 6.5 to 10.5 dB, with a median of 9.3 dB. Large-area shadowing thus reflects shadow fluctuations if the vehicle moves over many kilometers.

In contrast to this, in most papers on mobile propagation, only *small-area shadowing* is considered: log-normal fluctuations of the local-mean power over a distance of tens or hundreds of meters are measured. Marsan et al. [30] reported a median s_s of 3.7 dB for small-area shadowing. Preller and Koch [16] measured local-mean power levels at 10 m intervals and studied shadowing at 500 m intervals. The maximum standard deviation experienced was about $s_s = 7$ dB, but 50% of all experiments showed an s_s less than 4 dB.

Extending the distinction between large-area and small-area shadowing, the definition of shadowing is generalized here to cover any statistical fluctuation of the received local-mean power about a certain area-mean power, with the latter determined by predictable, large-scale mechanisms. Multipath propagation is separated from shadow fluctuations by considering the local-mean power. This definition implies that the standard deviation of the shadowing will depend on the geographical resolution of the estimate of the area-mean power. A propagation model that ignores specific terrain data produces about 12 dB of shadowing [9]. On the other hand, prediction methods using topographical data bases with unlimited resolution can, at least in theory, achieve an s_s of 0 dB. Thus, it can be argued that the standard deviation is a measure of the imprecision (and hence the economy!) of the terrain description. For generic system studies, the large-scale path loss is taken in the form of (2.13), so s_s will necessarily be large. On the other hand, for the planning of a practical network in a certain (known) environment, the accuracy of the large-scale propagation model may be refined. This may allow spectrally more efficient planning if the cellular layout is optimized for the propagation environment [31]. A further discussion of this issue is deferred until Section 3.2.1.

2.1.3 Multipath Fading

In mobile communications with relatively low antenna positions, the received signal is usually a superposition of a large number of reflected waves. Wave interference therefore leads to fast fluctuations of the field strength if the antenna position is varied. Only narrowband channels are considered: the interarrival times of the reflected waves are small compared to the time scale of the variations in the modulating signal. In other words, in narrowband channels, all reflections are assumed to add coherently (phasor addition).

2.1.3.1 Rician Fading

We now consider an unmodulated carrier transmitted by terminal i. In a typical Rician-fading channel, the received carrier is in the form of

$$v_i(t) = (C_i + \zeta_i) \cos \omega_c t + \xi_i \sin \omega_c t \qquad (2.17)$$

where the constant C_i represents the direct (line-of-sight) component, and the random variables ζ_i and ξ_i represent the in-phase component and quadrature component of the sum of the reflections. If the mobile terminal is in motion, ζ_i and ξ_i are functions of time, though this is not denoted explicitly. Rician fading occurs if the central limit theorem can be applied for the accumulations of in-phase components and quadrature components of each individual reflection. This occurs if the number of reflections is large and none of the reflections substantially dominates the joint reflected power. In this case, ζ_i and ξ_i are independently Gaussian-distributed random variables with identical pdfs, of the form $N(0, \bar{q}_{si})$ (i.e., with zero mean and variance equal to the local-mean reflected power $\bar{q}_{si}$). The received carrier $v_i(t)$ can also be expressed in terms of the amplitude ρ_i and the phase θ_i:

$$v_i(t) = \rho_i \cos(\omega_c t + \theta_i) \tag{2.18}$$

with

$$\rho_i = \sqrt{(C_i + \zeta_i)^2 + \xi_i^2}$$

$$\theta_i = \arctan\left(\frac{C_i + \zeta_i}{\xi_i}\right)$$

The instantaneous amplitude ρ_i has the Rician pdf [10]

$$f_{\rho_i}(\rho_i|\bar{q}_{si}, C_i) = \frac{\rho_i}{\bar{q}_{si}} \exp\left(-\frac{\rho_i^2 + C_i^2}{2\bar{q}_{si}}\right) I_0\left(\frac{\rho_i C_i}{\bar{q}_{si}}\right) \tag{2.19}$$

where $I_0(\cdot)$ is the modified Bessel function of the first kind of zero order. The local-mean power $\bar{p}_i$ is the sum of the power $\bar{q}_{di}$ in the dominant component[1] (with $\bar{q}_{di} = \frac{1}{2}C_i^2$) and the average power $\bar{q}_{si}$ in the scattered component, i.e., $\bar{p}_i = \bar{q}_{si} + \bar{q}_{di}$. The K-factor of the Rician distribution is defined as the radio of the direct power $\bar{q}_{di}$ and the scattered local-mean power $\bar{q}_{si}$ (so $K \triangleq \bar{q}_{di}/\bar{q}_{si}$). Substitution gives

$$\bar{q}_{si} = \frac{\bar{p}_i}{1 + K} \quad \text{and} \quad C_i = \sqrt{2\bar{q}_{di}} = \sqrt{\frac{2K\bar{p}_i}{1 + K}}$$

Hence, the pdf of the signal amplitude, expressed in the local-mean power $\bar{p}_i$ and the Rician K-factor becomes

$$f_{\rho_i}(\rho_i|\bar{p}_i, K) = (1 + K)e^{-K}\frac{\rho_i}{\bar{p}_i} \exp\left(-\frac{1 + K}{2\bar{p}_i}\rho_i^2\right) I_0\left(\sqrt{\frac{2K(1 + K)}{\bar{p}_i}}\rho_i\right) \tag{2.20}$$

The instantanous power p_i ($p_i = 1/2\rho_i^2$) has the noncentral chi-square pdf

$$f_{p_i}(p_i|\bar{p}_i, K) = f_{\rho_i}(\rho_i|\bar{p}_i, K)\left|\frac{d\rho_i}{dp_i}\right|$$

$$= \frac{(1 + K)e^{-K}}{\bar{p}_i} \exp\left(-\frac{1 + K}{\bar{p}_i}p_i\right) I_0\left(\sqrt{4K(1 + K)\frac{p_i}{\bar{p}_i}}\right) \tag{2.21}$$

---

[1]Since it is not subject to multipath fading, the power in the dominant component is not a stochastic variable. Nonetheless, we denote $\bar{q}_{di}$, rather than q_{di}, to make the notation consistent with the overlined notation for the local-mean scattered power $\bar{q}_{si}$.

Measurements have indicated that for the wanted signal in microcellular nets, the Rician factor may range from $K = 4$ to 1000 (6 to 30 dB) [10]. From Bultitude and Bedal [32] we conclude that a Rician factor of about 7 dB ($K \approx 5$) adequately describes most microcellular channels. In some specific cases, the Rician factor may be about 12 dB ($K \approx 16$).

2.1.3.2 Rayleigh Fading

In macrocells, with longer, often obstructed propagation paths, the power of the direct line-of-sight signal is small compared to the reflected signal power ($C_i \to 0$, $K \to 0$). This produces the special case of Rayleigh fading [1, 2, 33, 34]. In this case, the variance of ζ_i and ξ_i is equal to the local-mean power $\bar{p}_i$: the phase θ_i is uniformly distributed over [0, 2π], and the instantaneous amplitude ρ_i has the Rayleigh pdf [34]

$$f_{\rho_i}(\rho_i|\bar{p}_i, K = 0) = \frac{\rho_i}{\bar{p}_i} \exp\left(- \frac{\rho_i^2}{2\bar{p}_i} \right) \tag{2.22}$$

The corresponding total instantaneous power p_i ($p_i = (1/2)\rho_i^2 = (1/2)\zeta_i^2 + (1/2)\xi_i^2$) received from the ith mobile terminal is exponentially distributed about the mean power; that is,

$$f_{p_i}(p_i|\bar{p}_i, K = 0) = f_{\rho_i}(\rho_i|\bar{p}_i, K = 0) \left|\frac{d\rho_i}{d\bar{p}_i}\right| = \frac{1}{\bar{p}_i} \exp\left(- \frac{p_i}{\bar{p}_i} \right) \tag{2.23}$$

2.1.3.3 Nakagami Fading

Refined models for the pdf of a signal amplitude exposed to mobile fading have been suggested [33, 35, 36, 53]. In [35, 36], the pdf of the amplitude ρ_i of a mobile signal was described by the Nakagami m-distribution [37], namely,

$$f_{\rho_i}(\rho_i|\bar{p}_i = m\bar{p}) = \frac{\rho_i^{2m-1}}{\Gamma(m)2^{m-1}\bar{p}^m} \exp\left(- \frac{\rho_i^2}{2\bar{p}} \right) \tag{2.24}$$

where $\Gamma(m)$ is the gamma function [38], with $\Gamma(m + 1) = m!$ The local-mean power is $\bar{p}_i = m\bar{p}$. The corresponding instantaneous power p_i is gamma-distributed, with pdf

$$f_{p_i}(p_i|\bar{p}_i = m\bar{p}) = \frac{1}{\bar{p}\Gamma(m)} \left(\frac{p_i}{\bar{p}}\right)^{m-1} \exp\left(- \frac{p_i}{\bar{p}} \right) \tag{2.25}$$

The parameter m is called the *shape factor* of the Nakagami, or the *gamma distri-*

Figure 2.3 (left) Model of multipath-wave interference for Rayleigh-fading (narrowband) channels.
Figure 2.4 (right) Model of multipath-wave interference for Nakagami-fading (wideband) channels.

bution. In the special case $m = 1$, Rayleigh fading (2.22) is recovered, with an exponentially distributed instantaneous power (2.23); for larger m, the fluctuations of the signal strength are reduced compared to (2.22). The model behind (2.24) and (2.25) assumes multipath scattering with relatively large delay-time spreads with different clusters of reflected waves. Within any one cluster, the phases of individual reflected waves are random, but the delay times are approximately equal for all waves. As a result, the envelope of each accumulated cluster signal is Rayleigh-distributed. The average time delay is assumed to differ significantly between clusters. If the delay times also significantly exceed the bit time of a digital link, the different clusters produce serious intersymbol interference, and so the multipath self-interference then approximates the case of cochannel interference by multiple incoherent Rayleigh-fading signals.

Figures 2.3 and 2.4 illustrate the models of wave interference in Rayleigh and Nakagami multipath reception. Narrowband signals are addressed in the following. The corresponding model of Rician or Rayleigh fading for individual signals is retained, assuming delay time spreads are small compared to the rate of modulation of the phase of signal. The Nakagami model will be used in Chapters 3 and 4 to describe the interference accumulated from the multiple independent Rayleigh-fading sources.

2.1.3.4 Rate of Rayleigh Fading

In mobile radio channels with high terminal speeds, Rayleigh fading causes the signal to fluctuate rapidly. If the mobile antenna moves a small distance ϵ, the kth incident wave, arriving from the angle ϕ_k with respect to the instantaneous direction of motion, experiences a phase shift of

$$\Delta\theta_k = 2\pi \frac{\epsilon}{\lambda} \cos \phi_k$$

It is often assumed that the angle ϕ_k is uniformly distributed within $[0, 2\pi]$ [1]. Hence, if ϵ is on the order of half a wavelength λ or more, the phases of all incident waves become mutually uncorrelated. The amplitude of the total received signal also becomes uncorrelated with the amplitude at the point of departure. The normalized covariance $L(\epsilon)$ of the electric field strength for an antenna displacement ϵ is in the form of [1]

$$L(\epsilon) = J_0^2\left(2\pi \frac{\epsilon}{\lambda}\right) \tag{2.26}$$

with $J_0(.)$ the zero-order Bessel function of the first kind [38]. It can be seen in Figure 2.5 that the signal remains almost entirely correlated for a small displacement,

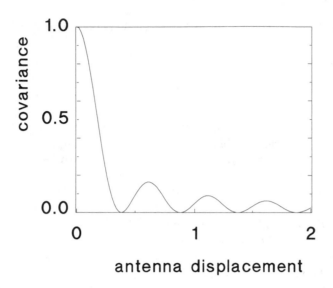

Figure 2.5 Auto-covariance $L(\epsilon)$ of the electric field strength in a Rayleigh-fading channel versus the normalized antenna displacement ϵ/λ (or $T_s f_m$) in the horizontal direction.

say, $\epsilon < \lambda/8$, but becomes rapidly independent for larger displacements, say, for $\epsilon > \lambda/2$.

The antenna displacement can also be expressed in the terminal velocity v and the time difference T_ϵ between the two samples ($\epsilon = vT_\epsilon$). So,

$$\frac{\epsilon}{\lambda} = vT_\epsilon \frac{f_c}{c} = T_\epsilon f_m \tag{2.27}$$

with f_m the maximum Doppler shift ($f_m = vf_c/c$). The rate of fading is considered further in Chapter 4.

2.1.3.5 Coherence Bandwidth

The foregoing properties of multipath fading hold for unmodulated carrier signals as well as for narrowband signals. Earlier, the property narrowband was defined in the time domain, considering interarrival times of reflections and the time scale of variations in the signal caused by modulation. Transformed into constraints in the frequency domain, this criterion is found to be satisfied if the transmission bandwidth does not substantially exceed the coherence bandwidth B_c of the channel, which is the bandwidth over which the channel transfer function remains virtually constant, with [1]

$$B_c = \frac{1}{2\pi\Delta_t} \qquad (2.28)$$

where Δ_t is the delay time spread of the significant reflections.

For a digital signal with high bit rate, dispersion is experienced as frequency-selective fading and *intersymbol interference* (ISI) [39]. No serious ISI is likely to occur if the symbol rate r_s is low, say, if $r_s << (2\Delta_t)^{-1}$.

In macrocellular mobile radio, delay spreads are mostly in the range of $\Delta_t \approx$ 100 ns to 10 μs [1]. A typical delay spread of 0.25 μs corresponds to a coherence bandwidth of $B_c \approx$ 640 kHz. Measurements made in the U.S. [40, 41] indicated that delay spreads are usually less than $\Delta_t \approx 0.2$ μs in open areas, about $\Delta_t \approx 0.5$ μs in suburban areas, and about $\Delta_t \approx 3$ μs in urban areas. Measurements in The Netherlands [42] showed that delay spreads are relatively large in European-style suburban areas, but rarely exceed $\Delta_t \approx 2$ μs. However, large distant buildings such as flats occasionally cause reflections with excess delays on the order of 25 μs. In densely and uniformly built European (historical) centers, delay spreads may be somewhat less than in urban (business) centers in the U.S., although no experimental evidence in the literature has been found by the author. In indoor and microcellular channels, the delay spread is usually smaller and rarely exceeds a few hundred nanoseconds [32, 43]. Seidel and Rappaport reported delay spreads in four European cities [43] of less than 8 μs in macrocellular channels, less than 2 μs in microcellular channels, and between 50 and 300 ns in picocellular channels.

2.1.4 Combined Shadowing and Rayleigh Fading

In the case of cellular VHF and UHF radio networks with propagation paths of $d =$ 1 to 60 km, the large-scale and small-scale mechanisms described in Sections 2.1.1 to 2.1.3 are generally assumed to be multiplicative (i.e., the narrowband propagation channel may be seen as a number of cascaded stochastical attenuators). Moreover, the attenuators (multipath fading, shadowing, and path loss) are mostly assumed to be mutually independent.

For a given area-mean power $\bar{p}_i$, the conditional pdf of the instantaneous power is [44, 35]

$$f_{p_i}(p_i|\bar{\bar{p}}_i) = \int_0^\infty \frac{1}{\bar{p}_i} \exp\left(-\frac{p_i}{\bar{p}_i}\right) \frac{1}{\sqrt{2\pi}\,\sigma_s\bar{p}_i} \exp\left(-\frac{\ln^2(\bar{p}_i/\bar{\bar{p}}_i^2)}{2\sigma_s^2}\right) d\bar{p}_i \quad (2.29)$$

This is sometimes called a Suzuki distribution [35]. No analytical solution for this integral is known to the author. In some studies, the large-area pdf of instantaneous power, thus conditional on $\bar{p}_i$, is described by a log-normal distribution rather than by (2.29). This has been confirmed by some experiments [39] and may lead to more convenient mathematical expressions [45].

In a number of network studies, the position of the terminal is a stochastic variable, too. If the pdf of the distance between the mobile terminal and the base station is denoted by $f_r(r)$, and if the area-mean is included in the form of (2.13a), the unconditional pdf of the received power becomes, after combining shadowing,

$$f_{p_i}(p_i) = \int_0^\infty \int_0^\infty \frac{1}{\bar{p}_i} \exp\left(-\frac{p_i}{\bar{p}_i}\right) \frac{f(r)}{\sqrt{2\pi}\,\sigma_s\bar{p}_i} \exp\left(-\frac{(\ln\bar{p}_i - \beta\ln r)^2}{2\sigma_s^2}\right) dr\, d\bar{p}_i \quad (2.30)$$

For some special cases of the $f_r(r)$, solutions for the integral over r can be found in closed form.

2.1.5 Laplace Transforms of Received Power

To facilitate the analysis in the following chapters, the Laplace image of the signal power received in a mobile channel with combined Rayleigh fading and shadowing is now discussed, following [46]. Laplace images of a pdf are particularly useful in the evaluation of the joint power received from multiple interfering signals. In general, the one-sided Laplace transform of a function $f(x)$ is defined as [38]

$$\mathcal{L}[f(x), s] \triangleq \int_{0-}^\infty \exp(-sx)f(x)dx \quad (2.31)$$

where s is the image variable. Moreover, Laplace images are closely related to "moment-generating functions" and "characteristic functions" used in probability theory, although mostly the kernel $\exp(-jsx)$ with $j = \sqrt{-1}$ is used for characteristic functions, rather than $\exp(-sx)$, as in (2.31).

The Suzuki pdf (2.29) is now inserted for $f(x)$ in (2.31). The Laplace image of the power received from the ith transmitting terminal, with area-mean power $\bar{p}_i$ is

$$\mathcal{L}(f_{p_i}, s) = \int_0^\infty \frac{1}{1 + s\bar{p}_i} \frac{1}{\sqrt{2\pi}\,\sigma_s\bar{p}_i} \exp\left(\frac{\ln^2(\bar{p}_i/\bar{\bar{p}}_i)}{2\sigma_s^2}\right) d\bar{p}_i \quad (2.32)$$

where the integral over x has been solved analytically. After substituting

$$x \overset{\Delta}{=} \frac{1}{\sqrt{2}\sigma_s} \ln\left(\frac{\bar{p}_i}{\bar{\bar{p}}_i}\right) \tag{2.33}$$

the Laplace transform of a Suzuki pdf is found to be

$$\mathscr{L}\{f_{p_i}(p_i), s\} = \frac{1}{\sqrt{\pi}} \int_{-\infty}^{\infty} \frac{\exp(-x^2)dx}{1 + s\bar{\bar{p}}_i \exp(\sqrt{2}x\sigma_s)} \overset{\Delta}{=} \phi_\sigma(s\bar{\bar{p}}_i) \tag{2.34}$$

where, for ease of notation, the function $\phi_\sigma(s)$ was defined as

$$\phi_\sigma(s) \overset{\Delta}{=} \frac{1}{\sqrt{\pi}} \int_{-\infty}^{\infty} \frac{\exp(-x^2)dx}{1 + s \exp(\sqrt{2}x\sigma_s)} \tag{2.35}$$

For channels without shadowing ($\sigma_s = 0$), this image becomes

$$\phi_0(s) = \frac{1}{1 + s} \tag{2.36}$$

This obviously corresponds to the Laplace image of the exponential pdf (2.22). Figure 2.6 illustrates the image $\phi_\sigma(s)$ for $\sigma_s = 0$, 1.36, and 2.72 ($s_s = 0$, 6, and 12 dB).

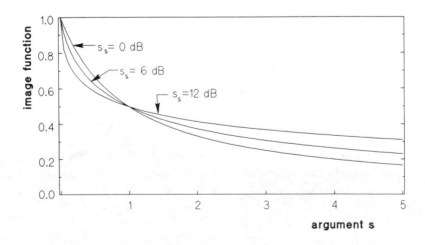

Figure 2.6 Laplace image of a Suzuki pdf for 0, 6, and 12 dB of shadowing, versus its argument s.

Further discussion of the mathematical properties of this image function and techniques for numerical evaluation are presented in Appendixes C and E, respectively.

2.1.6 Reciprocity

Two-way communication requires facilities for *inbound* (i.e., mobile-to-fixed), as well as *outbound* (i.e., fixed-to-mobile), communication. In circuit-switched mobile communication, such as cellular telephony, the inbound and outbound channels are also called the *uplink* and *downlink*, respectively. The propagation aspects described above are valid for inbound and outbound channels. This is understood from the reciprocity theorem: If, in a radio communication link the role of the receive and transmit antenna are functionally interchanged, the instantaneous transfer characteristics of the radio channel remain unchanged.

In mobile multiuser networks with fading channels, the reciprocity theorem does not imply that the inbound channel behaves identically to the outbound channel. Particular differences occur for a number of link aspects:

1. Man-made Noise Levels. The antenna of the base station is usually mounted on an appropriate antenna mast such that it does not suffer from attenuation caused by obstacles in its vicinity. The mobile antenna, on the other hand, is at most mounted a few meters above ground level. The manmade noise level, particularly automotive ignition noise, is likely to be substantially higher at the mobile antenna than at the base station antenna.

2. Effect of Antenna Diversity and Antenna Directivity. Multipath scatters mostly occur in the immediate vicinity of the mobile antenna. The base station more or less receives a transversal electromagnetic wave, whereas the mobile station receives a superposition of a set of reflected waves from random angles. Two antennas at the mobile terminal are likely to receive uncorrelated signal power levels if their separation is more than a wavelength (see (2.27)). At the base station site, however, all reflections arrive from almost identical directions. Therefore, diversity at the base station requires much larger antenna separation to ensure uncorrelated received signal power levels at the two antennas [1]. For the same reason, antenna directivity has different effects at the mobile and at the base station.

3. Correlation of Shadow Fading of Wanted and Interfering Signals. In a cellular network, shadow fading of the wanted signal received by the mobile station is likely to be correlated with the shadow fading of the interference caused by other base stations, or, in a spread-spectrum network [46], with the shadowing of simultaneously transmitted signals from the same base station. In contrast to this, at the base station, shadow fading of the wanted signal is presumably mostly statistically independent of shadow fading of the interference. However, experimental results for correlation of shadow attenuation are scarce.

4. Full-Duplex Channels. In full-duplex operation, multipath fading of inbound and outbound channels, which operate at widely different frequencies, may be uncorrelated. This will be the case particularly if the delay spread is large.

5. Multiplexing and Multiple Access. In a practical multiuser system with intermittent transmissions, inbound messages are sent via a *multiple-access* channel, whereas in an outbound channel, signals destined for different users can be *multiplexed*. In the latter case, the receiver in a mobile station can maintain carrier and bit synchronization of the continuous incoming bit stream from the base station, whereas the receiver in the base station has to acquire synchronization for each user slot. Moreover, in packet-switched data networks, the inbound channel has to accept randomly occurring transmissions by the terminals in the service area. Random-access protocols are required to organize the data traffic flow in the inbound channel, and access conflicts (*contention*) may occur.

In cellular networks with large traffic loads per base station, spread-spectrum modulation can be exploited in the downlink to combat multipath fading, whereas in the uplink, the signal power levels from the various mobile subscribers may differ too much to effectively apply spread-spectrum multiple access unless sophisticated adaptive power control techniques are employed [55].

6. Industrial Design. From a practical point of view, the downlink and the uplink will be designed under entirely different (cost) constraints, such as power consumption, size, weight and other ergonomic aspects, energy radiated into the human body, and consumer cost aspects.

7. Data Traffic Patterns. In packet data networks applied to traffic and transportation, the characteristics of the data traffic flows are known to differ between the uplink and the downlink. For instance, outbound messages from a fleet management center to the vehicles are likely to be more routine, of a more uniform length, and occur in a more regular pattern than messages in the opposite (inbound) direction.

2.1.7 Time Scales

Time variations of signals and system parameters play a key role in this study of mobile communication systems. This section summarizes the various time scales T_i ($i = 1, 2, \ldots, 10$) considered in the observation of the system. Each time scale indicates an order of magnitude of the rate of variations caused by a certain phenomenon.

T_1 Carrier Frequency

The frequency f_c of the carrier determines by far the fastest time scale $T_1 \approx 1/f_c$. The propagation models discussed here are appropriate in the frequency range of, say, $f_c = 150$ MHz to 2 GHz. So, the duration of a carrier cycle is on the order of $T_1 \approx 10^{-9}$ seconds. Only the interarrival time between the fictitious direct and ground-reflected waves in equation (2.2) is on the order of magnitude T_1.

T_2 Interarrival Times Between Various Obstacle-Reflected Waves

If obstacles are present near the mobile antenna, reflected waves are received with interarrival times of at most a few microseconds. T_2 indicates the order of magnitude of the delay time spread Δ_t (see Section 2.1.3.3).

T_3 Bandwidth of the Modulating Signal

A time scale T_3 is related to the bandwidth W of the modulating signal ($T_3 \approx 1/W$) or to the bit duration ($T_3 \approx T_b$). Only narrowband systems are considered, so modulation is assumed to be slow compared to the interarrival times of the various obstacle-reflected waves ($T_2 << T_3$). Spread-spectrum communication, in which the transmission bandwidth B_T is intentionally enlarged ($B_T >> W$), is not addressed. We assume narrowband channels with $B_T \approx W$. The effect of delayed reflections ($T_3 < T_2$) can be described as intersymbol interference in the time domain or as frequency selective fading in the frequency domain [47].

T_4 Propagation Delays Between Transmitter and Receiver

The order of magnitude of the propagation time between transmitter and receiver is denoted by T_4. In some systems, T_4 can be larger than the duration of a few bits. This delay may affect the performance in some full-duplex systems with feedback (e.g., feedback information on access conflicts between contending users in random-access networks). In addition, different distances of mobile users to the base station may require extension of time slots in slotted access systems.

T_5 Duration of a Data Packet

Blocks of L bits are considered ($T_5 \approx LT_3$, $L >> 1$). In our analysis of slotted packet radio networks, the duration of data packets is assumed fixed. The duration of a data packet is often taken as the unit of time for analyses concentrating on protocol issues.

T_6 Rate of Multipath Fading

The rate of fluctuations of the received signal amplitude determines the duration of intervals during which the channel performance is constant. Equation (2.27) gives some impression of T_6. The word "instantaneous" is used to denote a value averaged over a period substantially shorter than T_6. Gilbert [48] simplified the description of channel behavior into two states: periods of unsatisfactory performance called *fades* and periods of satisfactory performance called *nonfade periods*. The event in which the duration of a data block is short with respect to the average nonfade duration is called *slow fading*. *Fast fading* occurs when the packet duration is substantially longer than the average length of fade and nonfade intervals. This in contrast to a few scientific papers on modulation techniques, which address the case that the multipath fading is fast compared to the duration of a single bit ($T_6 \approx T_3$). To avoid confusion, we do not follow the common use of "fast" and "slow" fading to distinguish between multipath reception and shadowing; we use the properties "fast" and "slow" to refine the description of multipath fading.

T_7 Rate of Shadow Fading

Shadowing is much slower than multipath fading. With normal vehicle speeds, T_7 is on the order of a few seconds [30]. Local-mean averages are taken over an observation interval T with $T_6 \ll T \ll T_7$.

T_8 Waiting Times Between Retransmissions

A data block that cannot be decoded correctly by the receiver has to be retransmitted. The waiting time before a retransmission is performed is on the order of T_8. As in all other studies of packet radio over mobile radio channels known to the author, retransmitted blocks will be assumed to experience uncorrelated Rayleigh fading ($T_6 \ll T_8$).

In most systems for data communication, messages lost because of excessive interference are retransmitted. Under certain conditions [49–51], this can lead to nonergodic (unstable) system behavior: these retransmissions cause additional interference to other users, which, in its turn, increases the number of retransmissions of messages in the system. Unless stated otherwise, stable systems are addressed. This corresponds to the assumption that T_8 is sufficiently large to ensure an uncorrelated set of interfering signals during each retransmission.

T_9 Speed of Motion Through the Service Area (Near/Far Fading)

If the size of the service area is small (e.g. in microcellular nets), the position of the vehicle may change during a communication session. In most cases, however, it is reasonable to assume $T_7 \ll T_9$. Area-mean averages are taken over an observation interval $T_7 \ll T \ll T_9$.

T$_{10}$ *Data Traffic Intensities*

We assume that the (Poisson) arrival rate of attempted transmissions is constant (except in Section 8.6).

In the analyses, the following relations are always assumed to be satisfied:

Because of the nature of the multipath environment	$T_1 \ll T_2$
Because of the nature of land-mobile communication	$T_2 \ll T_6 \ll T_7 \ll T_9$
Because of error-detection coding and terminal addresses	$T_3 \ll T_5$
Because of the assumption of narrowband systems	$T_2 \ll T_3$

2.2 ASSESSMENT OF DETERMINISTIC PROPAGATION MODELS

Statistical propagation models, as discussed in the previous sections, are mainly used in generic system design. In contrast to this, deterministic models, considering specific terrain data, are used extensively in frequency management and during the planning of cell structures for mobile telephony networks. The effectiveness of the planning of cellular networks and selection of usable frequencies at various transmitter sites depend heavily on the availability of appropriate propagation models.

Narrowband propagation measurements in the VHF and UHF bands are analyzed in this and the next section. This investigation addresses radio propagation as encountered in point-to-point communication in rural areas, particularly in the ZODIAC network operated by the First Netherlands Army Corps in the North German Plain. The purpose of the investigation was to assess a model appropriate for the prediction of large-scale path losses and for frequency management. Operational links cover 5 to 30 km over the slightly hilly terrain: the interdecile range of the terrain, defined as the difference Δh between the highest and the lowest points in the terrain after the extreme highest and lowest 10% of terrain heights have been removed, is in the range of 20 to 100 m. The terrain is mainly agricultural, with scattered woodlands. Antenna heights range from 4 to 21 m, which is often insufficient to exceed the treetop level. Local propagation phenomena, both multipath reception and shadowing, introduce substantial fluctuations of the local field strength.

In frequency assignment for a complex radio relay network, evaluation of many path profiles is required in order to select the optimum location and frequencies for a relay node. This does not only concern the operational paths, because paths over which interference signals may travel have to be considered as well. Especially in transportable nets (e.g., for emergency aid or military applications), where relay nodes are deployed rapidly and may be relocated frequently, relatively simple algorithms are required for the prediction of propagation losses.

2.2.1 Propagation Forecasting Using Terrain Data

Forecasting path loss from terrain data can be seen as a three-step process (see Figure 2.7), namely, interpretation and classification of the path profile (step 1), evaluation of the effects of relevant mechanisms (step 2), and appropriately combining the effects of each mechanism (step 3). The major part of the theoretical work on propagation concerns step 2 only: given a set of path parameters, the theoretical losses due to isolated and idealized mechanisms can be calculated. Well-known examples, already discussed in Section 2.1, are the diffraction loss over an obstacle with a known knife-edge geometry and the loss caused by groundwave propagation over a smooth, plane, and perfectly conducting earth, given antenna heights and the path distance. Nonetheless, the reliability of the prediction of a field strength depends on the first and third steps. The first step is the interpretation of the terrain data by extracting parameters to be used in selected models. It is mainly in this step that many implementation decisions have to be made. Typical questions are "What is the effective height for diffraction of a hill covered by trees in winter?"; "When does a terrain irregularity act as one single obstacle and when should it be treated as two (or more) separate hills?"; and "What is the effective height of an antenna in an irregular, sloping, and forested terrain?" [17].

Step 3 describes the transition from different theories, and sometimes conflicting approximate models, to a reliable estimate of the path loss. Experience and expertise are required to select and combine the relevant effects. The extent to which mechanisms are present and mutually interact is an important consideration in the selection of a model. Since ample literature is available on the second step, the experiments reported here concentrate on the first and the third steps. For the application considered here, it was concluded from the technical literature that diffraction over single or multiple obstacles and ground-reflection loss were the most

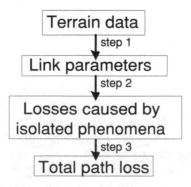

Figure 2.7 Forecasting propagation losses as a three-step process.

relevant mechanisms of propagation. Moreover, fluctuations of the received power level were likely to occur because of local effects.

2.2.2 Measurement Setup

The propagation measurements were carried out in the period between 1986 to 1988 by the Directie Materieel Koninklijke Landmacht (D.M.K.L./M.B.A.1) and the Physics and Electronics Laboratory (F.E.L.-T.N.O.) in The Hague. Initially, a set of stereotypical paths in The Netherlands was investigated to isolate propagation mechanisms such as diffraction (Haarlerberg) and propagation over a plane earth (Flevopolders). The measurements in the North German Plain contained typical mixed paths, as applied by the ZODIAC net. The transmitter was an Andret 740A signal generator with Ailtech 15100, 35512, and 5020 power amplifiers driven to a power level just below overload. Periodically, 4-second bursts of an unmodulated carrier were transmitted.

The transmit antenna was of a corner type (Siemens S42044-Q16-A2), positioned at a mast of $h_t = 21$ m. The antenna gain was 15 dB. Antenna positions were chosen in accordance with common practice during military operations. This often resulted in positions with relatively little antenna clearance and, probably, severe

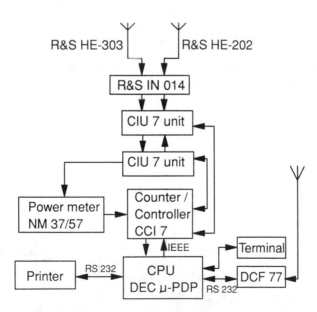

Figure 2.8 Receiver system setup for propagation measurements.

local multipath scattering. The receiving antenna was an active dipole (R&S HE-303 and HE-202) with horizontal polarization. At the receiver sites, antenna heights varied from $h_r = 4$ m to 22 m. Due to the low antenna directivity and the lower antenna heights, local scattering was presumably more severe at the receiver site than at the transmitter site. An Ailtech NM37/57 CIU7/CCI7 receiver was used to measure the signal power level during each burst (see Figure 2.8). To verify the reliability of these samples, the background noise level was measured between bursts. The equipment was controlled by an IBM PC at the transmitter site and by a micro PDP at the receiver end. Synchronization of transmitter and receiver was obtained from the DCF77 time reference transmitter in Frankfurt.

2.2.3 Initial, Dedicated Measurements in The Netherlands

Measurements in The Netherlands were conducted to get an insight into the relevance of the individual propagation mechanisms. Measurements in the flat, plane, and (in 1984 still) relatively bare Flevopolders closely agreed with theoretical propagation over a plane, perfectly conducting earth. Diffraction over an isolated hill was studied based on measurements at seven locations on a path crossing the Haarlerberg, which is located in a relatively smooth agricultural area near Zwolle and has a height of 55m relative to the surrounding area. The paths are characterized in Figure 2.9 and Table 2.2. Measurements were taken at $f_c = 240$ and 400 MHz. At location 1, the path loss closely agreed with the theoretical free-space loss. Observations during measurements in the North German Plain confirmed that, for short links, the measured path losses closely follow the free-space attenuation. Nulls caused by ground

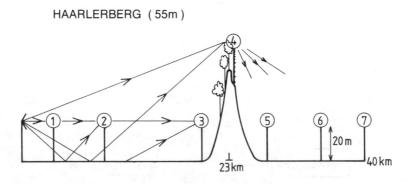

Figure 2.9 Measurement locations near Haarlerberg (1 to 7). Antenna height h_r, $h_t \approx 21$ m. Path length: (1) 3.4, (2) 9.4, (3) 21.2, (4) 23.9, (5) 28.3, (6) 34.0, and (7) 39.0 km.

Table 2.2
Behavior of Path Loss at Various Locations

Location	Path Loss	Height Gain
1	$A_m \approx A_{fs}$	None
2,3	$A_m \approx A_{fs} + A_R$	6 dB/oct
4	$A_m \approx A_{fs} + A_d$	1.5 dB/m
5,6,7	$A_{fs} + A_R + A_d < A_m < A_{fs} + A_R + A_d$	6 dB/oct

reflections have not been observed. Only for the longer paths (i.e., beyond the turn-over point $d = 4\pi h_t h_r / \lambda$), does the two-ray model (2.6) become appropriate, unless significant obstacles cause diffraction losses.

The Haarlerberg measurements in Figure 2.10 and the statistical results in Figure 2.2 suggest that SKE diffraction and ground reflection loss may not be independent, multiplicative mechanisms. Simple addition of logarithmic losses, as in the model $A = A_{fs} + A_d + A_R$, do not lead to an optimum model: in this case, interaction of both mechanisms is neglected. Moreover, the attenuation exponent β would exceed 4, which contradicts the measurements. On a path with dominant ground-reflection losses, isolated obstacles tend to temper reflection losses [19]. A more appropriate and accurate model for VHF/UHF propagation should change smoothly

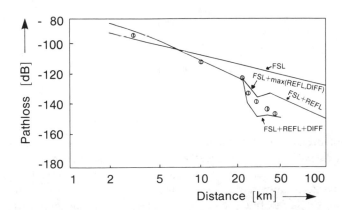

Figure 2.10 Path loss in decibels versus distance at 7 locations (①) and free-space loss (FSL), free-space loss plus the maximum of diffraction and ground-reflection loss (FSL + MAX (DIFF < REFL)) and free-space loss plus diffraction and ground-reflection loss (FSL + DIFF + REFL).

from a diffraction model (for paths mainly covered by vegetation) to a plane earth model (for paths over farmland). Antenna height gain tests near Haarlerberg further strengthened the impression that diffraction and reflection dominate different parts of the path.

2.2.4 Statistical Analysis of Measurements

2.2.4.1 Standard Deviation of Prediction Error

Whenever a measurement of a path loss A_m is conducted, a discrepancy e with the loss A_K predicted by a certain model K is experienced. This discrepancy is caused by a number of effects. The statistical evaluation conducted here is based on the assumption that these effects are mainly multiplicative (i.e., the error introduced by each effect is essentially an attenuation (or amplification) by a number of decibels). For A_m, A_K, and e expressed in decibels, the logarithmic standard deviation s_K of the error experienced when a propagation model A_K is verified is defined as

$$s_K^2 \triangleq E[e - E(e)] = E[(A_m - A_K - E[A_m - A_K])^2] \qquad (2.37)$$

where any systematical error $m_K \triangleq E[A_m - A_K]$ is removed from the variance. The variance s_K^2 is considered to occur due to four statistically independent effects: model deficiencies s_p, shadowing s_s, multipath scattering s_R, and measurement errors s_m. So, if attenuation mechanisms are assumed multiplicative, the logarithmic variance s_K is given by

$$s_K^2 = s_m^2 + s_p^2 + s_s^2 + s_R^2 \qquad (2.38)$$

Each term in (2.44) is discussed in more detail below.

Measurement Errors

Using state-of-the-art equipment, measurement inaccuracies, though present, should not be of principal concern. Typical errors are on the order of $s_m \approx 1$ to 2 dB. Nonetheless, erratic measurements can greatly influence the results. Faulty or loose connecters, damaged antennas, overloaded amplifiers, and so on are not always spotted immediately. Such effects tend to introduce some outliers in the set of measurement data, rather than smoothly increasing the apparent s_K^2 by a small extra term. Daily calibration measurements revealed that measurements over a few paths were unreliable; these have been removed. Similarly, samples taken with an insufficient *carrier-to-noise* (*C/N*) ratio were discarded.

It was realized that the selection of paths to be used for measurements may not be limited to paths already used by radio links and known to give satisfactory communication. Hence, equipment had to be significantly more sensitive than operational radio relay sets, otherwise any resulting absence of paths with heavy loss would bias the population of measurements, leading to an optimistic propagation model.

During the analysis, 3 out of 55 paths were suspected because losses largely disagreed with expected values. In the statistical evaluation, these samples have been maintained, except in a number of special experiments (see, for example, Section 2.3.5).

Local Attenuation Mechanisms

Shadowing and multipath reception introduce local fluctuations of the received power. As explained in Section 2.1.2, shadowing is considered to be an effect of an area description using a finite resolution for the terrain data. The terrain database contained the precise terrain heights at the antenna location, the maximum terrain heights in a 1- x 1-km grid along the path profile, and the type of terrain features. Consequently, shadowing imposes an insurmountable limit to the achievable accuracy of forecasts ($s_K > s_s$), presumably on the order of 6 to 7 dB. Although the rough resolution limits the assessment to very basic models, the accuracy that can be achieved is still substantially better than 12 dB, which would be expected if no terrain data at all were used [9].

Multipath fading is known to cause fluctuations of the received power for displacements over lengths on the order of a wavelength. In an environment with severe multipath reception, such as in urban mobile networks, multipath fluctuations are best removed by averaging over a number of samples. The averaging is to be performed over the received field strength in absolute values, rather than over decibel values [52]. In the ZODIAC radio relay network, multipath scattering was not initially assumed to be a major cause of fluctuations. Therefore, the analysis was performed based on individual samples.

During the assessment of a model for large-scale propagation attenuation, the effect of shadowing and frequency-nonselective multipath scattering can be effectively reduced by taking a sufficiently large number of samples at each location and an appropriate averaging technique. Nonetheless, during the planning of the network, the local fluctuations of received power have to be accounted for. A relevant consideration is whether the network is to be operated with randomly located antennas, as in most mobile and personal communication networks, or with antennas at carefully selected locations, as in point-to-point systems or (to some extent) in television broadcasting. Evidently, *fade margins*, deemed necessary in the planning procedures of a certain mobile network, can be reduced if more sophisticated measures are taken to ensure favorable antenna positions (e.g., by diversity techniques). The influence of local propagation mechanisms on the efficiency of spectrum use will be discussed further in Chapter 3 and the following chapters.

Model Deficiencies

The remaining term s_p represents model deficiencies. The aim of the assessment of an appropriate model is to effectively reduce s_p to zero. The dearth of appropriate theoretical results for the combined occurrence of multiple propagation mechanisms is one of the main limitations on the achievable reduction of s_p. The following sections deal with empirical results of the effects of interacting propagation mechanisms.

It can be asserted that, given the limited resolution of the terrain data and the presence of multipath scattering, the best achievable accuracy s_K is on the order of 6 dB. This is in the same range as the accuracy reported for propagation models in mobile cellular networks. Only very complex models perform better, provided that antennas can be placed with sufficient clearance ($s_R \to 0$ dB). The model by Okumura [15] for frequency planning of cellular networks [20] is reported to achieve an accuracy of about 6 to 7 dB. The accuracy of most models used for frequency management can be characterized by a standard deviation s_K of 8 to 10 dB.

2.2.4.2 Linear Regression

The linear regression R[e, X] between the residual forecast error e and a certain parameter X of the path profile is

$$R[e, X] = \frac{E[(A_m - A_k)X] - E[A_m - A_k]E[X]}{E[X^2] - E^2[X]} \tag{2.39}$$

This measure proves an effective method for generating insight into the statistical dependence of model imperfections on typical path characteristics. Not only X but many other mechanisms influence the total path loss, so the regression is performed for samples with severe statistical noise. To underscore this, the 80% confidence interval of R[e, X] was also computed. Regression has been performed with the following choices of the parameter X:

log d/1 km	The logarithm of path distance between transmitter and receiver.
log f_c/1 MHz	The logarithm of the carrier frequency.
v_e	The effective diffraction parameter of the path, defined in (2.7). The effective obstacle height h_m in (2.7) was computed according to the method by Bullington [12] (see Section 2.3.1).

Using theoretical models to predict the combined effect of predicting a number of the path parameters, the forecast error e can also be correlated with the predicted losses due to these mechanisms. This provides insight into the accuracy by which each individual propagation mechanism is represented in the model A_K and thus assists in the validation of step 3. The models considered were:

A_{mh} Diffraction loss over the main hill on the path.

A_{Dg} Theoretical multi-knife-edge diffraction loss according to the method by Deygout [13].

A_{Bu} Multi-knife-edge diffraction loss according to the method by Bullington, based on one equivalent obstacle [12].

A_R Theoretical (line-of-sight) reflection loss from a two-ray propagation model, above free-space attenuation. See (2.5), with $A_R = 0$ dB if $d < 4h_r h_t / \lambda$.

If an empirically corrected model A_N is defined with

$$A_N \triangleq A_K + R[e, X]X \qquad (2.40)$$

the new standard deviation

$$s_N^2 = E[\{A_m - A_k - R[e, X]X - E[A_m - A_k - R[e, X]X]\}^2] \qquad (2.41)$$
$$= s_K^2 - R^2[e, X] \, \mathrm{var}[X] < s_K^2$$

will be obtained for that particular set of measurements. The value of s_N is reported in the following evaluation to give some impression of the effect of improved modeling of the parameter X on the model K. It is not the author's intention to claim that the empirical model A_N is necessarily better than the theoretical model A_K.

In the following sections, a propagation model will be refined step by step, considering additional propagation mechanisms, based on insight gained through regression with the above parameters. It is our aim to find a model that combines existing theoretical models (step 3) and still reasonably agrees with the measured data, rather than to establish an optimum statistical match of the newly proposed model with the data, ignoring generally accepted theoretical models. This, however, may imply that the final model can exhibit a residual correlation with the aforementioned parameters.

2.2.4.3 Number of Samples

Theories from the combined occurrence of propagation mechanisms are scarce. As far as empirical results are reported (e.g., in [17]), the validity in a different setting is not guaranteed in advance, so verification is required. Dedicated experiments, such as the Haarlerberg experiment, can yield insight into selecting and combining the relevant mechanisms (step 3) if absence of local scattering and shadowing ensures the validity of this type of experiment considering a single path. Usually, however, a large set of samples is required to be able to draw statistical conclusions, since the statistical noise caused by local effects such as shadowing may well disguise the effect of the mechanism under study. In the analysis, it was often found that

$$R[e, X] \sqrt{\text{var}[X]} < s_s$$

That is, the variation of the received power caused by variations of X had a smaller standard deviation than the shadowing and multipath reception. This necessitates the analysis of hundreds of measurements to achieve reasonable reliability of the regression. For instance, in the case of correlation with distance, it was found from the measurements that

$$10\beta \sqrt{\text{var}[\log d]} \approx 7 \text{ dB}$$

That is, the variation of the received power caused by variations of the distance d had a standard deviation of about 7 dB. In this case, fluctuations caused by shadowing and multipath fading are in the same order of magnitude or a larger magnitude than the distance. The results reproduced in Figure 2.2 were obtained from approximately 1300 measurements.

In the analysis of *step 1* (interpretation of the profile), the problem appears even more complicated than that of linear regression. The degrees of freedom in defining effective heights and other compensation parameters would appear virtually unlimited. Implementation decisions often have a discontinuous effect on path loss. A small change in terrain profile (as changing the modeled effective diffraction height of trees) can exert a step-wise influence on the predicted diffraction loss if the implemented algorithm switches from a single knife-edge to a double knife-edge interpretation. Further, conclusions are sensitive to outliers or exceptional events. Because of such difficulties, no general method can yet be recommended from the experience gathered during the analysis. It does appear essential to take sufficient samples to smooth out discontinuities. Our measurements were taken from 55 paths, 7 antenna heights, and 4 UHF frequencies. Out of these 1540 samples about 200 samples were discarded because they had insufficient C/N. This set of data did not fully smooth out discontinuities. However, experiments revealed that the discontinuities sharply diminish with the accuracy of the implemented prediction model, even though a more detailed model generally has to take more decisions, each of which may contribute new discontinuities. For example, the Deygout MKE diffraction model consistently gave a significantly better-defined optimum than SKE methods.

2.3 EMPIRICAL RESULTS FOR DETERMINISTIC PROPAGATION MODEL

In this section, empirical results are presented for a number of propagation models. Initially, a set of 34 paths was used; in the final stage of the project, measurements over a total of 55 paths became available.

2.3.1. Modeling Single Knife-Edge (SKE) Diffraction

The dominant hill in a radio path is defined as the point along the profile with the highest value v_m of the diffraction parameter v. Model 1 is defined as free-space loss plus the knife-edge diffraction loss caused by the dominant hill, so

$$A_1 = A_{fs} + A_{mh} \qquad (2.42)$$

The effective height of trees was set to 20 m for coniferous woods and 10 m for other woods. This model appeared to give optimistic estimates of the path loss, with $m_1 \triangleq E[e] = 20.6$ dB. The standard deviation s_1 was found to be 11.7 dB. This method ignores the influence of any part of the terrain other than the dominant hill.

Table 2.3 shows that the increase of the path loss with distance is grossly underestimated and the frequency dependence is overestimated. Tables 2.3 to 2.7 and 2.9 to 2.11 present the regression $R[e, X]$ with a number of relevant path parameters X, the 80% confidence interval of $R[e, X]$, and the standard deviation of an empirically corrected model in the linear form of (2.40). Results in Tables 2.3 to 2.7 and 2.10 to 2.11 were obtained from 34 paths.

One of the propagation models suggested by Bullington includes a method in which the effect of irregular terrain, including various hills obstructing the direct line of sight, is combined into one effective obstacle [12]. The equivalent hill is found

Table 2.3
Empirical Corrections for Model 1: Diffraction Over the Dominant Hill in Free Space

$R[e, X]$	X	Confidence Interval	s_N dB
−1.3	v_e	−0.55 – 2.0	11.6
+0.45	A_{Dg}	0.38 – 0.52	11.2
+0.01	A_{mh}	−0.14 – 0.13	11.7
+0.37	A_{Bu}	0.26 – 0.47	11.5
18.8	$\log d$	16.2 – 21.4	11.1
−6.7	$\log f_c$	−8.9 – −4.5	11.6

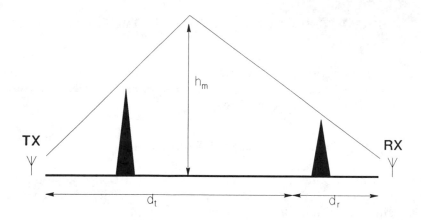

Figure 2.11 Construction of effective obstacle according to Bullington [12].

by drawing a fictitious line of sight from the transmitter antenna, touching its radio horizon and another line of sight from the receiver antenna, and touching the radio horizon of the latter. The top of the equivalent hill is found at the point where both lines cross (see Figure 2.11).

This model particularly applies to obstacles in free space (i.e., on paths where the terrain, except for a few hills, does not significantly enter the first Fresnel zone). The propagation loss is then obtained as

$$A_2 = A_{fs} + A_d(v_e) \tag{2.43}$$

No ground-reflection loss is taken into account. Table 2.4 summarizes the performance as assessed from the first set of measurements. The model accuracy is $m_2 \approx$ 19 dB and $s_2 \approx 11.5$ dB.

It can be seen that the distance dependence is still grossly underestimated and the frequency dependence overestimated. With model 2, a problem of implementation was encountered for forested terrain, because the equivalent obstacle is often only determined by woods or isolated trees in the immediate vicinity of the antennas. In such events, all other terrain irregularities on the propagation path are ignored.

2.3.2 Modeling Multiple Knife-Edge Diffraction

A significant improvement was found by applying multiple knife edges according to the method by Deygout [13]. The algorithm implemented takes account of up to three obstacles. First the dominant hill (v_{main}) is assessed. Secondary obstacles in the (fictitious) line of sight between the transmitter and the dominant hill, with diffrac-

Table 2.4
Empirical Corrections for Model 2: Diffraction Loss Over the Effective Hill in Free Space

R[e,X]	X	Confidence Interval	s_N (dB)
−0.3	v_e	−1.0 − 0.5	11.4
+0.4	A_{Dg}	0.33 − 0.47	11.1
+0.13	A_{mh}	0.00 − 0.26	11.5
+0.19	A_{Bu}	0.08 − 0.29	11.4
18.1	$\log d$	15.5 − 20.6	10.9
-6.6	$\log f_c$	−8.7 − −4.4	11.3

tion parameter v_1, and between the dominant hill and the receiver, with diffraction parameter v_2, are also taken into account. The propagation loss is then obtained as

$$A_3 = A_{fs} + A_d(v_{\mathrm{main}}) + A_d(v_1) + A_d(v_2) \qquad (2.44)$$

The technique for obtaining v_{main}, v_1, and v_2 is illustrated in Figure 2.12.

Results are in Table 2.5. The systematic error of this model is $m_3 \approx 15.3$ dB and the standard deviation $s_3 \approx 10.7$ dB. This model thus appears almost one decibel more accurate than the models considering diffraction losses over a single knife edge. The effective height of trees was empirically optimized and set to 25 m for coniferous woods and 15 m for other woods. This had a small effect on the accuracy achieved. No ground reflection loss is considered.

Model 3 seems to overestimate diffraction losses. A possible explanation is the lack of proper account of groundwave propagation. Because of ground reflections, the error is anti-correlated with frequency and positively correlated with distance.

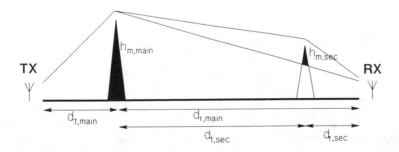

Figure 2.12 Technique for obtaining multiple knife-edge geometry according to Deygout [13].

Table 2.5

Empirical Corrections for Model 3: Deygout Triple Knife-Edge Diffraction Loss in Free Space

R[e, X]	X	Confidence Interval	s_N (dB)
−2.04	v_e	−2.72 – −1.37	10.6
−0.17	A_{Dg}	−0.24 – −0.11	10.6
−0.35	A_{mh}	−0.48 – −0.23	10.6
−0.18	A_{Bu}	−0.28 – −0.09	10.7
15.0	log d	12.6 – 17.5	10.3
−3.6	log f_c	−5.7 – −1.6	10.7

Since reflection losses are not yet taken into account and since these losses may create substantial statistical noise, the improvement in accuracy compared to SKE methods was deemed to be relevant.

2.3.3 Modeling Ground Reflection

The gross underestimate of the distance dependence of the path loss in models 1 to 3 can be explained by the ignorance concerning ground reflections. Height gain measurements confirmed the presence of the groundwave propagation mechanism: in the measurements in the North German Plain, the antenna height gain, as evaluated from measurements at h_r = 4, 7, 10, 13, 16, 19, and 21 meters, was assessed from the regression R[A_{fs}, log h_r]. Measurements over 34 paths showed a 4.5-dB/oct height gain, with an 80% confidence interval of 3.6 to 5.4 dB/oct. Perfect groundwave propagation would result in an antenna height gain of 6 dB/oct or, equivalently, 20 log h_r (see (2.6)). Possible explanations for the difference are that:

1. Effective antenna height when antenna sites are above the average ground level (e.g., on a hill) was not compensated for.
2. The average height gain could be substantially influenced by one outlier, where an obstacle was close to the receiving antenna.
3. At relative short ranges ($d < 4h_r h_t/\lambda$), the antenna height gain is not accurately approximated by a 6-dB/oct trend, since (2.6) becomes inappropriate if d is small.
4. Multipath reception caused by local reflections (e.g., from trees) might have influenced the height gain.

The height gain did not always exactly follow a smooth decibel-per-octave trend: multipath scattering produced peaks and dips with a standard deviation of 2.5 dB for f_c = 250 MHz, slowly increasing to 3.5 dB for f_c = 900 MHz. Diffraction by nearby obstacles (shadowing) only influenced the height gain if obstructions were very close to the antenna. For instance, an exponential height gain of 1 or 2 dB *per meter* was measured closely beyond Haarlerberg (Figure 2.9, location 4). Within small woods, an occasional 10 dB could be gained by placing the antenna between the trunks (at 7 to 10 m), rather than at the level of the highest leaves. Nonetheless, the general impression from the measurements was that ground reflections are the principal cause of the height gain.

For models including free-space loss and diffraction over multiple knife edges but ignoring ground reflections (such as model 3), the error e was found to contain a term of about 8 log h_r. This confirms that ground reflection and diffraction mechanisms interact.

The fourth model considers groundwave propagation and does not take any specific terrain profile into account. The model is based on theoretical results for propagation over a plane, perfectly conducting earth. An empirical term, 20 $\log(f_c/40 \text{ MHz})$, was postulated [9]. The standard deviation is $s_4 \approx 11.3$ dB for the set of 34 paths and 11.7 dB for the set of 55 paths, which closely agrees with $s_s \approx$ 12 dB, as reported in [9]. Interestingly, model 4 does not use any terrain data, but is nonetheless as accurate as models 1 and 2. Model 4 was found to overestimate path losses slightly ($m_4 \approx -2$ dB). Table 2.6 confirms that the empirical frequency correction is well within the confidence interval. The distribution of the error is in Figure 2.13. The ordinate is drawn such that any log-normal distribution of the error produces a straight line. The distribution departs somewhat from the log-normal distribution reported by Egli [9]. As Figure 2.13 illustrates, in 10% of the measurements a field strength higher than 10 dB above the area-mean is encountered, whereas 10% of the locations produces an excess attenuation of 20 dB below the area-mean field strength. This asymmetry may have been caused by multipath scattering in a number of measurements. In fact, Rician fading would produce more pronounced down-fades than up-fades.

Table 2.6
Empirical Corrections for Model 4, Proposed by Egli [9]

R[e, X]	X	Confidence Interval	s_N (dB)
4.2	v_e	3.6 – 4.8	11.3
+0.87	A_{Dg}	0.80 – 0.93	10.8
−0.06	A_R	−0.13 – 0.00	11.3
−4.3	log d	−7.0 – −0.17	11.3
−1.0	log f_c	−3.2 – 1.2	11.3

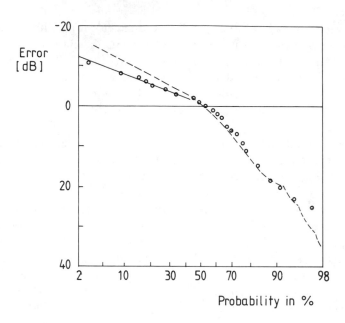

Figure 2.13 Distribution of the error (- -) of model 4, proposed by Egli, obtained from the unscreened set of measurements. Distribution of local-mean error, averaged per receiver location (o).

After averaging over all measurements at one receiver location (seven heights, four frequencies), the local-mean error was established and its distribution indicated by small circles (o) in Figure 2.13. Following the definition in Section 2.1.2, the local-mean error corresponds to large-area shadow attenuation.

A simple improvement of the model is made by distinguishing between paths over plane farmland and paths covered by vegetation. The empirical vegetation parameter Ψ ($0 < \Psi < 1$) is introduced, defined as the portion of samples along the path that contains woods or at least a number of trees. Each terrain sample describes the highest point in the terrain and the highest terrain feature (e.g., trees, in a 1- x 1-km square. For the paths considered, we found an average of 80% of all samples that contained woods of trees ($\mu_\Psi \triangleq E[\Psi] \approx 0.8$). The standard deviation was σ_Ψ $\triangleq E[(\Psi - \mu_\Psi)^2] \approx 0.12$. Empirically, the following regression was established:

$$A_5 = A_{fs} + A_R + 20 \log (f_c/40 \text{ MHz}) - 8 \log (d/20 \text{ km})$$
$$+ 30(\Psi - 0.8) - 50(\Psi - 0.8) \log (d/20 \text{ km}) \tag{2.45}$$

The standard deviation of this model was found to be $\sigma_5 \approx 11.2$ dB, which indicates some improvement over the model proposed by Egli (model 4).

The mean μ_Ψ and standard deviation σ_Ψ suggest that a typical path over a relatively open area has $\Psi \approx \mu_\Psi - \sigma_\Psi \approx 0.68$, whereas a typical woodland path has $\Psi \approx \mu_\Psi + \sigma_\Psi \approx 0.92$. Comparison of the Ψ-value with the character of the path on a detailed map generally confirmed the usefulness of the parameter Ψ. Figure 2.2 is obtained by inserting $\Psi = 0.68$ and 0.92 into (2.45). For paths over woodland ($\Psi \approx 0.92$), $\beta \approx 2.6$, and for paths over open areas ($\Psi \approx 0.62$), $\beta \approx 3.8$. This leads to the counter-intuitive conclusion that, in open areas, relatively shorter frequency reuse distances can be used, whereas in areas with more obstructions to propagation, longer reuse distance must be considered. An explanation is that, in open areas, interference power levels decay relatively rapidly because of ground reflections.

2.3.4 Modeling Combined Reflection and Diffraction Losses

The simplest way to approximate the effect of combined diffraction and reflection losses is to consider these mechanisms as independent and multiplicative (additive if decibel values are considered). Model 6 takes account of Deygout triple knife edges [13] and groundwave reflection losses, with

$$A_6 = A_{fs} + A_R + A_d(v_{main}) + A_d(v_1) + A_d(v_2) \tag{2.46}$$

This model gave, with 34 paths, a systematic error of $m_6 \approx 0.8$ dB and a standard deviation of $s_6 \approx 10.6$ dB.

It is evident from Table 2.7 that this model overestimates reflection losses. In particular, the frequency dependence of the reflection losses A_R in (2.46) impairs the accuracy. By adding the term $20 \log(f_c/40 \text{ MHz})$,

$$A_7 = A_{fs} + A_R + 20 \log(f_c/40 \text{ MHz}) + A_d(v_{main}) + A_d(v_1) + A_d(v_2) \tag{2.47}$$

Table 2.7
Empirical Corrections for Model 6

R[e, X]	X	Confidence Interval	s_N (dB)
−0.07	v_e	−0.69 – 0.54	10.6
−0.09	A_{Dg}	−0.16 – 0.02	10.6
−0.52	A_R	−0.58 – −0.46	9.8
−4.1	$\log d$	−6.6 – −1.7	10.6
+17.5	$\log f_c$	−15.6 – +19.3	9.7

the standard deviation can be reduced to $s_7 \approx 9.8$ dB. Moreover, the regression of the error with most path parameters becomes relatively small.

It was noticed that the terrain features near the antennas had a substantial effect. This was studied by splitting the paths into two sets. One set contained paths with both antennas located in open areas, the other file contained paths with at least one antenna located in an open area. For paths in the former set (wood-to-wood), the mean prediction error is $m_7 \approx -16$ dB (i.e., 16 dB too pessimistic). On the other hand, for paths in the latter set, the mean error appeared $m_7 \approx -25$ dB (i.e., 25 dB too pessimistic). This observation stimulated the introduction of local shadow losses in the model to compensate for heavier losses of antennas within woods.

2.3.5 Modeling Shadow and Scatter Losses

Scattering and shadowing in the vicinity of the antenna is an additional cause of propagation loss. Allsebrook and Parsons [53] draw attention to the clutter factor, which depends on terrain features near the antenna. The extent to which scattering and shadowing occur depends on the antenna clearance, and therefore also on the antenna height. Dedicated experiments in the North German Plain indicated that in a relatively open area or for an antenna a number of meters above the treetops, the influence of the antenna position is on the order of 2 to 3 dB. On the other hand, on wooded sites, fluctuation is substantially stronger. Occasionally, a 40-dB fluctuation was encountered within a few meters. According to [16], Rayleigh scattering may apply at UHF frequencies within woods; this could explain such large values. Practical circumstances often require that antenna clearance be much less than assumed in most theoretical propagation models for radio relay. In particular, models developed for civil radio relay were to be modified for rural communication or military point-to-point links with lower antenna positions. An antenna height of 21m is often insufficient to exceed treetop level in the North German Plain.

As discussed in Section 2.1, shadow attenuation is taken into account statistically by assuming a log-normal distribution of the local-mean power. Evidently, local terrain features affects not only the standard deviation, but also the area-mean of this distribution. Damosso and Lingua [54] proposed empirical corrections for the area-mean power. At $f_c = 450$ MHz the excess attenuation A_L was found to be on the order of [54]

9 dB	in urban areas
10 to 15 dB	in historical city centers
0 dB	in suburban areas,
-8 dB	in open areas
-10 dB	on highways

Shadow attenuation sharply increases with carrier frequency (see, for example, Figure 2.2 to 19 in [1]). For the propagation paths in the North German Plain, an empirically acceptable correction for local effects was found in the form of

$$A_{\text{loc}} = \begin{cases} c_1 \log (f_c/40 \text{ MHz}) & \text{in cluttered area (woods)} \\[2mm] \dfrac{c_1}{2} \log (f_c/40 \text{ MHz}) & \text{at the fringe of a wood} \\[2mm] 0 \text{ dB} & \text{in open area} \end{cases} \qquad (2.48)$$

At both the receiver and transmitter sites, a term in the form of (2.48) is considered. Model 8 considers

$$A_8 = A_{fs} + A_d(v_{\text{main}}) + A_d(v_1) + A_d(v_2) + c_2 A_R + A_{\text{loc},TX} + A_{\text{loc},RX} \qquad (2.49)$$

where c_1 and c_2 have been jointly optimized.

The optimum values of c_1 and c_2 were found near $c_1 \approx 6$ to 7 dB and $c_2 \approx$ 0.6 to 0.7. Figures 2.14 and 2.15 illustrate the effect of the choice of c_1 and c_2 on the standard deviation, respectively. Figures 2.14 and 2.15 have been obtained for

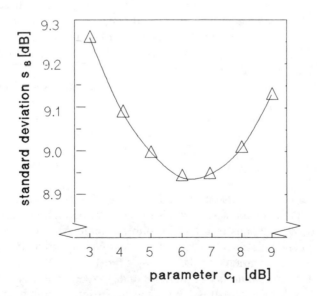

Figure 2.14 Standard deviation s_8 versus empirical parameter c_1 in decibels for shadow loss (52 paths).

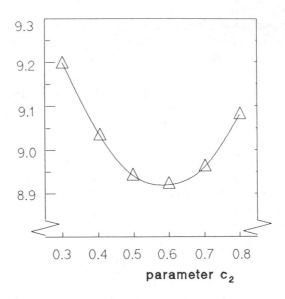

Figure 2.15 Standard deviation s_8 versus empirical parameter c_2 for ground reflections (52 paths).

a screened set of 52 paths; that is, three outliers were removed. These figures show that the effect of the choice of c_1 and c_2 on the standard deviation is limited (i.e., on the order of 0.1 dB for a relatively wide range of values). The distribution of the error e of model 8 is presented in Figure 2.16 for all samples ($\cdot$) and for path-average errors (x) in the unscreened set of 55 paths.

As illustrated by a straight line drawn in Figure 2.16, a log-normal distribution with $s_s \approx 6$ dB is appropriate for the central part of the distribution. Because of outliers, the tail of the distribution shows some deviation from the log-normal distribution, seen in Figure 2.16 as a departure from the straight line.

The standard deviation s_8 depended on the particular set of paths and the screening of the data (see Table 2.8), while the mean error appeared less sensitive to the selection of paths ($m_8 \approx -5.4$ dB). The standard deviation was about $s_8 \approx 9$ dB for the set of all measurements taken at relatively high receiver antenna heights ($h_r > 12$m), and $s_8 \approx 8$ dB for the screened set of 52 paths, which agrees with expected accuracies. For low antennas in a terrain with scattered woods, s_8 remained above 9 dB for the screened set of 52 paths. This is explained by the fact that multipath reception may cause deviations on the order of 6 dB.

To also verify the suitability of model 8 under exceptional cases, an unscreened set of measurements over 55 paths have been taken into account to compile Table 2.9. No major shortcomings are observed, though the effect of path length appears to be slightly overestimated.

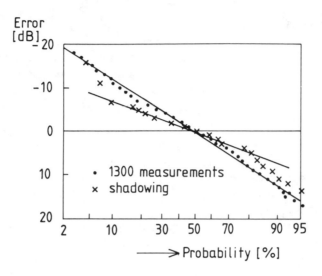

Figure 2.16 Distribution of the error (·) of model 8 obtained from the unscreened set of measurements. Distribution of local-mean error, averaged per receiver location (×).

Table 2.8
Standard Deviation s_8 for $c_1 = 6$ dB and $c_2 = 0.66$

Number of Paths	s_8 (dB)
33 (screened)	7.3
34	8.3
52 (screened)	9.0
55	9.9
52 (screened, $h_r > 12$m)	8.0
55 ($h_r > 12$m)	9.0

Table 2.9
Empirical Corrections for Model 8 With Reduced Reflection Loss and Local Shadowing (55 Paths)

E[e, X]	X	Confidence Interval	s_N (dB)
0.77	v	0.35 – 1.19	9.9
−0.15	A_{Dg}	−0.20 – −0.10	9.8
−0.11	A_R	−0.16 – −0.07	9.9
−3.6	log d	−5.3 – −2.0	9.9
0.7	log f_c	−0.8 – 2.1	9.9

2.3.6 Empirical Model by Blomquist

The model embodied in (2.49) is a linear combination of ground-reflection and diffraction losses. Blomquist [18] suggested a nonlinear function, namely,

$$A_9 = A_{fs} + \sqrt{A_d^2 + A_R^2} \qquad (2.50)$$

This empirical formula models multiple knife edges in free space if the diffraction loss is relatively high ($A_D \gg A_R$), as is the case on many wooded paths. On the other hand, for a relatively open area ($A_d \ll A_R$), groundwave propagation is recovered. As confirmed by other results (e.g., Figure 2.2 or the Haarlerberg experiments), this is in agreement with experience.

With the given set of measurements, it appeared that (2.9) is more accurate than linearly adding the three loss terms, as in (2.10) or model 6 described by formula (2.54). The model underestimates path losses by 8 dB and gives a standard deviation of $s_9 \approx 9.8$ dB (for 34 paths). Correlation of the error with A_d and A_R (see Table 2.10) did not reveal significant deficiencies of the model. Distance dependence is underestimated by 5 log d, while frequency dependence is underestimated by 7 log f_c.

It was found that adding (2.48) for local shadow losses at the transmitter and receiver sites, that is,

$$A_{10} = A_{fs} + \sqrt{A_D^2 + A_R^2} + A_{lcl,TX} + A_{lcl,RX} \qquad (2.51)$$

reduces the standard deviation of the error by almost 1 dB ($s_{10} \rightarrow 8.9$ dB, $m_{10} \approx -3.7$ dB with 34 paths). Table 2.11 shows that, nonetheless, some residual regression with A_R and A_{Dg} remains.

Table 2.10
Empirical Corrections for Model 9, by Blomquist [18] (34 Paths)

R[e, X]	X	Confidence Interval	s_N (dB)
0.58	v_e	0.01–1.15	9.7
+0.10	A_{Dg}	0.03–0.17	9.7
−0.05	A_R	−0.10–0.01	9.7
5.2	log d	2.9–7.5	9.7
7.1	log f_c	5.2–8.9	9.6

Table 2.11
Empirical Corrections for the Model by Blomquist Modified for Local Shadowing (34 Paths)

$R[e, X]$	X	Confidence Interval	s_N (dB)
0.92	v_e	0.04 – 1.43	8.8
+0.10	A_{Dg}	0.04 – 0.16	8.8
+0.14	A_R	0.09 – 0.19	8.8
3.1	$\log d$	1.0 – 5.2	8.9
−3.5	$\log f_c$	−5.2 – −1.8	8.8

2.4 CONCLUSION

The propagation effects most relevant to VHF/UHF land-mobile radio have been discussed. A statistical model for propagation attenuation and for the received signal power has been formulated for use in the following chapters in generic system design.

Field measurement of path losses in rural areas have been reported and used to evaluate the accuracy of a number of relatively simple propagation models. These idealized models take account of free-space loss, diffraction losses, ground-reflection losses, and shadowing. Linear regression of the forecast error with relevant path parameters and with losses predicted by idealized models proved an effective method for investigating the performance and possible shortcomings of propagation models.

The groundwave model proposed by Egli [9] gave the same accuracy as models considering diffraction loss over a single knife edge; that is, the standard deviation is about 11 or 12 dB. Their accuracy appeared equivalent, despite the fact that the former model does not take into account any specific terrain data, whereas knife-edge models require evaluation of the path profile. Hence, the model by Egli is much faster in terms of required computer time. For more accurate results, it appeared necessary to perform multiple knife-edge computations and carefully weigh the relative effects of groundwave and diffraction loss. The interaction of the effects of ground reflections and diffraction has been discussed by comparing models with field measurements. The empirical model proposed by Blomquist [18] was verified. A linear addition of losses may also be applied for the application studied. The measurements have also been used to verify the statistical models reviewed in the first part of this chapter. Because of ground reflections, the area-mean signal power decays faster in open areas than in areas with dense vegetation. Thus, larger separation of cochannel users in a wooded environment will be required to avoid harmful mutual interference. Moreover, because of shadowing and multipath scattering, received power levels are likely to fluctuate more in a wooded area than in an open area. This also necessitates the use of relatively large frequency reuse distances in wooded areas.

The error of relatively accurate models appeared to have a log-normal distribution. This confirms the log-normal distribution of small-area shadowing; however, for the model by Egli, the distribution of forecast error showed some deviations from the log-normal distribution.

REFERENCES

[1]   Jakes, W.C., *Microwave Mobile Communications*, New York: John Wiley and Sons, 1978.

[2]   Lee, W.C.Y., *Mobile Communications Design Fundamentals*, Indianapolis: Howard W. Sams & Co., 1986.

[3]   Parsons, D., *The Mobile Radio Propagation Channel*, New York: Halsted Press (John Wiley & Sons, Inc.), 1992.

[4]   Norton, K.A., "The Propagation of Radio Waves Over the Surface of the Earth and in the Upper Atmosphere," *Proc. of the IRE*, Part 1: Vol. 24, No. 10, Oct. 1936, pp. 1367–1387; Part 2: Vol. 25, No. 9, Sept. 1937, pp. 1203–1236.

[5]   Bullington, K., "Radio Propagation at Frequencies Above 30 Megacycles," *Proc. IRE*, Vol. 35, Oct. 1947, pp. 1122–1136.

[6]   Bullington, K., "Radio Propagation Fundamentals," *Bell Sys. Tech. Journal*, Vol. 36, May 1956, p. 593.

[7]   "Standard Definition of Terms for Radio Wave Propagation" Draft 4, January 1990, IEEE Wave Propagation Standards Subcommittee on Definitions of the IEEE Antennas and Propagation Society.

[8]   Longley, A.G., and P.L. Rice, "Prediction of Tropospheric Radio Transmission Over Irregular Terrain, a Computer Method," Environmental Science Service Administration, ESSA Tech. Rep. ERL 79-ITS 67 NTIS Acc. No. 676874, 1968.

[9]   Egli, J.J., "Radio Propagation Above 40 MC/s over Irregular Terrain," *Proc. IRE*, Oct. 1957, pp. 1383–1391.

[10]   Green, E., "Path Loss and Signal Variability Analysis for Microcells," 5th IEE International Conference on Mobile Radio and Personal Communication, Warwick, U.K., 11–14 December 1989, pp. 38–42.

[11]   Rice, P.L., *Transmission Loss Predictions for Tropospheric Communication Circuits*, Vols. I and II, National Bureau of Standards, Tech. Note 101.

[12]   Bullington, K., "Radio Propagation for Vehicular Communications," *IEEE Trans. on Veh. Tech.*, Vol. VT-26, No. 4, Nov. 1974, pp. 295–308.

[13]   Deygout, J., "Multiple Knife-Edge Diffraction of Microwaves," *IEEE Trans. on Ant. and Prop.*, Vol. AP-14, No. 4, July 1966, pp. 480–489.

[14]   Epstein, J., and D.W. Peterson, "An Experimental Study of Wave Propagation at 850 Mc," *Proceedings of the IRE*, Vol. 41, 1953, pp. 595–611.

[15]   Okumura, Y., E. Ohmori, T. Kawano, and K. Fukuda, "Field Strength and Its Variability in VHF and UHF Land Mobile Radio Service," *Review of the Elec. Comm. Lab*, Vol. 16, Nos. 9,10, pp. 825–873.

[16]   Preller, H.G., and W. Koch, "MATS-E, an Advanced 900 MHz Cellular Radio Telephone System: Description, Performance, Evaluation, and Field Measurements" *IEEE Comm. Mag.*, Vol. 24, No. 2, Feb. 1986, pp. 30–39.

[17]   Meeks, M.L., "A Propagation Experiment Combining Reflection and Diffraction," *IEEE Trans. on Ant. and Prop.*, Vol. AP-30, No. 2, March 1982, pp. 318–321.

[18]   Blomquist, A., and L. Ladell, "Prediction and Calculation of Transmission Loss in Different Types of Terrain," *AGARD Conf. Proc.*, No. 144, Electromagnetic wave propagation involving irregular surfaces and inhomogeneous media, paper 32.

[19]   Adams, B.W.P., "An Empirical Routine for Estimating Reflection Loss in Military Radio Paths in the VHF and UHF Bands," *IEE Int. Conf on Ant. and Prop.*, 28–30 Nov. 1978, IEE Conf. Publ. No. 169, p. 26.

[20]   Hata, M., "Empirical Formula for Propagation Loss in Land Mobile Radio Services," *IEEE Trans. on Veh. Tech.*, Vol. VT-29, No. 3, August 1980, pp. 317–325.

[21]   Harley, H., "Short Distance Attenuation Measurements at 900 MHz and 1.8 GHz Using Low Antenna Heights for Microcells," *IEEE J. Sel. Areas in Commun.*, Vol. JSAC-7, No. 1, 1989, pp. 5–10.

[22]   Chia, S.T.S., R. Steele, E. Green, and A. Baran, "Propagation and Bit Error Ratio Measurements for a Microcellular System," *J. Inst. of Electric and Radio Engineering*, Vol. 57, No. 6 (supplement), Nov./Dec. 1987, pp. S255–266.

[23]   Pickholtz, R.L., L.B. Milstein, D.L. Schilling, M. Kullback, D. Fishman, and W.H. Biederman, "Field Tests Designed to Demostrate Increased Spectral Efficiency for Personal Communications," *Proc. IEEE Globecom '91*, Phoenix, 1991, pp. 878–882.

[24]   Blackard, K.L., M.J. Feuerstein, T.S. Rappaport, S.Y. Seidel, and H.H. Xia, "Path Loss and Delay Spread Models as Functions of Antenna Height for Microcellular System Design," *Proc. 42nd IEEE Vehicular Technolgy Society*, Denver, Vol. 1, 10–13 May 1992, pp. 333–337.

[25]   Xia, H.H., H.L. Bertoni, L.R. Maciel, A. Lindsay-Stewart, R. Rowe, and L. Grindstaff, "Radio Propagation Measurements and Modelling for Line of Sight Micro-cellular Systems," *Proc. 42nd IEEE Vehicular Technolgy Society*, Denver, Vol. 1, 10–13 May 1992, pp. 349–354.

[26]   Goldsmith, A.J., and L.J. Greenstein, "An Empirical Model for Urban Microcells With Applicationa and Extensions," *Proc. 42nd IEEE Vehicular Technolgy Society*, Denver, Vol. 1, 10–13 May 1992, pp. 419–422.

[27]   Berg, J-E., R. Bownds, and F. Lotse, "Path Loss and Fading Models for Microcells at 900 MHz," *Proc. 42nd IEEE Vehicular Technolgy Society*, Denver, Vol. 2, 10–13 May 1992, pp. 666–671.

[28]   Fenton, L.F., "The Sum of Log-Normal Probability Distributions in Scatter Transmission Systems," *IRE Trans. Comm. Sys.*, Vol. CS-8, March 1960, pp. 57–67.

[29]   Schwartz, S.C., and Y.S. Yeh, "On the Distribution Function and Moments of Power Sums With Log-Normal Components," *Bell Sys. Tech. Journal*, Vol. 61, No. 7, Sept. 1982, pp. 1441–1462.

[30]   Marsan, M.J., G.C. Hess, and S.S. Gilbert, "Shadowing Variability in an Urban Land Mobile Environment at 900 MHz," *Electron. Lett.*, Vol. 26, No. 10, 10 May 1990, pp. 646–648.

[31]   Linnartz, J.P.M.G., and A.J.J. Meuleman, *Technical Aspects of an Improved Frequency Planning in the FM Broadcast Band* (in Dutch), Delft University Press, 1991, ISBN 90–6275–746–4.

[32]   Bultitude, R.J.C., and G.K. Bedal, "Propagation Characteristics on Microcellular Urban Mobile Radio Channels at 910 MHz," *IEEE J. Sel. Areas in Commun.*, Vol. JSAC-7, No. 1, 1989, pp. 31–39.

[33]   Vaughan, R.G., "Signals in Mobile Communications: A Review," *IEEE Trans. on Veh. Tech.*, Vol VT-35, No. 4, Nov. 1986, pp. 133–144.

[34]   Rayleigh, "On the Resultant of a Large Number of Vibrations of the Same Pitch and of Arbitrary Phase," *Phil. Mag.*, Vol. 10, Aug. 1880, pp. 73–78, and Vol. 27, June 1889, pp. 460–469.

[35]   Suzuki, H., "A Statistical Model for Urban Radio Propagation," *IEEE Trans. Comm.*, Vol. COM-25, No. 7, July 1977, pp. 673–680.

[36]   Charash, U., "Reception Through Nakagami Fading Multipath Channels With Random Delays,"
        IEEE Trans. on Comm., Vol. COM-27, No. 4, April 1979, pp. 657–670.
[37]   Nakagami, M., "The m-Distribution—A General Formula of Intensity Distribution of Rapid
        Fading," *Statistical Methods in Wave Propagation*," W. C. Hoffman, ed., Oxford: Pergamon,
        1960, pp. 3–36.
[38]   Abramowitz, M., and I.A. Stegun, eds., *Handbook of Mathematical Functions*, New York:
        Dover, 1965.
[39]   Turin, G.L., F.D. Clapp, T.L. Johnston, S.B. Fine, and L. Lavry, "A Statistical Model of
        Urban Multipath Propagation," *IEEE Trans. on Veh. Tech.*, Vol. VT-21, No. 1, Feb. 1972,
        pp. 1–9.
[40]   Cox, D.O., "Delay-Doppler Characteristics of Multipath Propagation at 910 MHz in a Suburban
        Mobile Radio Environment," *IEEE Trans. on Ant. and Prop.*, Vol. AP-20, No. 5, Sept. 1972,
        pp. 625–635.
[41]   Cox, D.O., and R.P. Leck, "Distribution of Multipath Delay Spread and Average Excess Delay
        for 910 MHz Urban Mobile Radio Path," *IEEE Trans. on Ant. and Prop.*, Vol. AP-23, No. 2,
        March 1975, pp. 206–213.
[42]   Van Rees, J., "Measurements of the Wideband Radio Channel Characteristics for Rural, Res-
        idential and Suburban Areas," *IEEE Trans. on Veh. Tech.*, Vol. VT-36, No. 1, Feb. 1987, pp.
        2–6.
[43]   Seidel, S.Y., and T.S. Rappaport, "Pathloss and Multipath Delay Statistics in Four European
        Cities for 900 MHz Cellular and Microcellular Communication," *IEE Electron. Lett.*, Vol. 26,
        No. 20, Sept. 1990, pp. 1713,1714.
[44]   Hansen, F., and F. Meno, "Mobile Fading—Rayleigh and Log-Normal Superimposed," *IEEE
        Trans. on Veh. Tech.*, Vol VT-26, No. 4, Nov. 1977, pp. 332–335.
[45]   Prasad, R., A. Kegel, and J.C. Arnbak, "Improved Assessment of Interference Limits in Cel-
        lular Radio Performance," *IEEE Trans. on Veh. Tech.*, Vol. 40, No. 2, May 1991, pp. 412–
        419.
[46]   Linnartz, J.P.M.G., "Site Diversity in Land-Mobile Cellular Telephony Network With Discon-
        tinuous Voice Transmission," *European Trans. on Telecomm. (ETT)*, Vol. 2, No. 5, Sept./Oct.
        1991, pp. 471–479.
[47]   Maseng, T., "On Selection of System Bit Rate in Mobile Multipath Channel," *IEEE Trans. on
        Veh. Tech.*, Vol. VT-36, No. 2, May 1987, pp. 51–54.
[48]   Gilbert, E.N., "Capacity of a Burst Noise Channel," *Bell Sys. Tech. Journal*, Vol. 39, 1960,
        pp. 1253–1266.
[49]   Gallager, R.G., "A Perspective on Multiaccess Channels," *IEEE Trans. Inf. Theory*, Vol. IT-
        31, No. 2, March 1985, pp. 124–141.
[50]   Carleial, A.B., and M.E. Hellman, "Bistable Behavior of ALOHA-Type Systems," *IEEE Trans.
        on Comm.*, Vol. COM-23, No. 4, April 1975, pp. 401–410.
[51]   Kleinrock, L., and S.S. Lam, "Packet Switching in a Multiaccess Broadcast Channel-Perfor-
        mance Evaluation," *IEEE Trans. on Comm.*, Vol. COM-23, No. 4, April 1975, pp. 410–423.
[52]   Peritsky, M.M., "Statistical Estimation of Mean Signal Strength in a Rayleigh Fading Envi-
        ronment," *IEEE Trans. on Comm.*, Vol. COM-21, No.11, Nov. 1973, pp. 1207–1213.
[53]   Allsebrook, K., and J.D. Parsons, "Mobile Radio Propagation in British Cities at Frequencies
        in the V.H.F. and U.H.F. Bands," *Proc. IEE*, Vol. 124, No. 2, 1977, pp. 95–102.
[54]   Damosso, E., and B. Lingua, "A Computer Prediction Technique for Land Mobile Propagation
        in VHF and UHF Bands," *Proc. Int. Comm. Conf. (ICC'83)*, Boston, 19–22 June 1983, Vol.
        1, pp. 59–63.
[55]   Eizenhöfer, A., "Application of Spread-Spectrum Technology in the Hybrid Mobile Radio Sys-
        tem MATS-D (in German)," *Frequenz*, Vol. 40, Nos. 9,10, 1986, pp. 255–259.

Chapter 3
Probability of Signal Outage

Cellular design of land-mobile radio networks meets the requirement of using the available spectrum in a highly efficient manner. Frequencies can be extensively reused by keeping spatial separations between cochannel cells as small as possible, compliant with the required system performance [1, 2]. However, determination of the minimum spacing between terminals transmitting with the same carrier frequency (i.e., the maximum acceptable interference power) is complicated by the fact that received signal power levels fluctuate because of multipath fading and shadowing. Not only the statistical behavior of any wanted signal power, but also that of all cochannel interferers, determines the performance of mobile radio nets. The *outage probability* is a useful performance measure, expressing the probability that satisfactory demodulation of the wanted signal fails because of excessive interference and noise power at the receiver input. In high-capacity cellular networks, it is interference, rather than noise, which imposes the main limits on the system performance. In this case, the outage probability is known as the *cochannel interference probability* [2] (i.e., the probability that the instantaneous *carrier-to-interference* (C/I) ratio drops below a certain margin required by the receiver design).

This chapter gives an analytical determination of outage probabilities in multiuser mobile networks with narrowband radio channels. Section 3.11 formulates the model for signal outage. The robustness of practical systems against noise and cochannel interference is discussed and a criterion for signal outage based on the C/I ratio is defined. Further, the statistical behavior of the joint interference signal is discussed. In Section 3.2, a mathematical method is presented to express the cochannel interference probability in Rician and Rayleigh-fading channels. A noise-free system is considered, and all cochannel cells are assumed to contain an active interferer. Results are found in the form of a multidimensional integral, and techniques for its numerical quadrature are discussed. This basic evaluation is generalized in Section 3.3 to include the effects of bursty interferers and noise impairments in Rayleigh-fading channels. Moreover, macrodiversity (i.e., reception of the wanted signal by cooperating receivers at multiple base station sites) is investigated. Relatively simple closed-form expressions are derived for the limiting case of large C/I

ratios. Numerical results derived from the general model are presented for the case of a cellular network with an idealized hexagonal repetition pattern.

3.1 MODEL FOR SIGNAL OUTAGES IN RADIO NETWORKS

In mobile telephony, the quality of service is often expressed in terms of the probability of outage experienced by subscribers near the boundary of the base station service area. Because of limited spectrum availability, radio networks become increasingly limited by mutual interference between users. Therefore, the outage probability is usually determined in terms of the probability that the signal-to-joint-interference ratio drops below a minimum required ratio z. If noise is also significant, an outage occurs when the signal-to-noise-plus-interference ratio $C/(I + N)$ drops below the threshold z.

The receiver parameter z appears in many studies throughout the technical literature. The parameter is called the *cochannel rejection ratio* [3], the *receiver threshold* [4], the *threshold ratio* [5], the *capture ratio* [6, 7], the *interference criterion* [8], or the *protection ratio* [9–11].

The protection ratio, however, is principally intended as a parameter in *system design* or *network planning* [9], describing the minimum required (average) C/I ratio, rather than merely a *receiver* parameter. For instance, as indicated by the CCIR in Report 358–5, fade margins may be implicitly included in the required protection ratio to allow for the effects of multipath fading or terrain irregularities [9]. Since the effect of channel fading will be addressed explicitly in the following sections, the term *protection ratio* is avoided here, and z will be considered a pure receiver parameter. Moreover, the protection ratio, as defined by the CCIR, is expressed in decibel values, while in our analyses, z will be expressed in absolute values.

The cochannel rejection ratio is defined as the maximum tolerable interference power level with respect to the power of the wanted signal; thus, the cochannel rejection ratio is z^{-1}. The terms *threshold* and *capture* were originally used in studies of the demodulation of FM broadcast transmissions, whereas the analysis is not necessarily restricted to analog exponential modulation. Nonetheless, in the light of the above observations, *receiver threshold* is considered the most appropriate name for z.

The choice of the value of z generally depends on the required quality of service and may therefore be somewhat arbitrary. The *outage criterion* can, for instance, be a certain figure of merit subjectively determined by a representative panel of listeners [12] or, in a digital sytem, an instantaneous bit error rate or digital-word erasure rate [13].

3.1.1 Receiver Threshold

In general, the receiver threshold z depends both on the required output performance (i.e., at baseband) and on the type of interference. If the characteristics of the interference resemble AWGN, the relationship between the required signal-to-noise ratio (SNR) at the detector output and z is well known. For a detailed review, we refer to text books such as [14]. For linear modulation, such as AM and SSB, and any Gaussian interference, the relationship between the SNR at the detector output and the RF C/I ratio is linear. However, in nonlinear modulation, such as phase modulation (PM) or frequency modulation (FM), the post-detection SNR can be greatly enhanced compared to baseband transmission or to linear modulation. This enhancement occurs as long as the received predetection SNR is above a certain minimum value, called the *threshold*. Typically, for FM signals, the threshold is in the range of 3 to 10 dB ($\gamma_r \approx 2$ to 10). This threshold fundamentally limits the noise immunity of various types of nonlinear modulation techniques [14]. Nonetheless, we adopt the word *threshold*, for exponential as well as for linear modulation, to describe the RF C/N or C/I ratio at which the quality of reception experienced by the user degrades below an acceptable value.

Gosling summarized typical receiver thresholds below which a voice baseband signal becomes almost unintelligible (see Table 3.1) [11].

Typical receiver thresholds for SSB, a modulation technique used in some North American mobile telephone systems, are on the order of 20 dB [12]. With exponential modulation, which is more commonly used in mobile radio nets, the receiver threshold z can be reduced but cannot be substantially lower than 5 to 10 dB. The CCIR indicates in Report 319 [12] that $z \approx 80$ (about 19 dB) for FM voice signals with an occupied bandwidth of $B_T \approx 12.5$ kHz (peak frequency deviation $f_\Delta \approx 4$ to 5 kHz) and $z \approx 7$ to 8 (8 to 9 dB) for FM voice signals with $B_T \approx 25$ to 30 kHz ($f_\Delta \approx 12$ to 15 kHz). A receiver threshold of 17 dB ($z \approx 50$) is used in the U.S. cellular radio network employing FM voice with a peak deviation of 12 kHz (and a channel spacing virtually identical to the occupied bandwidth of $B_T \approx 30$ kHz). For a C/I ratio higher than 17 dB, most subscribers experience the quality of voice reception as "good" or "excellent" [15]. An outage probability of 10% is tolerated for a Rayleigh-fading channel. The receiver threshold below which the received voice

Table 3.1
Receiver Threshold Range for Various Types of Analog Modulation [11]

Modulation Technique and Occupied Bandwidth	Receiver Threshold (dB)	
	Minimum	Maximum
FM 25 kHz	3	8
FM 12.5 kHz	6	15
SSB 5 kHz:	8	11
Heavy companding		
Moderate companding	11	15
No companding	15	20

becomes unintelligible is much lower and may be on the order of 6 to 10 dB, as suggested in Table 3.1. If, however, this outage criterion is used, tolerable outage probabilities will be significantly smaller, typically 1% or 0.1%.

For digital modulation, the receiver threshold z may be defined as the instantaneous C/I ratio at which the short-term average bit error probability P_b or the block erasure probability P_c exceeds a maximum tolerable value P_z, with P_z called the *outage criterion* [13, 16]. If interference has a Gaussian probability density and if the interference signal is independent from bit to bit, the relation between P_z and z is in the form of an error function if coherent detection is used [14]. On the other hand, for a single interferer with a peak-limited signal, such as a constant-envelope BPSK signal, the bit error rate is nearly a step function of the C/I ratio [13]. If the C/I ratio is above a certain threshold value z, error rates are negligible, whereas below this threshold, bit errors rapidly tend toward 50%. For the digital pan-European GSM system using Gaussian MSK modulation, a receiver threshold of 9.5 dB is considered to describe the performance of practical receivers appropriately [5].

3.1.2 Model for Joint Interference Signal

In order to determine the probability of outage, assessment of the statistical behavior of the cumulative interference is indispensable in multiuser systems. We distinguish between three cases of cumulation: the sum of multiple signals, each exposed to lognormal shadowing; the sum of multiple signals, each exposed to Rayleigh fading; and the sum of multiple signals, each exposed to combined shadowing and Rayleigh fading. In all three cases, the fading is assumed mutually independent for all interfering signals.

3.1.2.1 Multiple Log-Normal Signals

In the past, at least two different methods have been used to estimate the probability distribution of the joint interference power accumulated from several log-normal signals. Fenton [17] and Schwartz and Yeh [18] both proposed to approximate the pdf of the joint interference power by a log-normal pdf, yet neither could determine it exactly. The method by Fenton assesses the logarithmic mean and variance of the joint interference signal directly as a function of the logarithmic means and variances of the individual interference signals. This method is most accurate for small standard deviations of the shadow fading, say, for $s_s < 4$ dB. The technique proposed by Schwartz and Yeh is more accurate for s_s in the range of 4 to 12 dB, which corresponds to the case of land-mobile radio in the VHF and UHF bands. Their method first assesses the logarithmic mean and variance of a joint signal produced by cumulation of two signals. Recurrence is then used in the event of more than two interfering signals. A disadvantage of the latter method is that numerical computations become time-consuming. Table 3.2, compiled by Kegel in [19, 20], gives the logarithmic mean m_t in decibels ($\bar{p}_t = 10^{mt/10}$) and standard deviation s_t of the joint interference power for various numbers of contributing signals determined by Schwartz and Yeh's method.

Besides these methods by Fenton [17] and Schwartz and Yeh [18], a number of alternative (and often more simplified) techniques are used. For instance, in VHF radio broadcasting, signals fluctuate with location and time according to log-normal distributions. Techniques to compute the coverage of broadcast transmitters are in [21, 22].

Table 3.2
Mean m_t and Standard Deviation s_t (in decibels) of the Joint Power of n Signals With Uncorrelated Shadowing, Each With Mean $m_i = 0$ dB and With Identical Standard Deviation s_i [19]

n	$s_i = 0$ dB		$s_i = 6$ dB		$s_s = 8.3$ dB		$s_i = 12$ dB	
	m_t	s_t	m_t	s_t	m_t	s_t	m_t	s_t
1	0.00	0.00	0.00	6.00	0.00	8.30	0.00	12.00
2	3.00	0.00	4.58	4.58	5.61	6.49	7.45	9.58
3	4.50	0.00	6.90	3.93	8.45	5.62	11.20	8.40
4	6.00	0.00	8.43	3.54	10.29	5.08	13.62	7.66
5	7.00	0.00	9.57	3.26	11.64	4.70	15.37	7.13
6	7.50	0.00	10.48	3.04	12.69	4.41	16.74	6.74

3.1.2.2 Multiple Rayleigh-Fading Signals

Cumulation of multiple Rayleigh-fading signals requires investigation of the nature of the signals contributing to the interference. The joint interference signal $v(t)$ caused by n uncorrelated Rayleigh-fading narrowband signals with angular modulation is

$$v_t(t) = \sum_{i=1}^{n} \rho_i(t) \cos[\omega_{c_i} t + \theta_i(t) + \psi_i(t)]$$ (3.1)

Similar to (2.18), ρ_i are the random amplitudes and θ_i the random phases of the n Rayleigh-fading carriers. Angle modulation is represented by ψ_i. We now consider the signal behavior during an observation interval $\langle t_0, t_0 + T \rangle$, which is short compared to the rate of channel fading ($T \ll T_6$, see Section 2.1.7). This implies that ρ_i and θ_i are virtually constant. We distinguish between two extreme cases [7]:

1. *Coherent (or Phasor) Cumulation*: This occurs if, during the observation interval $\langle t_0, t_0 + T \rangle$, the phase fluctuations caused by the modulating signals are sufficiently small, and the carrier frequencies of the n signals are exactly equal; thus, if $\psi_i(t_0) \approx \psi_i(t_0 + \Delta t)$, $\theta_i(t_0) \approx \theta_i(t_0 + \Delta t)$ for any $\Delta t \in \langle 0, T \rangle$ and $\omega_{ci} \equiv \omega_c$. The power during interval T is found from

$$p_t = \frac{1}{T} \int_{t_0}^{t_0+T} v_t^2(t) \, dt$$

$$= \frac{1}{2} \left(\sum_{i=1}^n \rho_i \cos[\theta_i(t_0) + \psi_i(t_0)] \right)^2 + \frac{1}{2} \left(\sum_{i=1}^n \rho_i \sin[\theta_i(t_0) + \psi_i(t_0)] \right)^2 \quad (3.2)$$

The joint signal $v_t(t)$ behaves as a Rayleigh phasor, so the instantaneous power p_t is exponentially distributed. The local-mean power is equal to the sum of the local-mean power levels of the individual signals. Coherent cumulation can occur only if phase modulation with a very small deviation is applied. Moreover, if the observation interval is taken short with respect to the rate of modulation, coherent cumulation is appropriate. The latter case occurs for instances in digital systems if the joint interference signal is studied during one bit interval or during the lock-in of a carrier-recovery loop in a synchronous detector.

2. *Incoherent (or Power) Addition*: If the phases of each of the individual signals substantially fluctuate due to mutually independent modulation, the n signals add incoherently (see Figure 3.1). The interference power experienced during the observation interval is the power sum of the individual signals. Thus,

$$p_t = \frac{1}{T} \int_{t_0}^{t_0+T} v_t^2(t) \, dt = \frac{1}{T} \int_{t_0}^{t_0+T} \left(\sum_{i=1}^n \rho_i \cos[\omega_c t + \theta_i(t) + \psi_i(t)] \right)^2 dt$$

$$= \frac{1}{2} \sum_{i=1}^n \rho_i^2 + \int_{t_0}^{t_0+T} \text{crossterms} \, dt = \sum_{i=1}^n p_i \quad (3.3)$$

because incoherent cumulation ensures that the integrals over the crossterms vanish. With coherent addition, the joint interference signal may exhibit deep fades, caused by mutual canceling of phasors from the n signals. This cannot continue for a sustained period, because of the phase variations caused by angle modulation of each signal or by slightly different carrier frequencies due to Doppler shifts and free-running oscillators. Hence, in certain situations, the assumption of coherent interference cumulation may give optimistic results compared to incoherent cumulation [7, 20].

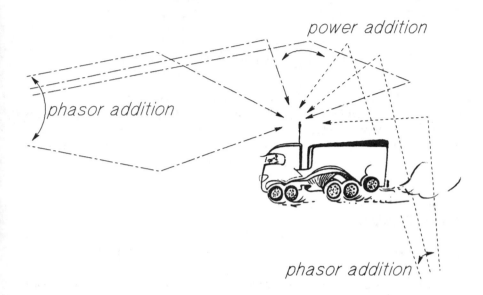

power addition

phasor addition

phasor addition

Figure 3.1 Model for incoherent cumulation of multiple interfering (narrowband) Rayleigh-fading signals.

With incoherent cumulation, the joint interference signal behaves as a band-limited Gaussian noise source if the number of components is sufficiently large. Moreover, any fade of one of the signals is likely to be hidden by the $n - 1$ other interfering signals. Hence, the joint interference signal tends to exhibit less multipath fluctuations per unit of time than the signal from one individual interferer. The special case of the power sum of a number of incoherent Rayleigh-fading signals with identical local-mean power levels will be discussed in the next section.

3.1.2.3 Multiple Incoherent Rayleigh-Fading Signals With Equal Mean Power

If the interference is caused by the power sum of n Rayleigh-fading signals, with identical local-mean power $\bar{p}$, the pdf of the joint interference power is the nth convolution of the exponential distribution (2.21) of the power of an individual interfering signal. Using the Laplace transform pair

$$f_{p_i}(p_i | \bar{p}_i = \bar{p}) = \frac{1}{\bar{p}} \exp\left(\frac{p_i}{\bar{p}}\right) \leftrightarrow \frac{1}{1 + s\bar{p}} \tag{3.4}$$

for $i = 1, 2, \ldots, n$, the pdf of the joint interference power is found [7, 23, 24]:

$$f_{p_t}(p_t|n) = \frac{p_t^{n-1}}{(n-1)! \, \bar{p}^n} \exp\left(\frac{p_t}{\bar{p}}\right) \leftrightarrow \left(\frac{1}{1+s\bar{p}}\right)^n \qquad (3.5)$$

That is, the joint interference power p_t has a gamma distribution (see Section 2.1.3.3). In Chapter 4, we will show that the pdf of the joint interference power caused by n interfering signals with different local-mean power levels can also be approximated by a gamma distribution. Even an approximation by a stationary (nonfading) band-limited Gaussian signal appears reasonable in certain analyses.

3.1.2.4 Multiple Incoherent Signals With Combined Shadowing and Rayleigh Fading

The early studies of signal outages and spectrum efficiency of cellular radio networks by Gosling [25] addressed channels with pure Rayleigh fading. As discussed in Section 2.1.2, neglecting shadowing is acceptable if the local-mean power levels of the signals are sufficiently known, for instance, from computation of path losses from a high-resolution terrain database. Moreover, on the downlink, the individual shadow attenuation of different arriving signals may be highly correlated. In these cases, shadowing has a limited effect on the fluctuations of the C/I ratio. French [26] studied the effect of combined Rayleigh fading and shadowing in the special case of a single interfering signal. The case of multiple interfering signals was assessed by Cox using Monte Carlo simulation [27]. Later, a number of mathematical investigations of the effect of multiple interfering signals with Rayleigh fading and shadowing were conducted [8, 10, 20, 28–30], though approximation techniques were used.

In the event of *coherent* cumulation of fading signals, the amplitude of the joint interference signal has a Rayleigh distribution. The local-mean power of the Rayleigh distributed envelope equals the sum of the power of the individual signals. If shadowing also occurs, the pdf of the local-mean power of the joint interference signal can be assessed by Fenton's [17] or Schwartz and Yeh's technique [18]. Muammar and Gupta [10] and Diakoku and Ohdate [28] extended the method by Fenton to evaluate cellular mobile radio systems with shadowing. The method by Schwartz and Yeh was extended in [29] and [30]. However, since the instantaneous interference power was assumed to be exponentially distributed about the local-mean power, these extensions were appropriate to coherent, rather than the more realistic incoherent, cumulation of the instantaneous interference power.

For *incoherent* cumulation of multiple Rayleigh fading signals, the distribution of the instantaneous joint interference power depends on the local-mean power of each contributing signal. This will be discussed further in Chapter 4. The joint interference signal cannot be characterized by the sum of the local-mean power levels only, because the variance of the joint signal power depends not only on the number of contributing signals but also on the relative power levels of the contribution signals [31]. This is in sharp contrast to the situation of coherent cumulation and substantially complicates the analysis. To evaluate incoherent cumulation, Prasad and Arnbak [32] proposed to approximate the Suzuki pdf of the instantaneous power of each individual interfering signal (2.29) with a log-normal pdf with area-mean power $\bar{\bar{p}}_{rs}$ and a logarithmic standard deviation s_{rs}. It was stated in [20, 32] that the first- and second-order *linear* moments of the approximate log-normal pdf and the exact Suzuki distribution match if one takes the logarithmic variance

$$\sigma_{rs}^2 = \sigma_s^2 + \ln 2 \tag{3.6}$$

and the logarithmic area mean

$$\bar{\bar{p}}_{rs} = \frac{1}{2}\sqrt{2}\,\bar{\bar{p}}_0 \tag{3.7}$$

Here, the area-mean is the absolute value corresponding to the logarithmic area-mean power $\bar{\bar{p}}_{rs} = \exp(m_{rs})$, with

$$m_{rs} = 10 \log_{10} \frac{\bar{\bar{p}}_0}{1\ \text{mW}} - 5 \log_{10} 2 \tag{3.8}$$

This suggests that Rayleigh fading reduces the logarithmic mean of the signal by about 1.5 dB and gives a standard deviation of $\sigma = \sqrt{\ln 2}$ ($s = 4.34\sqrt{\ln 2} \approx 3.6$ dB). In [20, 30], Schwartz and Yeh's method is applied for incoherent cumulation of a number of signals with a log-normal distribution with mean (3.7) and variance (3.6). For the wanted signal, the exact Suzuki distribution is considered in [20].

3.2 ANALYSIS OF COCHANNEL INTERFERENCE PROBABILITY

The instantaneous power received at receiver A from the ith mobile terminal is denoted by p_{A_i}. (This minor extension of the notation p_i will allow later extension of the analysis to networks with diversity reception.) Similarly, the local-mean power and the area-mean power are denoted by $\bar{p}_{A_i}$ and $\bar{\bar{p}}_{A_i}$, respectively. Throughout this analysis, the location of each mobile terminal and the corresponding area-mean power are assumed to be known, whereas the local-mean power and the instantaneous received power are considered to be stochastic variables. The signal from terminal 0 is taken to be the wanted signal.

The event of successful reception at a certain instant of the wanted signal at base station A is denoted by A_0, with

$$\Pr(A_0) \triangleq \Pr\left(\frac{p_{A_0}}{p_t} > z\right) = \int_{0-}^{\infty} f_{p_t}(x) \int_{zx}^{\infty} f_{p_{A_0}}(y)dydx \qquad (3.9)$$

Here, $f_{p_t}(\cdot)$ is the pdf of the instantaneous power p_t of the joint interference from n cochannel interfering signals. Defining $K(zx)$ as

$$K(xz) \triangleq \int_{xz}^{\infty} f_{p_{A_0}}(\lambda)d\lambda \qquad (3.10)$$

we may interpret expression (3.9) as an integral transform of the pdf of joint interference power. The kernel $K(zx)$ is the cumulative distribution (cdf) (3.10) of the power of the wanted signal. For microcellular networks with a Rician-fading wanted signal, the cdf of received power can be expressed as a Marcum Q-function [33]. For channels with Rayleigh fading, the cdf of received power is an exponential distribution and (3.9) becomes a Laplace transform. This will be worked out in more detail in the following sections.

If the joint interference signal shows significantly less severe fading than the wanted signal (e.g., because of incoherent cumulation), the local-mean outage probability can be approximated by [34]

$$\Pr(A_0) \approx K(z\bar{p}_t) \qquad (3.11)$$

3.2.1 Analysis for Rician-Fading Channel

In a microcellular network, the wanted signal is likely to be received with a dominant line-of-sight component. The pdf of the received signal amplitude is given by the Rician distribution (2.20), or, equivalently, the pdf (2.21) for received power. Interfering signals, however, arrive from more distant transmitters over obstructed, Rayleigh-fading propagation paths [35]. If shadowing is negligible, the joint interference power thus has the gamma distribution (3.5) [23, 24]. From the Rician pdf (2.19) and (3.10), it follows that

$$K(zx) = \int_{\sqrt{2zx}}^{\infty} f_{\rho_0}(\rho)d\rho = Q\left(\sqrt{2K}, \sqrt{\frac{(1+K)2zx}{\bar{p}_{A_0}}}\right) \qquad (3.12)$$

where $Q(a,b)$ is the Marcum Q-function [33], defined as

$$Q(a,b) \triangleq \int_{b}^{\infty} t \exp\left\{-\frac{a^2 + t^2}{2}\right\} I_0(at)\, dt \qquad (3.13)$$

In the special case of $b = 0$, $Q(a,0) = 1$. The probability (3.10) of successful reception becomes, after substituting $x = \lambda^2$,

$$\Pr(A_0|\bar{p}_{A_0}, \bar{p}, K, n) = \int_0^\infty \frac{2\lambda^{2n-1}}{(n-1)!\bar{p}^n} e^{-\lambda^2/\bar{p}} Q\left(\sqrt{2K}, \sqrt{\frac{2z(1+K)}{\bar{p}_{A_0}}}\lambda\right) d\lambda \quad (3.14)$$

This integral is solved in Appendix D and in [23] in terms of a sum of Laguerre polynomials. The local-mean power $\bar{p}$ of each interfering signal is related to the reuse distance using the path loss law (2.13b). Using approximation (3.11), we find [34]

$$\Pr(A_0|\bar{p}_{A_0}, \bar{p}, K, n) \approx Q\left(\sqrt{2K}, \sqrt{2(1+K)\frac{zn\bar{p}}{\bar{p}_{A_0}}}\right) \quad (3.15)$$

For interfering signals with strong shadowing ($\sigma_s \gg 6$ dB), (3.14) and (3.15) may become inaccurate. A log-normal distribution may be inserted for the joint interference in probability (3.9), rather than the gamma distribution [23].

3.2.2 Numerical Results for Microcellular Networks

In this section, an idealized cellular network with a hexagonal cell layout (honeycomb structure), as discussed in Chapter 1, is considered. In a cellular network, signal outages are caused mainly by the interference from the active users in the n nearest cochannel cells. Typically, $n = 6$ for idealized hexagonal cell layouts with base stations using omnidirectional antennas. In microcellular networks, the number of relevant interferers may be smaller if directional antennas are used, directed at a limited sector of each cell. Only interference from the nearest ring of cochannel cells is considered; interference from more distant users in second- and higher-order tiers of cochannel cells can usually be neglected, provided that β sufficiently exceeds 2.

Although an interfering user can be at any place inside his cell, it is assumed that all harmful interferers are at the reuse distance r_u ($r_i \equiv r_u$ for $i = 1, 2, \ldots, n$), where r_u is the distance between the centers of two cochannel cells. If $r_u \gg 1$, this approximation is reasonable. As a worst case, it can be considered that interfering terminals are near the fringe of their cells where they are closest to the cell of the terminal transmitting the wanted signal. In this case, $a_i > r_u - 1$ for $i = 1, 2, \ldots,$ n. Here, as in most papers on outage probabilities, we adopt the former assumption ($r_i \approx r_u$) for all first-order interferers. The affected user is considered to be at the cell boundary ($r_0 = 1$), which represents a worst case.

Figure 3.2 gives the outage probability versus the frequency reuse distance for a Rician-fading signal in the presence of $n = 6$ Rayleigh-fading interfering signals. The turnover distance of the path loss law (2.13b) is taken 0.67 times the cell radius

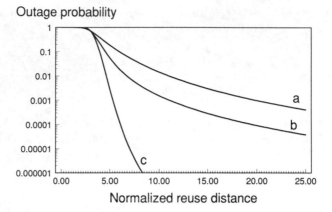

Outage probability

Normalized reuse distance

Figure 3.2 Outage probability versus normalized reuse distance r_u for $n = 6$ cochannel interferers in a microcellular net. Rician K-factor of wanted signal: (a) $K = 0$; (b) $K = 4$; (c) $K = 16$. Receiver threshold $z = 10$ (10 dB). Path loss turnover point $r_g = 0.67$.

$(r_g = 0.67\, r_0)$, with $\beta_1 = 2$ and $\beta_2 = 2$. Hence, the interference power caused outside the cell decreases approximately with r_u^{-4}. The Rician K-factor is 0 (Rayleigh fading), 4 (6 dB), 16 (12 dB). The receiver threshold is taken 10 dB ($z = 10$), which corresponds to the specifications of the digital GSM system [5]. Moreover, analog FM has a threshold of about 10 dB [13] (see also Table 3.1).

3.2.3 Exact Analysis for Rayleigh-Fading Channel[1]

In a Rayleigh-fading channel, the conditional pdf $f_{p_{A_0}}$ of the instantaneous power of the wanted signal is the exponential distribution (2.22). Hence, assuming the local-mean power of the wanted signal to be known [4, 36, 37],

$$\Pr(A_0|\bar{p}_{A_0}) = \int_{0-}^{\infty} \exp\left(-\frac{xz}{\bar{p}_{A_0}}\right) f_{p_t}(x)dx \triangleq \mathcal{L}\left\{f_{p_t}, \frac{z}{\bar{p}_{A_0}}\right\} \tag{3.16}$$

where $\mathcal{L}\{f, s\}$ denotes the one-sided Laplace transform (2.31) of the function f at the point s [38]. This observation will play a key role in the assessment of outage probabilities in narrowband macrocellular channels. One minus this conditional probability gives the *local-mean* outage probability.

If the joint interference power p_t is due to incoherent accumulation of n independently fading signals, its pdf is the n-fold convolution of the pdf of each individual signal power levels. Laplace transformation results in the multiplication of

---

[1]Section 3.2.3 and 3.2.4: Portions reprinted, with permission, from *IEEE Transactions in Communications*, Vol. 40, No. 1, January 1992, pp. 20–23. © 1992 IEEE.

n factors, each containing a Laplace image of the pdf of the received power from an individual signal of the form (3.10), namely,

$$\Pr(A_0|n,\bar{p}_{A_0}) = \mathcal{L}\left\{f_{p_t}, \frac{z}{\bar{p}_{A_0}}\right\} = \mathcal{L}\left\{f_{p_{A_1}} \otimes \cdots \otimes f_{p_{A_n}}, \frac{z}{\bar{p}_{A_0}}\right\} = \prod_{i=1}^{n} \mathcal{L}\left\{f_{p_{A_i}}, \frac{z}{\bar{p}_{A_0}}\right\} \quad (3.17)$$

where $\otimes$ denotes the convolution of two pdfs. This expression conveniently allows extension of the interference model by including extra multiplicative factors to account for the effect of a receiver noise floor, manmade burst noise, and, in a wideband channel, multipath self-interference caused by reflected waves arring with large delay.

For a noise-free narrowband channel, the local-mean interference probability $\bar{O}$ is found as [11, 26]

$$\bar{O} = 1 - \prod_{i=1}^{n} \frac{\bar{p}_{A_0}}{\bar{p}_{A_0} + z\bar{p}_{A_i}} \quad (3.18)$$

In order to calculate *area-mean* outage probabilities, we apply the Laplace image $\phi_\sigma(\cdot)$ of the Suzuki pdf. The image of the instantaneous power received from the *i*th interferer with area-mean power $\bar{\bar{p}}_{A_i}$ was denoted in (2.34) by

$$\mathcal{L}\left\{f_{p_{A_i}}, s\right\} = \phi_\sigma(s\bar{\bar{p}}_{A_i})$$

In the event that the *n* interferers transmit continuously, this image can be inserted in (3.17). The effects of intermittent transmission to avoid unnecessary interference caused during speech pauses will be discussed in Section 3.3 by appropriately adapting the expression for the Laplace image. Analogously, data networks with terminals randomly transmitting messages from unknown positions in the cell will be addressed in Chapters 7, 8, and 9.

By averaging (3.17) over the shadowing experienced by the desired signal, the probability becomes

$$\Pr(A_0|n) = \int_0^\infty f_{p_{A_0}}(\bar{p}_{A_0}|\bar{\bar{p}}_{A_0}) \prod_{i=1}^{n} \mathcal{L}\left\{f_{p_t}, \frac{z}{\bar{p}_{A_0}}\right\} d\bar{p}_{A_0}$$

$$= \int_0^\infty \frac{1}{\sqrt{2\pi}\,\sigma_s\bar{p}_{A_0}} \exp\left[-\frac{\ln^2(\bar{p}_{A_0}/\bar{\bar{p}}_{A_0})}{2\sigma_s^2}\right] \prod_{i=1}^{n} \phi_\sigma\left(z\frac{\bar{\bar{p}}_{A_i}}{\bar{p}_{A_0}}\right) d\bar{p}_{A_0} \quad (3.19)$$

Using the logarithmic local-mean power levels as integration variables (defined similarly to (2.33)), the probability is written in full as

$$\Pr(A_0|n) = \int_{-\infty}^{\infty} dy_0 \frac{\exp(-y_0^2)}{\sqrt{\pi}}$$

$$\cdot \left[\prod_{i=1}^{n} \frac{1}{\sqrt{\pi}} \int_{-\infty}^{\infty} \frac{\bar{\bar{p}}_0 \exp(\sqrt{2}y_0\sigma_s)}{\bar{\bar{p}}_0 \exp(\sqrt{2}y_0\sigma_s) + z\bar{\bar{p}}_i \exp(\sqrt{2}y_i\sigma_s)} \exp(-y_i^2)\, dy_i \right]$$

(3.20)

This expression gives the exact solution to the stated problem of the area-mean outage probability $\bar{O}$ in a macrocellular network with continuously transmitting interferers, since $\bar{O} = 1 - \Pr(A_0)$. For channels without shadowing effects ($\sigma_s = 0$, $\bar{\bar{p}}_i = \bar{p}_i = r_i^{-\beta}$), the probability (3.20) reduces to (3.17).

3.2.4 Numerical Results for Macrocellular Networks

To obtain numerical results in the presence of shadowing for $\sigma_s > 0$, the Hermite quadrature method has been applied (Appendix E). Figure 3.3 presents the resulting

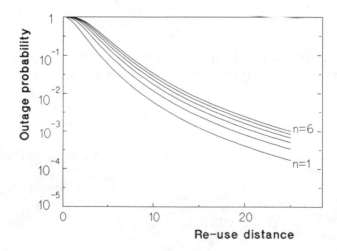

Figure 3.3 Outage probability versus normalized reuse distance r_u for 1, 2, ..., 6 cochannel interferers. Shadowing with $s_s = 6$ dB, receiver threshold $z = 10$ (10 dB). UHF groundwave propagation with $\beta = 4$.

outage probability versus the normalized reuse distance r_u. Shadowing with an intensity of $s_s = 6$ dB ($\sigma_s = 1.38$) is assumed. UHF groundwave propagation is considered with $\beta = 4$. For networks operating in irregular or urban terrain with $\beta \approx 3.3$, outage probabilities tend to be slightly larger.

Exact results can be found in closed form at particular points of the curve for a single interferer ($n = 1$): after substituting $y \triangleq y_0$ and $t \triangleq (1/2)\sqrt{2}(y_1 - y_0)$, (3.20) becomes

$$\Pr(A_0|n = 1) = \frac{1}{\pi} \int_{-\infty}^{\infty} \int_{-\infty}^{\infty} \frac{\exp(-2y^2 - 2\sqrt{2}ty - 2t^2)}{1 + zr_u^{-\beta} \exp(2\sigma_s t)} \, dy dt$$

$$= \frac{1}{\sqrt{\pi}} \int_{-\infty}^{\infty} \frac{e^{-t^2} dt}{1 + zr_u^{-\beta} \exp(2\sigma_s t)} = \phi_{\sqrt{2}\sigma}(zr_u^{-\beta}) \qquad (3.21)$$

Hence, for the special case $r_u = \sqrt[\beta]{z \exp(2k\sigma_s^2)}$, the outage probability $\bar{O}(r_u) \triangleq 1 - \Pr(A_0)$ can be found in closed form from expression (C.3) in Appendix C. For 6 dB of shadowing ($\sigma_s = 1.36$), $z = 10$ and $\beta = 4$, the following outage probabilities are found

Reuse distance r_u	Outage probability $\bar{O}(r_u)$
$k = 1$: $\quad r_u = \sqrt[\beta]{z \exp(2\sigma_s^2)} \approx 4.48$	$\dfrac{1}{2}\exp(-\sigma_s^2) \approx 0.07$
$k = 2$: $\quad r_u = \sqrt[\beta]{z \exp(4\sigma_s^2)} \approx 11.3$	$\exp(-3\sigma_s^2) - \dfrac{1}{2}\exp(-4\sigma_s^2) \approx 3.8 \cdot 10^{-3}$

The second example ($k = 2$) also closely agrees with the approximate expressions for large C/I ratios, which will be discussed further in Section 3.4.5.

3.2.5 Design of Real-World Networks

In the planning of a practical cellular voice network, its time-varying performance as experienced by the human user with a characteristic observation window T must satisfy minimum requirements. Presumably, the maximum tolerable outage probability is best specified for an observation window of at most a few seconds. This corresponds to a *small-area* mean outage probability $\bar{O}$, determined from more accurate large-scale propagation models than the simple path loss law (2.13). The effects of hills and other large obstacles are then accounted for in the large-scale path loss model, and $\bar{O}$ will vary from area to area.

In contrast to this, the *large-area* mean outage probability $\bar{\bar{O}}$, obtained from (3.20) with $s_s = 12$ dB and $\bar{p}_i \equiv r_i^{-\beta}$, thus averaged over the entire cell boundary, may not be very appropriate. The observation window may be too large to reasonably reflect the subjective performance experienced by a user. To ensure reliable communication on all parts of the cell boundary, $\bar{\bar{O}}$, is to be kept significantly smaller than the minimum required (small-area) outage probability. Network layouts based on a specified $\bar{\bar{O}}$, thus require a conservative approach: in most outer parts of the cell, the performance will be substantially better than the minimum required performance.

If small-area mean outage probabilities are computed for each small area on the cell boundary, this allows the design of cellular layouts customized for the particular propagation environment. This may significantly improve spectrum efficiency compared to the case of theoretical network planning without terrain data. However, a general, objective comparison between theoretical and customized cell planning cannot easily be given, because the spectrum efficiency depends a great deal on the creativity with which the practical cell layout is designed.

3.3 EXTENDED MODEL OF THE NETWORK

In this section, the analysis of Section 3.2 is now generalized to take into account:

1. Site diversity;
2. Discontinuous voice transmission;
3. A receiver noise floor.

First, the concept of diversity is introduced and a general discussion of discontinuous voice transmission in cellular telephone networks is given. The limiting effect of a receiver noise floor is discussed. Based on this, the analysis in the previous section is extended. The case of two cooperating receivers at different sites is addressed with this generalized model. The technique for numerical evaluation is discussed. Computational results are presented, both for the special case of channels without shadowing and for channels with shadowing. Approximate expressions for high average carrier-to-interference ratios are presented.

3.3.1 The Concept of Diversity

Major improvements in the quality of reception can be achieved if one exploits the fact that Rayleigh fading of the signal received at two antennas, sufficiently spaced apart, is uncorrelated (see (2.27)). The base station either selects the antenna that gives the best performance or appropriately processes and combines the signals from both antennas to enhance the quality of reception. This local technique is known as *antenna* (or *micro*) diversity [39–42]. The use of diversity reception allows higher local interference power levels for a certain required system performance. Hence, higher spectrum efficiency can be achieved. Using *site* (or *macro*) diversity [40, 41, 43, 44], signals from more base stations are combined to improve the signal received from a mobile terminal. This is in contrast to most existing cellular telephony nets, where a formal handover to a base station other than the primarily assigned one is required before it can participate in handling signals from a certain terminal. Macrodiversity combats outages more effectively than microdiversity, since not only Rayleigh fading but also shadowing and path loss can be uncorrelated at different receiver sites. In particular, site diversity is helpful for terminals near the boundary of a cell. If, in a conventional system, a terminal moves outside the range of one base station, a handover to an adjacent base station is performed. The optimum instant of handover is difficult to determine, because the mobile radio channel is of a stochastic nature [45]. A terminal moving near the boundary of two adjacent cells may experience degraded channel performance if the handover instant is chosen at an unfortunate moment. Site diversity can offer a "soft" handover for a terminal moving from one cell into an adjacent cell. Site diversity becomes of particular interest in future personal cordless telephone networks, since indoor wave propagation suffers from severe log-normal signal fluctuations caused by blocking or shadowing by walls [43].

The advantages of site diversity are obtained at the cost of increased complexity of the system hardware and of the access protocols. Signals arriving at different locations have to be compared and processed. This also requires an increased number of links in the fixed interconnection network. Moreover, if active terminals move over larger distances in a network with site diversity, determination of the involved group of tracking base stations is required. This may be more complicated than a straightforward handover in conventional systems.

In the following, it is assumed that each mobile terminal transmits a digitized voice channel in the form of frames (or segments) of bits with error detection coding. Whenever a base station receives an error-free voice segment, it is forwarded to the *mobile switch center* (MSC) and immediately scheduled on the appropriate telephone circuit. If a segment is received correctly at more than one base station simultaneously, the MSC may ignore all but one copy of the segment. Thus, a form of dynamic routing of subsequent voice segments occurs at this stage, violating the concept of circuit switching in the strict sense. However, we will consider a system that accepts an incidental loss of a voice segment due to an excessive burst of interference on the radio link (*loss system*). This is a major difference from packet radio networks where data packets lost on the radio channel are retransmitted (*delay system*), as, for example, are studied in Chapters 6 to 9. Communication between mobiles, base stations, and the MSC is assumed instantaneous. As a result, no re-sequencing of voice segments can occur. We do not address the design of the wired trunk network and the supporting protocols. The selection of one of the receiver outputs at the MSC is relatively easy in a digital system with error detection coding: a voice segment will be dropped if the number of bit errors exceeds a certain maximum of correctable errors. For ease of analysis, we do not evaluate exact block error rates, but shall assume a voice segment to be received correctly if its instantaneous power in at least one base station exceeds the joint interference power by at least a factor z. When the voice segment fails to capture any of the base stations, the user experiences a short outage of the telephone circuit. During the reception of each voice segment, the received power is assumed constant for all signals. A lower word erasure or undetected error rate could be achieved by combining the a posteriori information on detected bit sequences at the various receivers. This possibility is not addressed here.

This section derives generalized analytical results for outages in a telephony network with discontinuous voice transmission. The probability of loss of a voice segment is expressed for a mobile user on an end-to-end circuit basis. The analysis is limited to the uplink (i.e., the inbound channel from mobile terminals to the fixed *public switched telephone network* (PSTN)); multiplexing of signals in the downlink is not addressed here. The analysis includes interference and noise limitations.

3.3.2 Discontinuous Voice Transmission

In the basic analysis, it was assumed that in all first-order cochannel cells, interferers are active. This assumption is pessimistic for various reasons. First, if in a trunking network all available frequencies are in use, the blocking probability for newly arriving calls is unity. This is evidently unacceptable to the terminal. Secondly, if mobile terminals in cochannel cells happen to be close to the base station, the transmitting power levels may be reduced to minimize the interference caused to other terminals. Thirdly, the transmitter may be switched off during speech pauses, which also reduces the mutual interference [46]. If one investigates the link quality experienced, an observation interval has to be chosen appropriately. If small-area-mean outage probabilities are observed during a period longer than a few seconds but substantially shorter than the duration of an average telephone call, the channel occupancy and the required transmitting power levels are likely to remain fixed (unless slow frequency hopping is applied), but the channel fading, including multipath fading and shadowing, and the on/off pattern of the voice are stochastic variables. We now address the worst-case situation that each relevant cochannel cell contains a terminal at maximum power, which is only switched off during speech pauses of the voice.

Each mobile terminal i ($i = 0, 1, \ldots, N$, with N the number of interfering terminals) is assumed to transmit digitized voice segments (or *packets*) of fixed duration equal to the unit of time. The uplink channel is slotted; the time slots are perfectly synchronized throughout the entire network and known to all mobile and base stations. During a talkspurt of the ith mobile terminal (event i_{ON}), voice segments are generated and transmitted in each successive time slot. No voice segments are generated during gaps between talkspurts; the transmitter is switched off during these periods. This technique is known as *discontinuous voice transmission* (DTX) or *voice-excited transmission* (VOX) [46]. The probability of activity of terminal i during a certain time slot is $\Pr(i_{ON})$; the event of not transmitting is denoted by i_{OFF}, with $\Pr(i_{OFF}) = 1 - \Pr(i_{ON})$. The activity of terminals is assumed to be mutually independent. A typical value reported for speech activity is $\Pr(i_{ON}) \approx 0.43$ for a coding strategy using a slow voice on-off detector [47, 48]. For a fast coding strategy, detecting also short off periods (minigaps) within a principal speech burst, the activity is $\Pr(i_{ON}) \approx 0.36$.

If propagation delays can be neglected, the probability that a segment from terminal i causes interference to the wanted signal equals $\Pr(i_{ON})$. Taking into account periods of inactivity, the pdf of the instantaneous interference power at the receiver is

$$
f_{p_{A_i}}(x) = \Pr(i_{OFF})\delta(x)
$$

$$
+ \Pr(i_{ON}) \int_0^{\infty} \frac{1}{\bar{p}_{A_i}} \exp\left(-\frac{x}{\bar{p}_{A_i}}\right) \frac{1}{\sqrt{2\pi}\,\sigma \bar{p}_{A_i}} \exp\left(-\frac{1}{2\sigma^2} \ln^2\frac{\bar{p}_{A_i}}{\bar{\bar{p}}_{A_i}}\right) d\bar{p}_{A_i} \quad (3.22)
$$

with the Laplace transform

$$
\mathcal{L}\{f_{p_{A_i}}, s\} = \Pr(i_{OFF}) + \Pr(i_{ON})\phi_\sigma(s\bar{\bar{p}}_{A_i}) \quad (3.23)
$$

For ease of notation, an image function

$$
\phi(A, i) \triangleq \Pr\left(p_{A_0} < z p_{A_i} \middle| \bar{p}_{A_0}, \bar{\bar{p}}_{A_i}, i_{ON}\right) = 1 - \phi_\sigma\left(z\frac{\bar{\bar{p}}_{A_i}}{\bar{p}_{A_0}}\right) \quad (3.24)
$$

is introduced. This is a function of the *area-mean* power received from the interfering mobile i and the *local-mean* power received from the reference terminal 0, at base station A. It also depends on the standard deviation of the shadowing σ_s and on the receiver threshold z.

3.3.3 Receiver Noise Floor

For a receiver with RF-bandwidth B_T, the total received noise power is $N_A = N_0 B_T$, if N_0 is the one-sided spectral power density of the AWGN at the receiver input. Bandpass Gaussian noise is known to have a Rayleigh-distributed envelope, so the momentary noise power p_n (observed during $T \ll T_3 \ll T_6$, see Section 2.1.7) is exponentially distributed with pdf

$$f_{p_n}(x) = \frac{1}{N_A} \exp\left(-\frac{x}{N_A}\right) \tag{3.25}$$

and Laplace image

$$\mathcal{L}\{f_{p_n}, s\} = \frac{1}{1 + s N_A} \tag{3.26}$$

The momentary noise power may be considered constant only for very short time intervals, since the noise bandwidth is B_T. Signal impairments determined from (3.26), therefore, also tend to report the effect of short noise spikes rather than purely expressing harmful fading of the desired signal below the constant receiver noise floor. This is in contrast to the basic idea of an outage of the baseband signal during some period of time T (e.g., with T the duration of a voice packet). If we consider an observation time with $T \gg T_3$, the noise is wideband compared to the observation window. A more appropriate description is *constant* noise power, and therefore with wideband pdf

$$f_{p_n}(x) = \delta(x - N_A) \tag{3.27}$$

with Laplace image

$$\mathcal{L}\{f_{p_n}, s\} = \exp(-s N_A) \tag{3.28}$$

To find the outage probability, $s = z/\bar{p}_{A_0}$ is to be inserted (see (3.11)). For high SNRs (large $\bar{p}_{A_0}$), the behavior of the image function for small values of s is relevant. The zeroth and first-order terms of the Taylor expansion of (3.28) and (3.26) are identical, namely,

$$\mathcal{L}\{f_{p_n}, s\} \approx 1 - s N_A + \cdots \tag{3.29}$$

So, in practice, for high C/N ratios the models (3.25) and (3.27) are not substantially different. This suggests that the exact behavior of the pdf of the noise signal during the observation interval may only be of second-order relevance.

In the following, we use (3.28), corresponding to an observation interval that is large compared to the rate of change of the modulation of the wanted signal. This agrees with the assumption of incoherent cumulation of the signals from interfering terminals. The probability that the wanted signal with a known local-mean power is above the receiver noise floor at receiver A is

$$P_{NA} \triangleq \Pr(p_{A_0} > zN_A) = \exp\left(-z\frac{N_A}{\bar{p}_{A_0}}\right) \qquad (3.30)$$

For large C/N, $P_{NA} \rightarrow 1 - (zN_A/\bar{p}_{A_0})$. The noise N_A is now included as an additional uncorrelated interference signal; that is, P_{NA} is seen as an additional degrading factor in the product (3.11) describing the outage probability for a given local-mean power of the wanted signal. The above discussion addressed AWGN, band-limited by RF filters in the receiver.

In mobile radio, manmade (burst) noise further threatens the quality of reception [49, 50]. It is the impression of the author that by appropriately combining the wideband pdf (3.27) and the narrowband pdf (3.22), the effect of burst noise (i.e., noise with slower time constants than the modulating signal) can be addressed. This is left here as a recommendation for further study.

3.4 GENERALIZED EXACT ANALYSIS OF OUTAGE PROBABILITY[2]

Using (3.11), (3.22), (3.24), and (3.30), the probability of successful reception in the presence of noise, given the local-mean power of the wanted signal and the area-mean power of the interfering signals, becomes

$$\Pr(A_0|\bar{p}_{A_0}, \{\bar{\bar{p}}_{A_i}\}_{i=1}^N) = P_{NA} \prod_{i=1}^N 1 - \phi(A,i)\Pr(i_{ON}) \qquad (3.31)$$

The probability of correct reception, given the area-mean power levels of all $N + 1$ signals is found from averaging (3.31) over the range of the local-mean power of the desired signal; that is,

$$\Pr(A_0) = \Pr(A_0|\{\bar{\bar{p}}_{A_i}\}_{i=0}^N) = \int_0^\infty f_{\bar{p}_{A_0}}(\bar{p}_{A_0}|\bar{\bar{p}}_{A_0}) P_{NA} \left[\prod_{i=1}^N 1 - \phi(A,i)\Pr(i_{ON})\right] d\bar{p}_{A_0} \qquad (3.32)$$

[2]Section 3.4: Portion reprinted, with permission, from *European Transactions on Telecommunications (and Related Technologies)*, Vol. 2, No. 5, Sept/Oct 1991, pp. 471–479.

In the following discussion, conditionality on area-mean power levels will no longer be explicitly denoted. For $\Pr(i_{ON}) = 1$ and $N_A = 0$, (3.32) becomes identical to (3.19). It appears from (3.32) that the activity level $\Pr(i_{ON})$ of interferer i and the weight $\phi(A,i)$ of the vulnerability of the wanted signal to this interference are multiplicative.

Equation (3.32) gives the probability of loss of a wanted voice segment. An outage occurs if the user produces a talkspurt *and* the corresponding transmission is disturbed. If we assume that the system does not erroneously insert noise or crosstalk during speech pauses, the *time-average* outage probability is $\Pr(0_{ON}) \cdot (1 - \Pr(A_0))$. Transmitter identification codes can largely avoid disturbing signals in the baseband voice circuit when the desired carrier is switched off. Numerical results will be given for $\Pr(A_0)$ and the corresponding outage probability $1 - \Pr(A_0)$, and is thus conditional on the knowledge 0_{ON} that a wanted signal is transmitted.

3.4.1 A Posteriori Probability of Interference

We now explore the a posteriori information provided by the event A_0 about the activity of the interfering terminals. The probability $\Pr(k_{ON}|A_0)$ that a terminal k has transmitted an interfering voice segment, given that the wanted voice segment (of terminal 0) is received correctly by base station A, is found from Bayes rule:

$$\Pr(k_{ON}|A_0) = \frac{\Pr(A_0|k_{ON})}{\Pr(A_0)} \Pr(k_{ON}) \tag{3.33}$$

where $\Pr(A_0|k_{ON})$ is the probability that a wanted segment is received successfully, despite the knowledge that interferer k was active. Taking account of the noise floor, event k_{ON}, and possible activity of $N - 1$ other interferers with probability $\Pr(i_{ON})$,

$$\Pr(A_0|k_{ON}, \bar{p}_{A_0}) = P_{NA} \{1 - \phi(A,k)\} \prod_{i=1, i \neq k}^{N} 1 - \phi(A,i)\Pr(i_{ON}) \tag{3.34}$$

where we assumed the local-mean power of the wanted signal to be known. The conditional probability of activity of k can be expressed using (3.33), namely,

$$\Pr(k_{ON}|A_0, \bar{p}_{A_0}) = \frac{P_{NA}\{1 - \phi(A,k)\}\Pr(k_{ON}) \displaystyle\prod_{i=1, i \neq k}^{N} 1 - \phi(A,i)\Pr(i_{ON})}{P_{NA} \displaystyle\prod_{i=1}^{N} 1 - \phi(A,i)\Pr(i_{ON})}$$

$$= \frac{1 - \phi(A,k)}{1 - \phi(A,k)\Pr(k_{ON})} \Pr(k_{ON}) \tag{3.35}$$

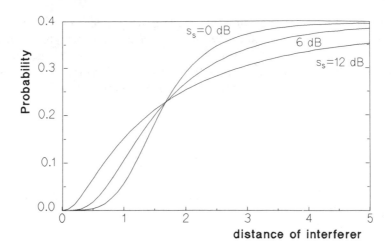

Figure 3.4 A posteriori probability $\Pr(k_{ON}|A_0)$ of activity of interferer k versus distance of the interferer a_k, given that the wanted signal from distance $a_0 = 1$ was received successfully. Voice activity $\Pr(k_{ON}) = 0.4$. Receiver threshold $z = 4$. Shadowing $s_s = 0$, 6, and 12 dB.

Remarkably, the probability that interferer k was active during a slot in which the wanted signal was received successfully at receiver A is defined uniquely by $\Pr(k_{ON})$ and the image $\phi(A,k)$: only variables concerned with terminal 0 and k appear in (3.35). The a posteriori probability thus does not appear to be dependent either on the noise level or on further interfering terminals.

For $\Pr(k_{ON}) < 1$, the a posteriori probability $\Pr(k_{ON}|A_0)$ is always less than the a priori probability $\Pr(k_{ON})$. Successful reception of the wanted signal at receiver A gives the information that during that particular time slot the number of interfering signals is presumably relatively low.

Only normalized reuse distances up to $a_k = r_u = 5$ (including the cases $C = 1, 3, 4$, and 7), are shown in Figure 3.4 because relatively dense reuse patterns are possible if site diversity is used. For relatively severe shadowing, say $s_s > 6$ dB, the a posteriori probability (3.35) significantly differs from the a priori probability $\Pr(k_{ON})$, particularly for short reuse distances. For larger reuse distances ($r_u > 5$), the differences become small for reasonable shadowing ($s_s < 12$ dB).

Probability (3.35) is now used to express the probability of capture at receiver B, given the event A_0.

3.4.2 Signal Outage With Macrodiversity

Assume that the positions of the $N + 1$ terminals are given. Terminal i is at a normalized distance a_i from base station A and at a normalized distance b_i from base station *B* (see Figure 3.5). Both Rayleigh fading and shadowing are assumed to be uncorrelated at the two base stations. We now consider the probability that the wanted signal captures the receiver at *B*, given the *local-mean* power of the wanted signal at receivers *A* and *B*, the *area-mean* power levels of the interference at receivers *A* and *B*, and on the condition that the test signal is received correctly at receiver *A* (event A_0).

Similar to (3.31), we write

$$\Pr(B_0|A_0, \bar{p}_{A_0}, \bar{p}_{B_0}) = P_{NB} \prod_{i=1}^{N} 1 - \phi(B,i)\Pr(i_{\text{ON}}|A_0, \bar{p}_{A_0}) \qquad (3.36)$$

where P_{NB} is the probability that the wanted signal is above the noise floor at receiver *B*. Inserting (3.35) gives

$$\Pr(B_0|A_0, \bar{p}_{A_0}, \bar{p}_{B_0}) = p_{NB} \prod_{i=1}^{N} 1 - \frac{1 - \phi(A,i)}{1 - \phi(A,i)\Pr(i_{\text{ON}})} \phi(B,i)\Pr(i_{\text{ON}}) \qquad (3.37)$$

Since $1 - \phi(A,i)$ is smaller than $1 - \phi(A,i)\Pr(i_{\text{ON}})$ if $\Pr(i_{\text{ON}}) < 1$, the probability (3.37) is smaller than the unconditional probability (3.31) (if $P_{NA} = P_{NB}$). Hence, a voice segment that captures receiver *A* has an enhanced probability of capturing receiver *B*, too. This is explained by the fact that with relatively high a posteriori

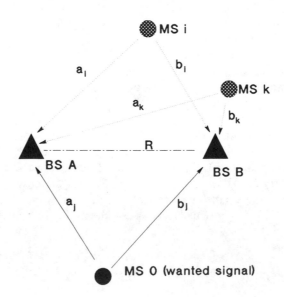

Figure 3.5 Model for macrodiversity: receivers are located in base stations *A* and *B*; the wanted signal arrives from mobile terminal 0; interferers are *i* and *k*.

probability the interference level during that particular time slot is low. The probability that the wanted voice segment captures at least one of the two base stations (i.e., the event—A_0 or B_0 or both), equals

$$\Pr(A_0 \vee B_0) = \Pr(A_0) + \Pr(B_0) - \Pr(A_0)\Pr(B_0|A_0) \qquad (3.38)$$

Conditional on the local-mean power of the wanted signal at receivers A and B, and by inserting (3.37) and (3.31), this probability becomes

$$\Pr(A_0 \vee B_0|\bar{p}_{A_0}, \bar{p}_{B_0}) = p_{NA}\left[\prod_{i=1}^{N} 1 - \phi(A,i)\Pr(i_{ON})\right]$$

$$+ p_{NB}\left[\prod_{i=1}^{N} 1 - \phi(B,i)\Pr(i_{ON})\right]$$

$$- p_{NA}p_{NB}\left[\prod_{i=1}^{N} 1 - \psi(A,B,i)\Pr(i_{ON})\right] \qquad (3.39)$$

where we introduced the function

$$\psi(A,B,i) \triangleq \phi(A,i) + \phi(B,i) - \phi(A,i)\phi(B,i) \qquad (3.40)$$

Assuming local-mean power levels of the wanted signal at the two base stations are mutually independent, averaging gives

$$\Pr(A_0 \vee B_0) = \int_0^\infty \int_0^\infty \Pr(A_0 \vee B_0|\bar{p}_{A_0}, \bar{p}_{B_0}) f_{\bar{p}_{A_0}}(\bar{p}_{A_0}) f_{\bar{p}_{B_0}}(\bar{p}_{B_0}) \, d\bar{p}_{A_0} \, d\bar{p}_{B_0} \qquad (3.41)$$

where the conditional probability (3.39) and log-normal pdfs (2.29) are to be inserted. In a noise-free system with the activity level $\Pr(i_{ON}) = 1$ for all terminals (i.e., if continuous voice transmission is employed), the outage probability corresponding to (3.41) with (3.39) can be rewritten as

$$O \triangleq 1 - \Pr(A_0 \vee B_0) = [1 - \Pr(A_0)][1 - \Pr(B_0)] \qquad (3.42)$$

where each of the two outage probabilities for single base stations on the right-hand side of (3.42) has the form of (3.32) with $\Pr(i_{ON})$ equal to unity. This agrees with the fact that, for mutually independent shadowing and multipath fading on each signal path, outage probabilities at the two receivers may be multiplied to assess the overall probability of circuit outage.

3.4.3 Computational Results for Rayleigh-Fading Channels Without Shadowing

The distances from terminal i to receivers A and B are denoted by a_i and b_i, respectively. We normalize $\bar{\bar{p}}_{A_0} = a_0^{-\beta} = 1$ and $\bar{\bar{p}}_{B_0} = b_0^{-\beta} = 1$. If shadowing is absent ($\sigma_s = 0$), the image function (3.24) reduces to the simple analytical form of (2.36), so

$$\phi(A,i) = \frac{z\bar{p}_{A_i}}{z\bar{p}_{A_i} + \bar{p}_{A_0}} = \frac{za_0^{\beta}}{za_0^{\beta} + a_i^{\beta}} \tag{3.43}$$

For this case, the probability of reception at the two base stations simultaneously (the event $A_0 \vee B_0$) can be expressed in closed form [37]. For Rayleigh-fading channels without shadowing, macrodiversity gives a slight improvement over microdiversity. This appeared to be caused by different path losses to the two receivers. For small reuse distances, the mean interference power at the two base stations becomes substantially different for each interfering terminal. On the other hand, for increasing reuse distances ($r_u \to \infty$), the interference situations experienced at A and B eventually become identical, since $a_i \approx b_i \approx r_u$, so for $r_u \to \infty$ and σ_s, macrodiversity does not offer enhanced performance compared to microdiversity [37].

3.4.4 Computational Results for Macrocellular Networks

Many proposals exist for the grid of receiver locations to support macrodiversity in cellular networks. Base stations with omnidirectional or sector antennas can be placed in the center or in certain corners of each cell [44]. In this section, we consider a wanted signal arriving from a point in the cell with the most adverse propagation conditions. In most concepts, this worst-case distance is also equal to the cell radius ($a_0 = b_0 = 1$). The distances from any interfering terminal to the two receivers A and B are assumed to be identical and equal to the reuse distance. Thus, $a_i = b_i = r_u$ for $i = 1, 2, \ldots, N$, so $\bar{p}_{A_i} = \bar{p}_{B_i} = r_u^{-\beta}$. Six equally interfering terminals are considered ($N = 6$), corresponding to a network with omnidirectional antennas.

Equation (3.39) will be used, implying that shadow fading of a signal from a terminal is assumed to be uncorrelated at receivers A and B. Figures 3.6 and 3.7 give the outage probability for two-branch macrodiversity for shadowing of $s_s = 6$ and 12 dB, respectively. Results are compared with the case of no diversity described by (3.32). The curves have been computed from the Hermite polynomial method (see Appendix E) using a 20-point integration. For the outage probability with macrodiversity, a three-dimensional integration is required, but in the special case of Rayleigh-fading channels without shadowing, the probability can be expressed in closed form (see (3.43) and (3.39)).

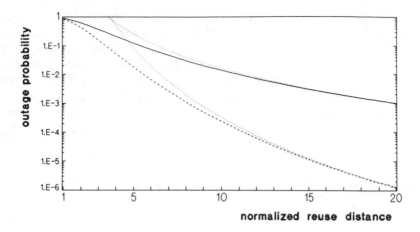

Figure 3.6 Outage probability versus reuse distance. Shadowing of s_s = 6 dB. Receiver threshold z = 10 (10 dB). UHF path loss law β = 4, 6 cochannel interferers, voice activity $\Pr(i_{ON})$ = 0.4. (——) single receiver, (- - -) macrodiversity, ($\cdots$) approximate expressions for large C/I ratios.

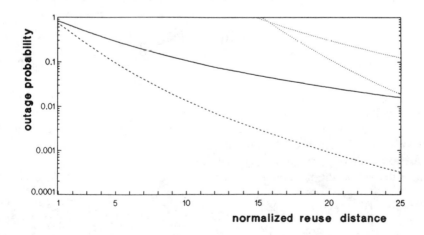

Figure 3.7 Outage probability versus reuse distance. Shadowing of s_s = 12 dB. Receiver threshold z = 10 (10 dB). UHF path loss law β = 4, 6 cochannel interferers, voice activity $\Pr(i_{ON})$ = 0.4. (——) Single receiver, (- - -) macrodiversity, ($\cdots$) approximate expressions.

It can be seen that macrodiversity substantially diminishes the outage probability as compared to the case of one receiver: the curves suggest that, for large r_u, the outage probability with two-branch macrodiversity is almost equal to the square of the outage probability at a single receiver. This agrees with results depicted in Figure 3.4, which show that the probabilities for the events A_0 and B_0 are not significantly correlated if large reuse distances are employed. However, to achieve a certain prescribed outage probability (e.g., 10^{-2} or 10^{-3}), macrodiversity is seen to allow relatively short reuse distances. Since $1 - \Pr(A_0 \vee B_0)$ departs from $\{1 - \Pr(A_0)\} \cdot \{1 - \Pr(A_0)\}$, correlation appears to be relevant.

3.4.5 Limiting Case for High Carrier-to-Interference Ratio

Although the formal expansion of $\phi_\sigma(s)$ was seen to diverge (Chapter 2), the second-order approximation of the image,

$$\phi(A,i) \approx z \frac{\bar{p}_{A_i}}{\bar{p}_{A_0}} \exp(\tfrac{1}{2}\sigma_s^2) - z^2 \frac{\bar{p}_{A_i}^2}{\bar{p}_{A_0}^2} \exp(2\sigma_s^2) \tag{3.44}$$

is useful for relatively weak interference ($z\bar{p}_{A_i} << \bar{p}_{A_0}$). We use (3.44) to focus on the asymptotic behavior for large C/I ratios. If the noise is neglected ($p_{NA} = p_{NB} = 1$), after inserting this in (3.32), one finds

$$\Pr(A_0) \approx \int_0^\infty d\bar{p}_{A_0} f_{\bar{p}_{A_0}}(\bar{p}_{A_0}) \left[\prod_{i=1}^N 1 - z\frac{\bar{p}_{A_i}}{\bar{p}_{A_0}} \exp(\tfrac{1}{2}\sigma_s^2) \Pr(i_{ON}) \right.$$
$$\left. + z^2 \frac{\bar{p}_{A_i}^2}{\bar{p}_{A_0}^2} \exp(2\sigma_s^2) \Pr(i_{ON}) \right] \tag{3.45}$$

Note that the approximation will not be valid for $\bar{p}_{A_0} \to 0$ and $\bar{p}_{A_i} \to \infty$ (i.e., in the far tails of the log-normal distribution). Nonetheless, in the interval that provides the dominant contribution to the integral, the approximation seems reasonable. A formal proof of the accuracy of these approximations is not presented, but the approximate expressions will be compared with the exact results.

After substituting the integration variable according to (2.33), (3.45) becomes

$$\Pr(A_0) \approx \int_{-\infty}^\infty dy \frac{\exp(-y^2)}{\sqrt{\pi}} \prod_{i=1}^N \left[1 - z\frac{a_0^\beta}{a_i^\beta} \Pr(i_{ON}) \exp(-\sqrt{2}\sigma_s y + \tfrac{1}{2}\sigma_s^2) \right.$$
$$\left. + z^2 \frac{a_0^{2\beta}}{a_i^{2\beta}} \Pr(i_{ON}) \exp(-2\sqrt{2}\sigma_s y + 2\sigma_s^2) \right] \tag{3.46}$$

Since the integrand contains a factor $\exp(-y^2)$, the effect of the approximations in the tails remains small. The product over N leads to 3^N terms. For $a_0 \ll a_i$, terms with $(a_0/a_i)^{M\beta}$ rapidly vanish with increasing M. Particularly for moderate values of shadowing, only low-order terms contribute significantly to the integral. Hence,

$$\Pr(A_0) \approx 1 - \sum_{i=1}^{N} z \frac{a_0^\beta}{a_i^\beta} \frac{\Pr(i_{ON})}{\sqrt{\pi}} \int_{-\infty}^{\infty} \exp(-y^2 - \sqrt{2}\sigma_s y + \tfrac{1}{2}\sigma_s^2)dy$$

$$+ \sum_{i=1}^{N}\sum_{k=1}^{N} z^2 \frac{a_0^{2\beta}}{a_i^\beta a_k^\beta} \Pr(i_{ON})\Pr(k_{ON}) \int_{-\infty}^{\infty} \exp(-y^2 - 2\sqrt{2}\sigma_s y + 2\sigma_s^2)dy$$

$$= 1 - z a_0^\beta \exp(\sigma_s^2) \sum_{i=1}^{N} a_i^{-\beta}\Pr(i_{ON})$$

$$+ z^2 a_0^{2\beta} \exp(4\sigma_s^2) \sum_{i=1}^{N}\sum_{k=1}^{N} a_i^{-\beta}a_k^{-\beta}\Pr(i_{ON})\Pr(k_{ON}) \tag{3.47}$$

In the first-order term of (3.47), a factor $z \cdot \exp(\sigma_s^2)$ is multiplied by the sum of the area-mean power levels of each interfering signal. This result is surprising, since the method by Schwartz and Yeh (see Table 3.2) indicates that, if multiple log-normal signals are added, the area-mean power of the joint signal tends to be significantly larger than the sum of the individual area-mean power levels [30]; that is, simple addition of area-mean signal power levels, as in the first-order term of (3.47), would not be a good approximation. But in the special case of very large C/I ratios, (3.47) suggests that (1) N weak interfering signals cause the same outage probability as one signal with N times stronger area-mean power, and (2) Schwartz and Yeh's method may overestimate outage probabilities. Presumably, the method by Schwartz and Yeh focuses on finding a good approximation for the main body of the pdf, whereas in the case of large C/I ratios, the tails of the pdf become increasingly important.

The mathematical appearance of this approximate outage probability would also suggest that the receiver threshold z can be multiplied by a shadow fade margin $\exp(\sigma_s^2)$ to obtain a *protection ratio* suitable as a rule of the thumb for the planning of practical systems. As discussed in Chapter 2, this shadow fade margin depends on the resolution of the terrain data used. Suppose that network planning with exact estimates of the local-mean power ($\sigma_s = 0$) requires a certain protection ratio determined by the threshold z. Then with modest shadowing ($s_s = 6$ dB, $\sigma_s = 1.36$) arising from less accurate terrain data, an extra protection margin of $\exp\{(1.36)^2\} \approx 6.36$ (or 8 dB) would be required to ensure the same (first-order) outage probability. Without terrain data at all, as in generic network design with $s_s = 12$ dB ($\sigma_s = 2.72$), the shadow fade margin should be even larger. However, it will be seen later (e.g., in Figure 3.8) that (3.47) has lost its accuracy in this range, so a simple factor $\exp\{(2.72)^2\} \approx 1633$ (or 32 dB) would be unrealistically large. Nonetheless, one may conclude that if, in a practical system with a prescribed maximum acceptable outage probability, crude terrain data is used, this might be at the cost of the spectrum efficiency of the system.

In Figures 3.6 through 3.8, only the first-order approximation is shown. Second-order terms were required to allow investigation of the approximate expressions for site diversity, because all first-order terms in (3.39) cancel for two-branch site diversity. Most second-order terms also cancel; only the terms that arise from multiplication of first-order terms in the third member of the right-hand side of (3.39) remain. Including these terms, the second-order approximate expression for the probability of successful reception at high C/I ratios becomes

$$\Pr(A_0 \vee B_0) \approx 1 - z^2\, a_0^\beta b_0^\beta \exp(2\sigma_s^2) \cdot \left[\sum_{\substack{i=1 \\ k \neq i}}^{N} \sum_{k=1}^{N} a_i^{-\beta} b_k^{-\beta}\, \Pr(i_{\mathrm{ON}})\Pr(k_{\mathrm{ON}}) + \sum_{i=1}^{N} a_i^{-\beta} b_i^{-\beta}\, \Pr(i_{\mathrm{ON}}) \right]$$

$$(3.48)$$

In the special case where all interfering terminals are at the reuse distance ($a_i = b_i = r_u$) and their voice activity is identical [$\Pr(i_{\mathrm{ON}}) \equiv P_{\mathrm{ON}}$], (3.48) becomes

$$\Pr(A_0 \vee B_0) \approx 1 - z^2 \left(\frac{a_0 b_0}{r_u^2}\right)^\beta \exp(2\sigma_s^2)[N(N-1)P_{\mathrm{ON}}^2 + NP_{\mathrm{ON}}] \qquad (3.49)$$

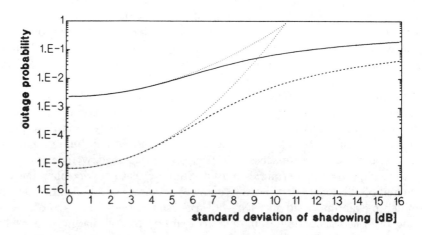

Figure 3.8 Outage probability versus standard deviation of the shadowing. (——). Single receiver, (- - -) macrodiversity, ($\cdots$) approximate expressions for large C/I ratios. Reuse distance r_u = 10. Receiver threshold z = 10 (10 dB).

Figure 3.8 depicts the outage probability and the asymptotes as a function of the standard deviation of the shadowing. The accuracy of the asymptotes decreases with increasing shadowing (σ_s) or decreasing (r_u). Reasonably accurate results occur for $s_s < 6$ dB and $r_u > 10$. This suggests that the simple approximate expressions can be of interest if practical cell layouts are planned with the aid of a high-resolution topographical database. On the other hand, when the standard deviation of the shadowing is as large as $s_s = 12$ dB, the approximate results appear to be inaccurate by more than an order of magnitude. For low C/I ratios, the approximations tend to overestimate the outage probability, so they can therefore be used in conservative designs.

3.5 CONCLUSION

The probability of a signal outage is a relevant measure of the performance of a mobile communication network with fading channels. In an interference-limited network, the fluctuations not only of the wanted signal, but also of the interference signals must be taken into account.

Assessment of outage probabilities began to be more developed during the 1980s, and a large number of technical papers appeared on this topic. By and large, these studies employed approximate techniques. An exact analytical technique has been presented in this chapter for the computation of outage probabilities in narrowband multiuser mobile radio channels, such as those for cellular telephony or trunking networks with closed user groups. The analysis includes the effects of both discontinuous voice transmission and receiver noise. Radio wave propagation in a land-mobile radio channel has been modeled by taking account of uncorrelated Rayleigh fading and shadowing for each contributing signal path. Relevant parameters are the area-mean power levels of the wanted signal and each interfering signal, the receiver threshold, the groundwave path loss exponent, the standard deviation of the shadow fading, and the voice activity of the interfering terminals. The analysis demonstrates that the standard deviation of the shadowing has a substantial effect on the outage probability. In the operational range of cellular networks, the outage probability tends to increase exponentially with increasing shadow fading.

If low-resolution terrain data is used for network planning, large shadow fluctuations ($s_s \approx 12$ dB) require large reuse distances to ensure reliable communication. Moreover, because of the large-area and small-area properties of shadowing, the resulting average performance is likely to be nonuniformly distributed over the boundary of the cells. On the other hand, for highly accurate terrain data, all shadowing fluctuations are included in the estimates of the area-mean power levels $\bar{p}_i$. This allows the design of cellular structures customized for the particular propagation environment, which is likely to yield improved spectrum efficiency. The analysis was extended to two-branch site (or macro) diversity in a macrocellular telephone network.

Also, approximate expressions for large reuse distances and moderate shadow fading ($s_s < 6$ dB) have been derived. In channels with shadow fading, site diversity gives substantial enhancement of the quality of the end-to-end circuit. In channels without shadowing, site diversity is still slightly more effective in combating signal outage than microdiversity, because the average power levels of the interference signals differ due to the different losses for the two paths to the receivers. Numerical results have been given to assist in generic design of systems based on hexagonal cells with a uniform propagation model. The expressions for outage probability are in a form that would also allow more detailed computer-aided investigation of a "real-life" cellular network, giving highly accurate estimates of the area-mean received power levels (e.g., as can be obtained by using a topological database).

The special case of uncorrelated shadowing was considered. Correlation of the shadow attenuation reduces the differences in signal power received from each transmitter at the various sites; this is likely to reduce the effectiveness of macrodiversity (see, for example, [51]).

REFERENCES

[1]   MacDonald, V.H., "The Cellular Concept," *Bell Syst. Tech. Journal*, Vol. 58, No. 1, Jan. 1979, pp. 15–41.

[2]   Report 740–2, "General Aspects of Cellular Systems," *Recommendations and Reports of the CCIR, 1986, Vol. VIII-1: Land Mobile Service, Amateur Service, Amateur Satellite Services*, XVIth Plenary Assembly, Dubrovnik, 1986.

[3]   Technical Characteristics and Test Conditions for Non-speech and Combined Analogue Speech/Non-speech Radio Equipment With Internal or External Antenna Connector, Intended for the Transmission of Data, for Use in the Land-Mobile Service (DRAFT), European Telecommunications Standards Institute, I-ETS [A], Document 300113, Valbonne, France, 1990.

[4]   Verhulst, D., M. Mouly, and J. Szpirglas, "Slow Frequency Hopping Multiple Access for Digital Cellular Radiotelephone," *IEEE J. Sel. Areas Comm.*, Vol. SAC-2, No. 4, July 1984, pp. 563–574.

[5]   Walker, J., ed., "Mobile Information Systems," Artech House, Inc., Norwood, MA, 1990.

[6]   Lundquist, L., and M. P. Peritsky, "Co-channel Interference Rejection in a Mobile Radio Space Diversity System," *IEEE Trans. on Veh. Tech.*, Vol. VT-20, No. 3, Aug. 1971, pp. 68–75.

[7]   Arnbak, J.C., and W. van Blitterswijk, "Capacity of Slotted-ALOHA in a Rayleigh Fading Channel," *IEEE J. Sel. Areas Comm.*, Vol. SAC-5, No. 2, Feb. 1987, pp. 261–269.

[8]   Nagata, Y., and Y. Akaiwa, "Analysis for Spectrum Efficiency in Single Cell Trunked and Cellular Mobile Radio," *IEEE Trans. on Veh. Tech.*, Vol. VT-35, No. 3, Aug. 1987, pp. 100–113. (See also [28].)

[9]   Report 358–5 "Protection Ratio and Minimum Field Strength Requirements in Mobile Services," *Recommendations and Reports of the CCIR, 1986, Volume VIII-1: Land Mobile Service, Amateur Service, Amateur Satellite Services*, XVIth Plenary Assembly, Dubrovnik, 1986.

[10]   Muammar, R., and S.C. Gupta, "Cochannel Interference in High-Capacity Mobile Radio Systems," *IEEE Trans. on Comm.*, Vol. COM-30, No. 8, Aug. 1982, pp. 1973–1978.

[11]   Gosling, W., "Protection Ratio and Economy of Spectrum Use in Land Mobile Radio," *IEE Proceedings*, Vol. 127, Pt. F, June 1980, pp. 174–178.

[12]   Report 319–6, "Characteristics of Equipment and Principles Governing the Assignment of Frequency Channels Between 25 and 1000 MHz for Land Mobile Services," *Recommendations and Reports of the CCIR, 1986, Volume VIII-1: Land Mobile Service, Amateur Service, Amateur Satellite Services*, XVIth Plenary Assembly, Dubrovnik, 1986.

[13]   Chuang, J.C.-I., "Comparison of Coherent and Differential Detection of BPSK and QPSK in a Quasi-static Fading Channel," *IEEE Trans. on Comm.*, Vol. COM-38, No. 5, May 1990, pp. 565–567.

[14]   Carlson, A.B., *Communication Systems*, McGraw-Hill, 1968.

[15]   Arredondo, G.A., J.C. Feggeler, and J.F. Smith, "Voice and Data Transmission," *Bell Sys. Tech. Journal*, Vol. 58, No. 1, Jan. 1979, pp. 97–122.

[16]   Oetting, J.D., "A Comparison of Modulation Techniques for Digital Radio," *IEEE Trans. on Comm.*, Vol. COM-27, No. 12, Dec. 1979, pp. 1752–1762.

[17]   Fenton, L.F., "The Sum of Log-Normal Probability Distributions in Scatter Transmission Systems," *IRE Trans. Comm. Syst.*, Vol. CS-8, Mar. 1960, pp. 57–67.

[18]   Schwartz, S.C., and Y.S. Yeh, "On the Distribution Function and Moments of Power Sums with Log-Normal Ccomponents," *Bell Sys. Tech. Journal*, Vol. 61, No.7, Sept. 1982, pp. 1441–1462.

[19]   Kegel, A., "Mobile Networks" (in Dutch), *Lecture Notes Communication Systems (Selected Topics)* (L77), Delft University of Technology, Jan. 1991.

[20]   Prasad, R., A. Kegel, and J.C. Arnbak, "Improved Assessment of Interference Limits in Cellular Radio Performance," *IEEE Trans. on Veh. Tech.*, Vol. 40, No. 2, May 1991, pp. 412–419.

[21]   O'Leary, T., and J. Rutkowski, "Combining Multiple Interfering Field Strengths: the Simplified Multiplication Method and the Physical and Mathematical Basis," *EBU Telecommunication Journal*, Vol. 49, XII, 1982, pp. 823–831.

[22]   Report 945–1, "Methods of Assessment for Multiple Interference," and Report 944, "Theoretical Network Planning," *Recommendations and Reports of the CCIR, Broadcasting Service (Sound), Vol. X-1*, XVIth Plenary Assembly (Dubrovnik) International Radio Consultative Committee (CCIR), I.T.U.

[23]   Prasad, R., and A. Kegel, "Effects of Rician Faded and Log-Normal Shadowed Signals on Spectrum Efficiency in Micro-Cellular Radio," *Proc. 40th IEEE Veh. Tech. Conf.*, St. Louis, MO, May 1991, extended version submitted for journal publication, 1992.

[24]   Prasad, R., A. Kegel, and J. Olsthoorn, "Spectrum Efficiency Analysis for Microcellular Mobile Radio Systems," *Electron. Lett.*, Vol. 27, No. 5, 28 Feb. 1991, pp. 423–424.

[25]   Gosling, W., "A Simple Mathematical Model of Co-channel and Adjacent Channel Interference in Land-Mobile Radio," *IEEE Trans. on Veh. Tech.*, Vol. VT-29, No. 4, Nov. 1980, pp. 361–364.

[26]   French, R.C., "The Effect of Fading and Shadowing on Co-channel Reuse in Mobile Radio," *IEEE Trans. on Veh. Tech.*, Vol. VT-28, No. 3, Aug. 1979, pp. 171–181.

[27]   Cox, D.C., "Cochannel Interference Considerations in Frequency Reuse Small-Coverage-Area Radio Systems," *IEEE Trans. on Comm.*, Vol. COM-30, No. 1, Jan. 1982, pp. 135–142.

[28]   Daikoku, K., and H. Ohdate, "Optimal Channel Reuse in Cellular Land Mobile Radio Systems," *IEEE Trans. on Veh. Tech.*, Vol. VT-32, No. 3, Aug. 1983, pp. 217–224.

[29]   Prasad, R., A. Kegel, and J.C. Arnbak, "Analysis of System Performance of High-Capacity Mobile Radio," *Proc. IEEE Veh. Tech. Conf. 1989*, San Francisco, 3–5 May 1989, pp. 306–309.

[30]   Prasad, R., and J.C. Arnbak, "Comments on Analysis for Spectrum Efficiency in Single Cell Trunked and Cellular Mobile Radio," *IEEE Trans. on Veh. Tech.*, Vol. VT-36, No. 4, Nov. 1988, pp. 220–222.

[31]   Sowerby, K.W., and A.G. Williamson, "Outage Probability Calculations for Multiple Cochannel Interferers in Cellular Mobile Radio Systems," *IEE Proceedings*, Vol. 135, Pt. F, No. 3, June 1988, pp. 208–215.

[32]   Prasad, R., and J.C. Arnbak, "Effects of Rayleigh Fading on Packet Radio Channels With Shadowing," *Proc. IEEE Tencon 1889*, Bombay, Nov. 1989, pp. 27.4.1–27.4.3.

[33]   Van Trees, H.L., "Detection, Estimation and Modulation Theory," John Wiley, 1968.

[34]   Merani, M.L., "A Comparison Between Outage Probability Evaluations in Microcellular Systems," *Proc. 42nd IEEE Vehicular Technolgy Society*, Denver, Vol. 1, 10–13 May 1992, pp. 443–446.

[35]   Yao, Y.-D., and A.U.H. Sheikh, "Outage Probability Analysis for Microcell Mobile Radio Systems With Cochannel Interferers in Rician/Rayleigh Fading Environment," *Electron. H5 Letters*, Vol. 26, No. 13, 21 June 1990, pp. 864–866.

[36]   Linnartz, J.P.M.G., "Exact Analysis of the Outage Probability in Multiple-User Mobile Radio," *IEEE Trans. on Comm.*, Vol. COM-40, No. 1, Jan. 1992, pp. 20–23.

[37]   Linnartz, J.P.M.G., "Site Diversity in Land-Mobile Cellular Telephony Network With Discontinuous Voice Transmission," *European Transactions on Telecommunications*, Vol. 2., No. 5, Sep./Oct. 1991, pp. 471–480.

[38]   Abramowitz, M., and I.A. Stegun, eds., *Handbook of Mathematical Functions*, New York: Dover, 1965.

[39]   Jakes, W.C., Jr., ed., "Microwave Mobile Communications," New York: John Wiley and Sons, 1974.

[40]   Lee, W.C.Y., "Mobile Communications Design Fundamentals," Indianapolis: Howard W. Sams & Co., 1986.

[41]   Chang, L.F., and C.-I. Chuang, "Diversity Selection Using Coding in a Portable Radio Communications Channels With Frequency-Selective Fading," *IEEE Journal on Sel. Areas in Comm.*, Vol SAC-7, No. 1, Jan. 1989, pp. 89–98.

[42]   Sowerby, K.W., and A.G. Williamson, "Estimating Reception Reliability in Cellular Mobile Radio Systems" *Proc. 42nd IEEE Vehicular Technolgy Society*, Denver, Vol. 1, 10–13 May 1992, pp. 151–154.

[43]   Cox, D.C., H.W. Arnold, and P.T. Porter, "Universal digital portable communications: A system perspective," IEEE Journal on Sel. Areas in Comm. Vol. SAC-5, No. 5, June 1987, pp. 764–773.

[44]   Bernhardt, R.C., "Macroscopic diversity in frequency reuse radio systems," IEEE Journal on Sel. Areas in Comm., Vol. SAC-5, No. 5, June 1987, pp. 862–870.

[45]   Bye, K.J., "Handover Criteria and Control in Cellular and Microcellular Systems," *Proc. 5th IEE Int. Conf. on Mobile Radio and Personal Communications*, Warwick, UK, 11–14 Dec. 1989, pp. 94–98.

[46]   Braun, H.J., G. Cosier, D. Freeman, A. Gilloire, D. Sereno, C.B. Southcott, and A. van der Krogt, "Voice Control of the Pan-European Digital Mobile Radio System," *CSELT Tech. Rep.*, Vol. 18, No. 3, June 1990, pp. 183–187.

[47]   Brady, P.T., "A Model for Generating On-Off Speech Patterns in Two-Way Conversation," *Bell Sys. Tech. Journal*, Vol. 48, No. 7, Sept. 1969, pp. 2445–2472.

[48]   Goodman, D.J., and S.X. Wei, "Factors Affecting the Bandwidth Efficiency of Packet Reservation Multiple Access," *Proc. IEEE Veh. Tech. Conf. 1989*, San Francisco, 3–5 May 1989, pp. 292–298.

[49]   Prasad, R., A. Kegel, and W. Hollemans, "Performance Analysis of Digital Mobile Radio Cellular System in Presence of Co-channel Interference, Thermal and Impulsive Noise," *IEEE Conf. on Personal, Indoor and Mobile Comm.*, London, 23–25 Sep. 1991, pp. 154–159.

[50]   Hagn, G.H., *Manmade Radio Noise and Interference* (preprint), AGARD EPP Meeting, Lisbon, 24–30 Oct. 1987.

[51]   Safak, A., and R. Prasad, "Effects of Correlated Shadowing Signals on Channel Re-use in Mobile Radio Systems" *IEEE Trans. on Veh. Tech.*, Vol. 40, No. 4, Nov. 1991, pp. 708–713.

Chapter 4
Threshold Crossing Rate and Average Fade/Nonfade Duration

The outage probability discussed in Chapter 3 describes the percentage of time that the C/I ratio on a mobile link is below a certain threshold. A further relevant performance measure is the duration of the outages. The effect of short but frequently occurring outages is entirely different from the effect of rare but lengthy outages. To understand the temporal behavior of link outages better, this chapter studies the threshold crossing rate and the average fade/nonfade duration in a Rayleigh-fading channel.

During a link fade in an *analog* cellular network for voice communication, the user experiences not only an absence of the intended signal, but also a burst of noise or crosstalk. Even if link outages are very short, they collectively degrade the system performance, although they may not be individually recognized. Generally, only outages lasting longer than some tens of milliseconds are recognized as a dropout of the telephony circuit. In *digital* CW communication, the fade duration determines the length of error bursts. If an error-correcting code with interleaving is employed to combat bursts of bit errors, the interleaving interval should be chosen substantially longer than the expected duration of a signal fade. However, in land-mobile radio, codes with short block lengths are most often used, rather than convolutional codes or codes with long block lengths and interleaving. A combination of forward error correction and error detection with retransmission is commonly applied. In voice communication, blocks are dropped if received with too many bit errors, whereas in computer communication, retransmission of lost data is required. In both cases, the probability that a block of a certain length is error-free or received with only a small number of bit errors critically determines the performance of the system. Hence, an appropriate optimum block length ensures that the probability that a block fits inside a nonfade interval is sufficiently high. Shorter blocks suffer from relatively larger overheads, whereas longer blocks are more likely to experience one or more signal fades. It may be more efficient to retransmit an incidental lost packet than to employ a heavily redundant code on a channel with a low bit error probability most of the time.

The duration of the fades in a mobile radio channel has been discussed in several text books, such as [1, 2]. Most results for the level crossing rates and fade durations apply only for a single fading signal in the presence of stationary noise or a fixed receiver threshold. In this chapter, the classical theory developed by Rice for single-user channels [3–5] is extended for multiuser networks where the mutual interference between users limits the quality of the channel. The probability that the C/I ratio drops below a certain threshold is determined, taking into account fluctuations of not only the wanted signal, but also of the joint interference signal. In Sections 4.1 and 4.2, respectively, the models for the wanted signal and for the joint interference signal are formulated. In Section 4.3, the sum of multiple incoherent Rayleigh-fading narrowband signals is approximated as a joint interference signal with Nakagami fading. Outage probabilities for a Rayleigh-fading signal in the presence of Nakagami-fading interference are derived in Section 4.4. In Section 4.5, expressions for the rate of crossing a specified C/I threshold are derived. The average fade and nonfade durations are found in Section 4.6. It is shown in Section 4.7 that the expressions for the threshold crossing rate and the average nonfade duration for a single fading signal with respect to a stationary (nonfading) threshold level (e.g., determined by a receiver noise floor) occur as limiting cases of the results obtained in Sections 4.5 and 4.6. Section 4.8 discusses the distribution of the duration of nonfade periods. An approximate model is developed in Section 4.9 to compute the probability that a signal outage lasts longer than a specified time. The probability that the C/I ratio remains above the threshold for a prescribed duration (e.g., during reception of a data block) is studied in Section 4.10. Conclusions are summarized in Section 4.11.

4.1 SIGNAL MODEL[1]

A sum of n uncorrelated Rayleigh-fading narrowband signals

$$v_t(t) = \sum_{i=1}^{n} \rho_i(t) \cos[\omega_{c_i} t + \theta_i(t) + \psi_i(t)] \qquad (4.1)$$

interferes with a wanted signal $v_0(t)$ of the form

$$v_0(t) = \rho_0(t) \cos[\omega_{c_0} t + \theta_0(t) + \psi_0(t)] \qquad (4.2)$$

---

[1]Sections 4.1–4.7: Portions reprinted, with permission, from *Electronics and Communication (AEÜ)*, Vol. 43, No. 6, pp. 345–349, November/December 1989. © 1989 AEÜ.

Analogously with (3.1), the carrier phases $\theta_i(t)$ and $\psi_i(t)$ describe any Doppler modulation with maximum frequency shift $f_m = \omega_m/2\pi$ arising from terminal mobility and angle modulation, respectively. This phase modulation and slightly different carrier frequencies are assumed to result in incoherent (power) addition of the signals. All $n + 1$ signals present in the radio channel are subject to independent Rayleigh fading. The amplitude ρ_i ($i = 0, 1, \ldots, n$) and its derivative $\dot{\rho}_i$ are statistically independent stochastic variables, with joint pdf [1, 5, 6]

$$f_{\rho_i, \dot{\rho}_i}(\rho_i, \dot{\rho}_i) = \frac{1}{\sqrt{2\pi}\sigma_{\rho_i}} \exp(-\dot{\rho}_i^2/2\sigma_{\rho i}^2) \cdot \frac{\rho_i}{\bar{p}_i} \exp(-\rho_i^2/2\bar{p}_i) \tag{4.3}$$

where $\bar{p}_i$ is the local-mean power of the ith signal, and $\sigma_{\rho i}$ is the standard deviation of the time derivative of the amplitude, calculated from

$$\sigma_{\rho i}^2 = \frac{1}{2}\omega_m^2 \bar{p}_i = 2\pi^2 \frac{v_i^2}{\lambda^2}\bar{p}_i \tag{4.4}$$

with v_i the velocity of terminal i and λ the wavelength. In contrast to the notation in Chapters 2 and 3, μ and σ denote *linear* moments in this chapter.

4.2 INTERFERENCE MODEL

We assume that the bit rate r_b of the angle-modulated data largely exceeds the Doppler modulation $f_m(T_3 << T_6)$. So fluctuations in the phase $\theta_i(t)$ are relatively fast compared to amplitude fluctuations from mobility. Incoherence of the signals then allows power addition of the interference signals. Exact expressions for the joint interference power pdf are found by convolving n pdfs of the individual interference powers, using the Laplace transformation of each pdf. The image function of the pdf of the instantaneous power of the ith Rayleigh-fading signal with local-mean $\bar{p}_i$ is $\phi_i(s) = (\bar{p}_i s + 1)^{-1}$. The Laplace image of the joint instantaneous interference power is found by multiplication; that is,

$$\phi_t(s) = \prod_{i=1}^{n} \phi_i(s) = \prod_{i=1}^{n} \frac{1}{\bar{p}_i s + 1} \tag{4.5}$$

For a number of signals, say $n = 1, 2,$ or 3, reverse transformation appears possible [7, 8]. For larger n, this approach becomes impractical. Another disadvantage is that the analytic form of the joit pdf abruptly changes if two (or more) local-mean power levels happen to be equal. For instance, in the event of two interfering signals, the pdf of instantaneous interference power p_t is

$$f_{p_t}(p_t) = \begin{cases} \dfrac{1}{\bar{p}_1 - \bar{p}_2} \left[\exp(-p_t/\bar{p}_1) - \exp(-p_t/\bar{p}_2) \right] & \text{if} \quad \bar{p}_1 \neq \bar{p}_2 \\[2ex] \dfrac{p_t}{\bar{p}_1} \exp(-p_t/\bar{p}_1) & \text{if} \quad \bar{p}_1 = \bar{p}_2 \end{cases} \tag{4.6}$$

Intuitively, however, the fading experienced should not be highly sensitive to small mutual differences in the mean power levels of the n components. Table 4.1 and Figure 4.1 give the pdf (4.6) of the joint signal power if the local-mean power of the minor signal is ∞, 10, 6, 3, 0.5, and 0 dB below the local-mean power of the major interferer in curves (a) to (f), respectively.

The probability density for low interference power p_t follows from the *initial-value theorem* of the one-sided Laplace transform [8, page 20]

$$f_{p_t}(0) = \lim_{s \to \infty} s\phi_t(s) = \lim_{s \to \infty} \frac{s}{\displaystyle\prod_{i=1}^{n} \bar{p}_i s + 1} \tag{4.7}$$

which is zero for $n = 2, 3, \ldots$, but equals $\bar{p}_l^{-1}$ for $n = 1$. Thus, deep fades occur occasionally with one signal, but become very rare for larger n. An explanation is that with more signals, a deep fade of one signal is likely to be hidden by the signal power received from other signals.

Tractable analytical expressions for the pdf $f_{\rho,\dot{\rho}}(\rho, \dot{\rho})$, for a joint interference signal accumulated from several fading signals with arbitrary local-mean power levels have not been found. In the following, we therefore approximate the pdf of the joint signal with a Nakagami-fading signal, with the same mean and variance as the exact pdf.

Table 4.1
Mean $\bar{p}$ and m-Parameter of the Joint Signal for a Major Interferer Plus a Minor Interferer With a Local-Mean Power A dB Below the Major Interferer

A	∞	10	6	3	0.5	0	dB
$\bar{p}$	1	0.92	0.85	0.83	0.95	1	$\bar{p}_1$
m	1	1.2	1.5	1.8	1.99	2	
Curve:	(a)	(b)	(c)	(d)	(e)	(f)	

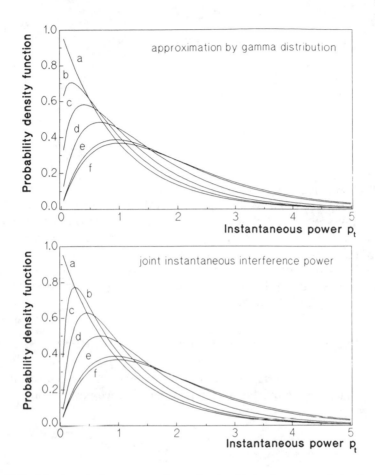

Figure 4.1 PDF of the instantaneous power of two signals with incoherent cumulation (exact, below) and an approximation by a gamma distribution, assuming a Nakagami fading envelope (above). Curves (a) to (f) are discussed in Table 4.1.

The kth order moment μ_k of the pdf of the joint interference power, defined as

$$\mu_k \triangleq E[p_t^k] = \int_0^\infty p_t^k f_{p_t}(p_t) dp_t \qquad (4.8)$$

is found simply from the kth derivative of the image function (4.5). The mean power of the joint interference power is identical to the sum of the mean power of the individual signals, namely,

$$\bar{p}_t = \mu_1 = -\phi'(0) = \sum_{i=1}^{n} \bar{p}_i \qquad (4.9)$$

Similarly, the variance is found from

$$\sigma^2 = \mu_2 - \mu_1^2 = \phi_t''(0) - \{\phi_t'(0)\}^2 = \sum_{i=1}^{n} \bar{p}_i^2 \qquad (4.10)$$

In Section 3.1.2.3, we addressed the special case that all n interference signals have equal mean power ($\bar{p}_i \equiv \bar{p}$). The Laplace image then reduces to $(\bar{p}s + 1)^{-n}$. Inverse transformation (3.5) shows that the instantaneous power p_t of the cumulated interference is then gamma distributed, with mean $\mu_1 = n\bar{p}$ and variance $\sigma^2 = n\bar{p}^2$. The pdf of the amplitude of the joint interference ρ_t is the Nakagami m-distribution (2.24) with $m = n$. Consequently, in the special case of n interfering signals with equal mean power, the Nakagami distribution accurately describes the pdf of the amplitude of the joint interference signal.

4.3 APPROXIMATE MODEL FOR JOINT INTERFERENCE SIGNAL

For the general case of interference caused by n signals with mutually different mean power, the characteristic function of the joint interference power is approximated by

$$\phi_t(s) \approx \frac{1}{(\bar{p}s + 1)^m} \qquad (4.11)$$

This is equivalent to approximating the pdf of the joint power by the gamma distribution (3.5) or (2.25). Both the means and the standard deviations of the instantaneous signal power p_t in the exact and approximated gamma pdf match each other if we insert

$$\bar{p} = \frac{\sum\limits_{k=1}^{n} \bar{p}_k^2}{\sum\limits_{k=1}^{n} \bar{p}_k} \quad \text{and} \quad m = \frac{\left[\sum\limits_{k=1}^{n} \bar{p}_k\right]^2}{\sum\limits_{k=1}^{n} \bar{p}_k^2} \qquad (4.12)$$

in (3.5), where m is a real continuation of the integer number n, with $1 \leq m \leq n$.

As an example, the exact joint pdf of the instantaneous interference power p_t of two signals obtained by using the exact expression (4.6) is compared with the approximation by a gamma function using (4.12) for several ratios of the local-mean power levels in Figure 4.1. The estimates for $\bar{p}$ and m are given in Table 4.1. It is asserted from Figure 4.1 that the discrepancy is acceptable, even for the critical case $p_t \rightarrow 0$, especially if the interferers have similar mean power levels.

In the event of many interferers ($n >> 1$), the joint interference signal does have a constant envelope, even though each indivudual component may be a constant envelope FM or BPSK signals. As known from the central limit theorem, the cumulated signal tends toward band-limited Gaussian noise. Nonetheless, we define the *instantaneous* amplitude; that is, averaged over a time scale T longer than the time constants of variations caused by modulation, but smaller than the time constants of the multipath fading ($T_3 << T << T_6$). The amplitude can then only be related to the instantaneous power p_t. No such phenomenon as an envelope can be defined. This is in contrast to the constant envelope of a single phase-modulated signal ($n = 1$). In any observation window $T_1 << T << T_3$, the joint interference power is exponentially distributed, as in Rayleigh fading, irrespective of n. The influence of n is recognized only after averaging the signal power over a suitable time T long enough to experience incoherent addition ($T >> T_3$), but short enough to avoid fluctuations due to mobility ($T << T_6$). Remember that the influence of latter fluctuations on the cumulated signal decreases with increasing n: a temporary fade of any one of the n interfering signals is likely to be hidden by the interference received from other transmitters.

4.4 OUTAGE PROBABILITY

As in Chapter 3, we assume that the wanted signal is correctly received as long as its instantaneous p_0 power exceeds the joint interference power by at least a ratio z. Then the signal amplitude $\rho_0(t)$ exceeds the amplitude $\rho_t(t)$ of the joint interference by at least a ratio $\sqrt{z}$. We approximate the exact result (3.20) using (4.11). So,

$$\Pr\{\rho_0 > \sqrt{z}\rho_t\} = \phi_t\left(\frac{z}{\bar{p}_0}\right) \approx \left(\frac{z\bar{p}}{\bar{p}_0} + 1\right)^{-m} \qquad (4.13)$$

which is an exact expression for n interfering signals with identical mean power levels. This simplification is adopted here to generate results that can be combined with results for the threshold crossing rate with Nakagami-fading interference.

4.5 THRESHOLD CROSSING RATE

The instantaneous amplitude $\rho_t(t)$ of the joint interference is found from the power sum

$$\rho_t^2(t) = \sum_{i=1}^{n} \rho_i^2(t) \tag{4.14}$$

The first derivative of $\rho_t(t)$ with respect to the time t gives

$$\dot{\rho}_t(t) = \frac{\sum\limits_{i=1}^{n} \rho_i(t)\, \dot{\rho}_i(t)}{\rho_t(t)} \tag{4.15}$$

According to (4.3), the derivatives of all individual amplitudes are zero-mean Gaussian distributed with variance (4.4). Because of the linear form of (4.15), the derivative $\dot{\rho}_t(t)$ of the amplitude of the cumulated signal, given all individual amplitudes $\{\rho_1, \ldots, \rho_n\}$, has a Gaussian distribution with zero mean and variance

$$\sigma_{\rho t}^2 = \frac{\dfrac{1}{2}\sum\limits_{i=1}^{n} \rho_j^2(t)\omega_{m_i}\bar{p}_i}{\rho_t^2(t)} \tag{4.16}$$

Analytical results can be obtained if all interference signals can be assumed to experience uncorrelated fading and have equal mean power ($\bar{p}_i \equiv \bar{p}$) and equal Doppler spectra ($\omega_{mi} \equiv \omega_m$). The latter assumption also implies that all transmitters move with equal speed. Under these conditions, equation (4.16) simply reduces to

$$\sigma_{\rho t}^2 = \frac{1}{2}\omega_m^2\, \bar{p} \tag{4.17}$$

Remarkably, this is independent of n, but the relative standard deviation $\sigma_{\rho t}$ decreases with $1/\sqrt{n}$, since the mean joint interference power $\bar{p}_t$ is equal to $n\bar{p}$.

In an interference-limited network, the C/I ratio determines the quality of communication. For ease of notation, in this chapter we consider the ratio $\dot{y}$ between the *amplitudes* of wanted signal and the joint interference. The joint pdf of this ratio and its derivative is found by substituting [6]

$$\rho_0 = \rho_t y \tag{4.18}$$

and

$$\dot{\rho}_0 = \dot{y}\rho_t + \dot{\rho}_t y \tag{4.19}$$

The pdf of $\dot{y}$ and results from the integration [6]

$$f_{y,\dot{y}}(y,\dot{y}) = \int_0^\infty d\rho_t \int_0^\infty d\dot{\rho}_t \; \rho_t f_{y0}(\rho_t y) f_{\dot{y}0}(\dot{y}\rho_t + \dot{\rho}_t y) f_{\rho_t}(\rho_t) f_{\dot{\rho}_t}(\dot{\rho}_t) \tag{4.20}$$

where the four-dimensional pdf can be separated because the instantaneous amplitude and its derivative are statistically independent for both the wanted and the interference signals. The pdfs (4.3), (2.24), and a Gaussian pdf of $\dot{\rho}_t$, with variance (4.17), are inserted into (4.20). This gives

$$f_{y,\dot{y}}(y,\dot{y}) = \int_0^\infty \int_0^\infty d\rho_t d\dot{\rho}_t \frac{\rho_t^{2n+2} y \exp(-a_1 \dot{\rho}_t^2 - a_2 \rho_t \dot{\rho}_t - a_3 \rho_t^2)}{2\pi\sigma_{\rho 0}\sigma_{\rho t}\,\bar{p}^n \bar{p}_0 (n-1)!\, 2^{n-1}} \tag{4.21}$$

with

$$a_1 \triangleq \frac{1}{2\sigma_{\rho t}^2} + \frac{y^2}{2\sigma_{\rho 0}^2}$$

$$a_2 \triangleq \frac{2y\dot{y}}{2\sigma_{\rho 0}^2} \tag{4.22}$$

$$a_3 \triangleq \frac{\dot{y}^2}{2\sigma_{\rho 0}^2} + \frac{y^2}{2\bar{p}_0} + \frac{1}{2\bar{p}}$$

The integration over $\dot{\rho}_t$ of an exponent of a second-order polynomial can be performed [9]:

$$f_{y,\dot{y}}(y,\dot{y}) = \int_0^\infty \sqrt{\frac{2}{\pi}} \frac{\rho_t^{2n+2} y}{\bar{p}^n \bar{p}_0 (n-1)! 2^n \sqrt{\sigma_{\rho 0}^2 + \sigma_{\rho t}^2 y^2}} \exp\left(\frac{a_2^2 - 4a_1 a_3}{4a_1}\rho_t^2\right) d\rho_t \tag{4.23}$$

This integral over ρ_t can be solved analytically using equation (15.77) in [9]. The joint pdf of the C/I amplitude ratio and its derivative with respect to time is found to be in the form of

$$f_{y,\dot{y}}(y,\dot{y}) = \frac{1*3*\ldots*(2n + 1)y(y^2\bar{p} + \bar{p}_0)^{n+1}\omega_m^{2n+2}}{\bar{p}^n\bar{p}_0(n - 1)!2^{2n+1}\left[\dot{y}^2 + \dfrac{(y^2\bar{p} + \bar{p}_0)^2\omega_m^2}{2\bar{p}\bar{p}_0}\right]^{n+3/2}} \tag{4.24}$$

This pdf can be used to determine the threshold crossing rate by applying the method used by Rice in [3–5]. The rate $N(z)$ at which the threshold z is crossed in the positive direction is found from

$$N(z) = \int_0^\infty \dot{y}\, f_{y,\dot{y}}(\sqrt{z},\dot{y})d\dot{y} \tag{4.25}$$

This integral is solved as

$$N(z) = \frac{\omega_m}{\sqrt{2\pi}} \frac{\Gamma(n + \frac{1}{2})}{\sqrt{n}\Gamma(n)} \frac{1}{\eta}\left(1 + \frac{1}{\eta^2 n}\right)^{-n} \tag{4.26}$$

where the fade margin η is defined as

$$\eta^2 \triangleq \frac{\bar{p}_0}{z\bar{p}_t} = \frac{\bar{p}_0}{zn\bar{p}} \tag{4.27}$$

The fade margin η is thus expressed as a ratio of amplitudes, rather than a ratio of power levels. In the following sections, we will consider not only the case of large fade margins ($\eta > 1$), which corresponds to the case of a strong signal with weak interference, but also the case of $\eta < 1$. In the latter situation, the interference is relatively strong. This case is highly relevant to contention-limited random-access channels in which many signals compete for access to a common receiver.

The threshold crossing rate is seen to be a function of the parameters ω_m, η, and n only. Figure 4.2 gives the threshold crossing rate as a function of fade margin η for various n.

If the fade margin η is less than unity (0 dB), the signal has an average power $\bar{p}_0$ below the average minimum power $zn\bar{p}$ required for reliable reception. In this event, nonfade periods (with $p_0 > zp_t$) will be rare and short. It is noted that the threshold crossing rate heavily depends on the number of components. With many

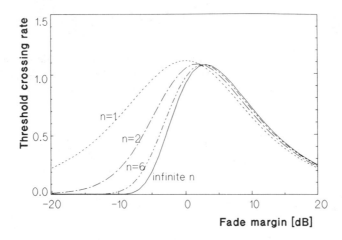

Figure 4.2 The normalized threshold crossing rate $N(z)/f_m$ versus fade margin in decibels for n interfering signals with equal mean power.

components ($n \gg 1$), the interference power is nearly constant. Crossing the threshold will be caused mainly by any temporary increase of the power of the wanted signal, which is rare; the corresponding threshold crossing rate increases sharply with η if η approaches unity (if $\sqrt{0.1} < \eta < 1$). On the other hand, for small n, the interference may exhibit fading, so a higher C/I can occasionally occur due to downfading of the joint interference. The threshold crossing rate only increases smoothly with increasing η. Figure 4.2 verifies that for one interferer ($n = 1$) the threshold crossing rate is correctly a symmetric function about $\eta = 1$ (0 dB): if the mean power levels of the wanted signal and the interferer are interchanged, the threshold crossing rate should remain unchanged (if $z = 1$).

The case of conventional radio communication corresponds to relatively low joint interference power ($n\bar{p} \ll \bar{p}_0$) and a threshold z of a few decibels, and, hence, a fade margin η much higher than unity ($\bar{p} \gg 1$). The threshold z will be crossed only when the wanted signal experiences a deep fade. Fading of the joint interference does not play a major role. Accordingly, the influence of the number of interferences is small.

This leads to the overall conclusion that, in the case of mobile telephony with its requirement of relatively high signal-to-interference most of the time (high η), any number of interfering signals may well be approximated by a nonfading noise-type signal with joint local-mean power $\bar{p}_t \approx n\bar{p}$. In contention-limited mobile packet radio nets, however, severe interference is allowed to occur (low η), since we accept collisions of contending packets which can be retransmitted. In the latter case, downfading of the interference signal can enhance the probability of correct packet reception if $n = 1$. This will be addressed in more detail in Section 4.10.

4.6 AVERAGE FADE/NONFADE DURATION

The probability of a signal outage ($C/I < z$) should be equal to the threshold crossing rate multiplied by the average duration of a fade. Similarly, the probability that $C/I > z$ equals the threshold crossing rate multiplied by the average nonfade duration.

The average nonfade duration is thus

$$\bar{\tau}_{NF}(z) = \frac{\Pr(\rho_0 > \sqrt{z}\rho_t)}{N(z)} = \frac{\Pr(y^2 > z)}{N(z)} \qquad (4.28)$$

Inserting (4.13) with (4.27) and (4.26) gives

$$\bar{\tau}_{NF}(z) = \eta \cdot \frac{\sqrt{2\pi}}{\omega_m} \cdot \frac{\sqrt{m}\Gamma(m)}{\Gamma(m + \frac{1}{2})} \qquad (4.29)$$

with $m = n$. This is a relatively simple expression, since all factors in (4.13) cancel with factors in (4.26). Equation (4.29) is in the form of a product of a constant, the fade margin η and a factor only depending on the m-parameter (i.e., on the number of interferers n). This is in contrast to the mathematical expression for the threshold crossing rate (4.26) or the outage probability (4.13), where the parameters η and m are interacting.

Figure 4.3 gives the average duration of a nonfade interval normalized to the maximum Doppler shift caused by vehicle mobility. It is seen that the influence of the number of interfering signals (or the m-parameter) is small, particularly if $m \geq 2$. This is understood from the asymptotic expansion [7]:

$$\chi \overset{\Delta}{=} \frac{\sqrt{m}\Gamma(m)}{\Gamma(m + \frac{1}{2})} \approx 1 + \frac{1}{8m} + \dots \qquad (4.30)$$

The value of χ for $m = 1$ is $\chi = 2/\sqrt{\pi} \approx 1.13$ [7]. The average nonfade interval decreases slightly with m. A possible explanation for this seemingly conflicting result is that in the case of one interferer ($m = n = 1$), the downfades of the wanted signal may coincide with the fades in the interference, and so prolong the capture period. Significant upfades of the interference, reducing the nonfade period, are not expected.

For practical receivers, the threshold z, and hence the fade margin η, may well be a function of the type of interference, and thus of m. In practice, the nonfade duration might be more dependent on the character of the interference signal than suggested in Figure 4.3.

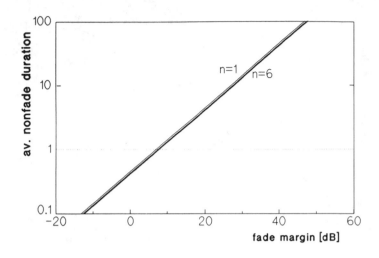

Figure 4.3 Normalized average nonfade duration $_{NF}f_m$ versus the fade margin η for $n = 1, 2, \ldots, 6$ interfering signals.

The average *fade* duration is found from

$$\bar{\tau}_F(z) = \frac{\Pr(\rho_0 < \sqrt{z}\rho_t)}{N(z)} = \frac{\Pr(y^2 < z)}{N(z)} \tag{4.31}$$

Inserting (4.13) with (4.27) and (4.26) gives

$$\bar{\tau}_F(z) = \eta \frac{\sqrt{2\pi}}{\omega_m} \left[\left(1 + \frac{1}{\eta^2 n}\right)^n - 1 \right] \frac{\sqrt{n}\Gamma(n)}{\Gamma(n + \frac{1}{2})} \tag{4.32}$$

For $\eta > 1$ (0 dB), the effect of the shape factor m (or n) of the interference is almost negligible (see Figure 4.4). However, for $\eta < 1$, m has a substantial effect on the average fade duration: for small η and large n, the wanted signal is almost continuously in a fade ($\bar{\tau}_F \to \infty$), whereas for $n = 1$, the fade of the wanted signal is occasionally interrupted by a fade of the interference signal, which causes $p_0 > zp_1$.

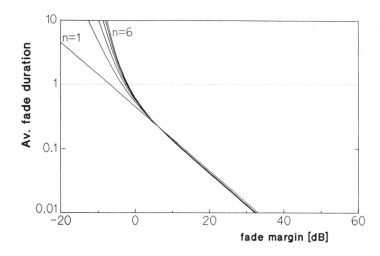

Figure 4.4 Normalized average fade duration $\bar{\tau}_F f_m$ versus the fade margin η in decibels for $n = 1, 2,$ ..., 6 interfering signals.

4.7 NOISE-LIMITED CHANNELS

Results for the rate of crossing a fixed (nonfading) threshold can be verified from (4.26) and (4.29). To this end, the number of interferers is increased without limit ($n \to \infty$, $m \to \infty$) under the condition that the total power $n\bar{p}$ is constant. This inherently implies that $\bar{p} \to 0$ and, thus, $\sigma_{pt} \to 0$. For large n the interference behaves like wide-sense stationary (nonfading) Gaussian noise. Any temporary fade of the amplitude of one signal out of the n interferers is likely to be hidden by $n - 1$ other transmissions.

We define η as it is in (4.27). Since $\Gamma(m + 1/2) / (\sqrt{m} \cdot \Gamma(m)) \to 1$ for $m \to \infty$ and $(1 + xm^{-1})^m \to e^x$ for $m \to \infty$, the limiting threshold crossing rate is

$$\lim_{n \to \infty} N(z) = \frac{\omega_m}{\sqrt{2\pi}} \frac{1}{\eta} \exp(-1/\eta^2) \tag{4.33}$$

Similarly, the average nonfade duration (4.29) and the average fade duration (4.32) tend toward the limits

$$\lim_{m \to \infty} \bar{\tau}_{NF}(z) = \lim_{m \to \infty} \eta \frac{\sqrt{2\pi m}\ \Gamma(m)}{\omega_m \Gamma(m + \frac{1}{2})} = \eta \frac{\sqrt{2\pi}}{\omega_m} \tag{4.34}$$

and

$$\lim_{n \to \infty} \bar{\tau}_F(z) = \eta[\exp(\eta^2) - 1] \frac{\sqrt{2\pi}}{\omega_m} \tag{4.35}$$

respectively.

4.8 DISTRIBUTION OF FADE/NONFADE DURATION

Rice [5] presented a method for computing the pdf of the duration of fades and illustrated the method with an analysis of reception of a Rayleigh-fading signal with a Gaussian fading spectrum. A Gaussian spectrum applies, for instance, to the event of fading over ionospheric shortwave (HF) channels. In land-mobile radio communication at VHF or UHF, the Doppler spectrum of the multipath fading caused vehicle motion with constant speed, is certainly not a Gaussian spectrum. Spectral components at $f_c + f_m \cos\phi_i$ exist because waves are arriving from an angle ϕ_i (see Section 2.3.1.2) with a certain (e.g., uniform) probability density. The spectrum is thus band-limited to $[f_c - f_m, f_c + f_m]$. If reflections arrive from random angles uniformly distributed over $[0, 2\pi\rangle$, a high concentration of power occurs near the boundaries of the Doppler band [1], irrespective of the direction of motion of the mobile antenna with respect to the fixed base station antenna. This leads to the relatively frequent occurrence of nonfade durations corresponding to the time required for a movement over one-half of a wavelength.

In urban environments with dominant reflections from particular directions, the Doppler spectrum may diverge from the theoretical models but is still band-limited within $[f_c - f_m, f_c + f_m]$. Analysis and experiments [10] showed that the distribution of the durations of deep fades (10 dB or more below the local-mean signal level) is not very sensitive to the exact shape of the Doppler spectrum [1, 10]. This is in contrast to the distribution of shallower fades, which is influenced more by the specific shape of the Doppler spectrum [10–12].

In an urban environment, the different vehicle speeds significantly influence the statistics of the fade/nonfade durations. Because of the typical urban traffic speed statistics, the durations of fade/nonfade periods are further spread about the dominant values, and the pdfs of fade/nonfade durations may tend toward an exponential one. However, a substantial percentage of the time an operating terminal may have zero speed. During these static periods, no contributions to threshold crossings are made.

4.9 RATE OF DROPOUTS

Short signal outages are experienced as clicks. The clicks cumulatively degrade the circuit, but appear too rapidly to be individually distinguishable [13]. Only longer outages are experienced as temporary "clipping" or a "dropout" of the circuit. The analysis of Chapter 3 addressed the probability of outage, irrespective of whether this outage would be experienced as a click or as a dropout. However, Gruber and Strawczynski [14] reported that the subjective rating of outages depends greatly on the duration of the speech clips. In this section, the rate of occurrence of outages that exceed a specified time T_0 is determined.

First, the significant time characteristics of speech are reviewed. Human speech signals contain periods of activity and of silence [15]. (This fact was already exploited in Section 3.2.2.) A number of models have been proposed for voice activity, many of which were based on a Markov Chain. For example, Goodman and Wei [16] used a speech activity model to evaluate the performance of a system with packetized voice, employing a random-access protocol (*packet reservation multiple access* (PRMA)) to accommodate multiple users. A rough model of voice activity is obtained by considering principal talkspurts and gaps in the speech as a two-state Markov Chain. As a refinement, minispurts and minigaps can be distinguished in each principal talkspurt (see Table 4.2).

Typically, a voice waveform can be described as a stationary statistical process during intervals on the order of 20 to 40 ms. This interval is substantially shorter than the duration of a syllable. Consonant-vowel-consonant syllables were analyzed and the following typical durations were observed by Richards [17]:

Initial consonants:	70 to 250 ms, average about 120 ms
Vowels:	150 to 500 ms, average about 250 ms
Final consonants:	100 to 350 ms, average about 190 ms

Experiments on satellite links revealed that speech clips longer than $T_0 = 50$ ms at the beginning of a talkspurt (called a *freeze-out* or *front end clipping* (FEC)) cause perceptible mutilation of initial plosive, stop, fricative, and nasal consonants [22]. For *time-assigned speech interpolation* (TASI, *digital speech interpolation* (DSI)) on satellite links, freeze-outs are caused by the competition among signals from various subscribers due to statistical multiplexing of multiple voice signals. Statistical multiplexing of voice signals in personal communication networks has been proposed by Goodman and Wei [16]. A typical requirement is that the probability of a clip longer than 50 ms at the beginning of a talkspurt should be less than 2%.

In mobile radio, clips caused by outages of the RF-signal occur at random instants (they are thus not necessarily at the beginning of talkspurts, as in TASI), and are called *midspeech burst clipping* (MSC) [14]. Little or no impairment results

Table 4.2
Average Duration of Events in Human Speech [16]

Condition	Av. Duration (sec)
Principal talkspurt	1.00
Principal gap	1.35
Minispur	0.275
Minigap	0.050

for less than 2% MSC with durations less than 4 ms. Clipping durations from 16 to 64 ms produce noticeable quality degradations unless the percentage of speech clipping is less than 0.2%. Although the subjective quality degrades for MSC with durations less than 64 ms, the intelligibility is not expected to be seriously degraded. Speech clips larger than 64 ms cause quality degradation as well as reduced intelligibility.

In cellular networks with CW transmission, MSC clipping occurs because of signal fades. If many voice signals are statistically multiplexed in common radio channels, as in PRMA [16], FEC can also occur, but this situation is not addressed here. In order to investigate the probability of MSC of excessive length, exponentially distributed fade durations are postulated with memoryless threshold crossings. Thus, the distribution of the fade duration is taken

$$f_{\tau_F}(\tau_F) = \frac{1}{\bar{\tau}_F} \exp(-\tau_F/\bar{\tau}_F) \tag{4.36}$$

The average number of dropouts per second, denoted by $N(z, T_0)$, is simply found as the number of threshold crossings per second multiplied by the probability that the fade lasts longer than T_0, so

$$N(z,T_0) = N(z) \cdot \exp(-T_0/\bar{\tau}_F) \tag{4.37}$$

Inserting (4.26) and (4.32) gives

$$N(z,T_0) = \frac{\omega_m \chi}{\sqrt{2\pi}} \frac{1}{\eta} \left(1 + \frac{1}{n\eta^2}\right)^{-n} \exp\left\{\frac{-\omega_m T_0}{\eta \sqrt{2\pi}} \chi \left[\left(1 + \frac{1}{\eta^2 n}\right)^n - 1\right]^{-1}\right\} \tag{4.38}$$

with χ defined in (4.30). Numerical results are in Figure 4.5 for the dropout rate, normalized to the maximum Doppler shift f_m, versus the fade margin η. A typical value of the maximum Doppler shift is $f_m \approx 60$ Hz, (e.g., with a carrier frequency of $f_c = 900$ MHz and vehicle speed of 72 km/h ($v = 20$ m/s)). For $T_0 = 16$ ms, we find $T = 1$.

For very small fade margins ($\eta < 1$), the signal is almost continuously in a fade. Since the nonfade duration tends toward zero if $\eta \to 0$, the dropout rate approaches the threshold crossing rate. For high fade margins, the dropout rate is almost independent of n. Long dropouts are relatively rare, particularly if the fade margin is higher than, say, 10 dB. However, at lower vehicle speeds and relatively small fade margins, some signal outages will be experienced as dropouts.

In digital transmission systems with protection against burst errors (e.g., by error-correcting coding), the dropout rate (4.38) is relevant for determining the probability of an error burst with excessive run length larger than the error-correcting capability of the code.

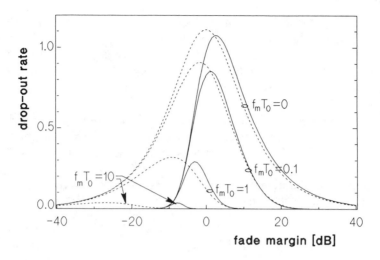

Figure 4.5 Normalized rate of dropouts, $N(z, T_0)/f_m$, with duration longer than the normalized criterion $T_0 f_m = 0$, 0.1, 1, and 10, versus the fade margin η in decibels. One interferer (- - -), multiple interferers (———), ($n = 6$).

4.10 PROBABILITY OF FADE-FREE RECEPTION OF A DATA BLOCK

In digital communication using blocks of L bits to support a data rate r_b, the duration of each block is $T_s = L r_b$. Successful reception of a data block is assumed to occur if the received signal power exceeds the interference power for the entire duration of the block. The case of Rayleigh-fading signals with known local-mean power levels is addressed. Exponentially distributed nonfade durations are postulated with memoryless threshold crossings. Hence, analogous to (4.36),

$$f_{\tau_{NF}}(\tau_{NF}) = \frac{1}{\bar{\tau}_{NF}} \exp\left(-\frac{\tau_{NF}}{\bar{\tau}_{NF}}\right) \tag{4.39}$$

The probability that a nonfade period continues for a duration of at least T_s is found by the integration of (4.39) from T_s to ∞. The probability of successful reception (event A_0) becomes

$$\Pr(A_0) = \Pr(c_0) \Pr(\tau_{NF} > T_s)$$

$$= \Pr(c_0) \exp\left(-\frac{T_s}{\bar{\tau}_{NF}}\right) \tag{4.40}$$

where c_0 denotes the event that the C/I ratio is above the threshold. We now compare the case of interference caused by a single Rayleigh-fading signal with the event of cumulation of many weak interfering signals. The wanted signal experiences Rayleigh fading with exponentially distributed received power. So, in the event of many interferers,

$$\Pr(c_0|n \to \infty) = \exp\left(-\frac{z\bar{p}_t}{\bar{p}_0}\right)$$

(4.41)

For one Rayleigh-fading interferer, it was shown in Chapter 3 (equation (3.18)) that

$$\Pr(c_0|n = 1) = \frac{\bar{p}_0}{\bar{p}_0 + z\bar{p}_1}$$

(4.42)

The asymptotic behavior for strong interference of (4.41) and (4.42) is seen to be essentially different. With many interferers, the probability c_0 decreases exponentially if the interference power is increased. This is in contrast to the case of one single signal, where the probability is only inversely proportional to the interference power. In the event of $n = 1$, the signal can be received correctly for the short duration of a deep fade of the interferer. However, we now show that this relatively slow degradation of $\Pr(c_0)$ holds only for signal segments of infinitely short duration ($T_s = 0$). The probability of correct reception of a segment of finite duration vanishes exponentially if the interference power is nonzero, even when the interference is caused by a single Rayleigh-fading signal.

For a single interferer, the probability of successful reception is

$$\Pr(A_0|n = 1) = \frac{\eta^2}{\eta^2 + 1} \exp\left\{-\frac{T_s \omega_m}{\sqrt{2\pi\eta}} \frac{\Gamma(1\frac{1}{2})}{\Gamma(1)}\right\}$$

(4.43)

where $\Gamma(3/2) = 1/2 \cdot \sqrt{\pi}$ and $\Gamma(1) = 1$. In the event of interference composed of many weak signals, one obtains

$$\Pr(A_0|n \to \infty) = \exp\left(-\frac{1}{\eta^2} - \frac{T_s \omega_m}{\sqrt{2\pi\eta}}\right)$$

(4.44)

Figure 4.6 illustrates the corresponding probabilities as a function of the fade margin η for various normalized packet durations T_n ($T_n = T_s f_m$).

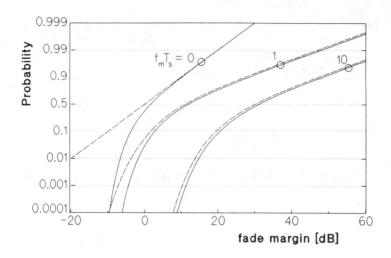

Figure 4.6 Probability of reception of a data block without fade as a function of the fade margin, (- - -) for one interferer and (——) for many interferers. Normalized block duration $T_n = T_s f_m = 0$, 1, and 10.

4.10.1 Effective Throughput

The above analysis showed a decreasing probability of successful reception for increasing packet duration. However, longer packets contain more user data and relatively less overhead.

The effective throughput (i.e., $\Pr(A_0) \cdot L_u / (L_H + L_u)$ with L_u the number of user-data bits and L_H the number of overhead bits) is depicted in Figure 4.7. The duration of the data block (or the data packet) is normalized to the Doppler modulation according to $T_n = T_s f_m$. The packet duration T_s in seconds, of course, depends on both the bit rate and the data format [18–21]. We give two examples:

HDLC: In the High-Level Data Link Control (HDLC) protocol [18,19], a frame contains at least $L_H = 48$ overhead bits to accommodate a start flag, address and control bytes, and CRC block error detection. At a bit rate of $r_b = 1200$ b/s, this corresponds to $T_s = 0.04$ sec ($T_n = 2.4$ if $f_m = 60$ Hz) and at $r_b = 16$ kb/s to $T_s = 3$ ms ($T_n = 0.18$ if $f_m = 60$ Hz).

MOBITEX: In the first generation of the Mobitex public packet-switched mobile network [20], the bit rate is $r_b = 1200$ b/s. For a data message of minimum duration ($T_s \approx 13.3$ ms, $L = 16$ bits), $T_n = 0.8$ is found. The maximum packet duration ($T_s = 3.315$ sec for $L = 255$ bits) corresponds to $T_n \approx 200$, though in this case the packet is split into smaller blocks and a selective repeat algorithm is used to enhance the throughput efficiency if blocks are lost in a signal fade.

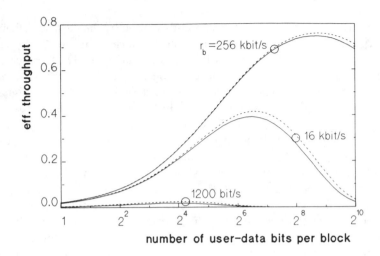

number of user–data bits per block

Figure 4.7 Effective throughput versus the number of user-data bits L_u. Total block duration $T_s = (L_H + L_u)/r_b$ with $L_H = 48$ header bits. Maximum Doppler shift $f_m = 60$ Hz. Bit rates $r_b = 1200$, 16k, and 256 kb/s. Many interferers (———), one interferer (- - -). Fade margin $\eta = \sqrt{10}$ (10 dB).

   In systems with higher data rates, T_n is found to be substantially smaller. In the GSM-system, a time slot has duration $T_s \approx 0.577$ ms, so $T_n \approx 0.035$ if $f_m = 60$ Hz. Results in (4.43), (4.44), and Figure 4.7 have been derived for narrowband Rayleigh fading. For high bit rates ($T_3 << T_2$), however, reception may be impaired by intersymbol interference. In this case, the effect of increasing the bit rate is less favorable than depicted in Figure 4.7.

4.11 CONCLUSION

In mobile radio, not only the probability of signal outage but also the duration of outages determine the quality of communication. The threshold crossing rate and the average nonfade duration have been studied analytically for the case of a Rayleigh-fading signal in the presence of n interfering signals. The fading of the interfering signals has been assumed to be mutually independent, with identical mean power levels and Doppler modulation widths. The mathematical analysis presented here is a good approximation of the outbound (base-to-mobile) channel in a cellular network. Since only the receiver is in motion and multipath scattering occurs in the vicinity of the receiving antenna, the width of the Doppler spectrum is equal for the wanted signal and all interfering signals. For signals arriving from different sources, Rayleigh fading is nearly independent. Shadowing caused by features in the vicinity of mobile is correlated and may thus be less relevant in an interference-limited design of the downlink. Further, interference is mainly caused by the closest cochannel base stations (typically six), which are all at about the reuse distance. Under such conditions, the average fade duration can be expressed analytically, as described here.

The expressions for the threshold crossing rate and the average nonfade duration have a more general application whenever the joint interference signal can be approximated by a Nakagami-fading signal. In the inbound random-access channel of a packet-radio network, the assumption of equal mean power for all interferers is less realistic. For this case, it has been demonstrated that a Nakagami pdf closely approximates the pdf of the joint interference power. The results also suggested that mainly the rate of fading of the wanted signal will determine the average fade duration. A nonfade duration is most likely to be terminated by a downfade of the wanted signal, whereas an upfade of the joint interference signal is very unlikely. As far as the average nonfade duration is concerned, it has been shown that a further approximation may be reasonable. The exact distribution of the received power from the different fading interfering signals appeared less relevant. Nonfade durations may therefore be assessed by considering (multiple) interfering signals to be a noise-type signal with constant power.

REFERENCES

[1]   Jakes, W.C., Jr., ed., *Microwave Mobile Communications*, New York: John Wiley and Sons, 1974.

[2]   Lee, W.C.Y., *Mobile Communications Design Fundamentals*, Indianapolis: Howard W. Sams & Co, 1986.

[3]   Rice, S.O., "Mathematical Analysis of Random Noise," *Bell Sys. Tech. Journal*, Vol. 24, No. 1, Jan. 1945, pp. 46–156.

[4]   Rice, S.O., "Statistical Properties of a Sine Wave Plus Random Noise," *Bell Sys. Tech. Journal*, Vol. 27, No. 1, Jan. 1948, pp. 109–157.

[5]   Rice, S.O., "Distribution of the Duration of Fades in Radio Transmission: Gaussian Noise Model," *Bell Sys. Tech. Journal*, Vol. 37, No. 1, May 1958, pp. 581–635.

[6]   Lank, G.W., and I.S. Reed, "Average Time to Loss of Lock for an Automatic Frequency Control Loop With Two Fading Signals and a Related Probability Density Function," *IEEE Trans. Inf. Theory*, Vol. IT-12, No. 1, Jan. 1966, pp. 73–75.

[7]   Abramowitz, M., and I.A. Stegun, *Handbook of Mathematical Functions*, New York: Dover, 1965.

[8]   Spiegel, M.R., *Laplace Transforms*, Schaum's Outline Series, New York: McGraw Hill, 1965.

[9]   Spiegel, M.R., *Mathematical Handbook of Formulas and Tables*, Schaum's Outline Series, New York: McGraw-Hill, 1968.

[10]   Bodtmann, W.F., and H.W. Arnold, "Fade Duration Statistics of a Rayleigh Distributed Wave," *IEEE Trans. on Comm.*, Vol. COM-30, No. 3, March 1982, pp. 549–553.

[11]   Swarts, F., H.C. Ferreira, and D.R. Oosthuizen, "Renewal Channel Model for PSK on Slowly Fading Rayleigh Channel," *Electron. Lett.*, Vol. 25, No. 22, 26 Oct. 1989, pp. 1514–1515.

[12]   Krantzik, A., and D. Wolf, "Statistical Properties of Fading Processes Describing a Land Mobile Radio Channel" (in German), *Frequenz*, Vol. 44, No. 6, 1990, pp. 174–182.

[13]   *Recommendations and Reports of the CCIR, 1986, Volume VIII-1: Land Mobile Service, Amateur Service, Amateur Satellite Services*, XVIth Plenary Assembly, Dubrovnik, 1986.

[14]   Gruber, J., and L. Strawczynski, "Subjective Effects of Variable Delay and Speech Clipping in Dynamically Managed Voice Systems," *IEEE Trans. on Comm.*, Vol. COM-33, No. 8, August 1985, pp. 801–808.

[15]   Brady, P.T., "A Model for Generating On-Off Speech Patterns in Two-Way Conversation," *Bell Sys. Tech. Journal*, Vol. 48, No. 7, Sept. 1969, pp. 2445–2472.

[16]   Goodman, D.J., and S.X. Wei, "Factors Affecting the Bandwidth Efficiency of Packet Reservation Multiple Access," *Proc. IEEE Veh. Techn. Conf. 1989*, San Francisco, 3–5 May 1989, pp. 292–298.

[17]   D.L., Richards, *Communications by Speech*, London: Butterworth & Co., 1973.

[18]   Schwartz, M., *Telecommunication Networks Protocols, Modelling and Analysis*, Reading, MA: Addison-Wesley, 1987, pp. 135–156.

[19]   "Technical Characteristics and Test Conditions for Non-speech and Combined Analogue Speech/Non-speech Radio Equipment With Internal or External Antenna Connector, Intended for the Transmission of Data, for Use in the Land-Mobile Service" (DRAFT), European Telecommunications Standards Institute, I-ETS [A] version 3.3.2, Valbonne, France, 1990.

[20]   *Mobitex Terminal Specification LZBA 703 1001*, Swedish Telecommunications Administration, Televerket Radio, Farsta, Sweden, Oct. 1986.

[21]   Siew, S.K., and D.J. Goodman, "Packet Data Transmission Over Mobile Radio Channels," *IEEE Trans. on Veh. Tech.*, Vol. COM-38, No. 2, May 1989, pp. 95–101.

[22]   Campanella, S.J., "Digital Speech Interpolation," *Comsat Tech. Rev.*, Vol. 6, No. 1, Spring 1976, pp. 127–158.

Chapter 5
Average Bit Error Rate in CW Communication

The performance of generic receivers in narrowband multiuser radio networks has been addressed in Chapters 3 and 4, with outage probability used as the performance measure. In the specific case of digital communication, the probability of bit error or block erasure provides another relevant measure of the performance of a radio link. Most studies of the average bit error rate (BER) in a fading channel (e.g., [1]) only consider AWGN without interference. On the other hand, studies with cochannel interference mostly consider channels without fading (e.g., [2]). The combined effect of multipath fading and cochannel interference is rapidly becoming an area of active research (e.g., [3–6]). Recently, Steele et al. [4, 5] studied the BER for an MSK signal in mobile networks with Rayleigh fading and multiple interfering signals. For our purpose, models as proposed in [5] appear prohibitively complicated. The motivation for this chapter was the need for a mathematically tractable, yet reasonably accurate expression to study the performance of digital communication systems. To this end, the bit error rate for channels with Rayleigh fading, noise and multiple interferers, and using standard receivers optimized for AWGN channels [1, 7] are considered with a number of simplifying assumptions, allowing us to compute the long-term average BER in microcellular and macrocellular networks, the probability of successful reception of a block of data, and (in Chapter 7) to calculate the throughput of random-access networks. Since the model proposed here is possibly too crude to compare the performance of slightly different digital modulation techniques, we confine ourselves to BPSK.

This chapter is organized as follows: Section 5.1 briefly reviews standard BPSK receivers and associated models proposed in the literature for cochannel interference in fading channels. Section 5.2 deals with the average BER for a desired BPSK signal with given (instantaneous) power, received from a mobile transmitter in the presence of AWGN and multiple interfering BPSK signals with Rayleigh fading. The local-mean BER is determined in Section 5.3, and area-mean BERs in cellular networks are calculated in Sections 5.4 and 5.5 for microcellular and macrocellular networks, respectively. Approximate BER expressions are derived at high average signal-to-interference ratio and negligible noise level. Section 5.6 removes the (pessimistic) approximation that the bits of each interfering signal have precisely aligned bit timing. Section 5.7 addresses the probability of successful reception of a block of bits. Conclusions are in Section 5.8.

5.1 BRIEF REVIEW OF MODELS FOR BER IN MOBILE CHANNELS WITH COCHANNEL INTERFERENCE

Optimum receivers for BPSK are well known for the AWGN channel, and their bit error rates have been studied extensively [1, 7, 8]. An optimum receiver can be implemented by correlating (e.g., by multiplying and integrating) the received signal with a locally generated perfect copy of the carrier. A block diagram of this receiver is shown in Figure 5.1. Bandpass filters, present in any realistic receiver for practical reasons (e.g., to remove adjacent-channel interference) are not drawn in Figure 5.1. From a theoretical point of view, they are obsolete [7], and imperfect bandpass filters may even degrade performance.

Ideally, the BER for coherent detection of BPSK in time-invariant AWGN channels is [1, 7]

$$P_b(e) = \frac{1}{2} \text{erfc} \sqrt{\frac{E_b}{N_0}} \qquad (5.1)$$

where N_0 is the (one-sided) spectral power density of the AWGN, E_b is the (constant) energy per bit ($E_b = p_0 T_b$) and erfc denotes the complementary error function [9].[1]

Extension of these results to analysis of the BER in realistic mobile fading channels with interference requires consideration of a number of possible issues, such as:

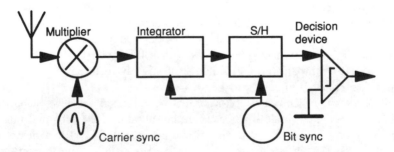

Figure 5.1 Block diagram of a correlation receiver for coherent detection of BPSK signals.

---

[1]The erfc-function is defined as $1 - \text{erf}(\cdot)$, with the erf-function described in Section 7.2. Expressions in terms of the $Q(\cdot)$-function [8] are obtained from $\text{erfc}(k) = 2Q(\sqrt{2}\,k)$.

1. Additive noise (manmade or thermal);
2. Amplitude (or energy per bit) of the received signal;
3. Fluctuations of the amplitude and phase of the wanted signal during detection of a bit (particularly if $T_3 > T_6$);
4. Delay spread of the wanted signal, causing ISI (particularly if $T_3 < T_2$);
5. Synchronization impairments caused by fading (particularly phase fluctuations) of the wanted signal during previous bits;
6. Parameters of the interfering signal(s), particularly:
    (a) Carrier phase offsets;
    (b) Bit timing (phase) offsets;
    (c) Amplitudes of the interfering signals;
    (d) Carrier frequency and bit (clock) frequency offsets;
    (e) Fading of the amplitude of interfering signals during detection of a bit.
7. Synchronization impairments of the receiver caused by the interfering signals;
8. Delay spread of interfering signals.

Since only narrowband systems with frequency-nonselective ("flat") fading channels are addressed here ($T_2 << T_3$), we ignore effects 4 and 8. The time constants of the fading [10] are assumed to be large compared to the bit duration ($T_3 << T_6$). This simplifies 3 and 3(e). Assuming receivers perfectly synchronizing to the wanted signal, impairments 5 and 7 are neglected.

A number of models and expressions have been proposed for the BER in the presence of cochannel interference; for instance:

1. Sheikh, Yao, and Wu [11] assumed bit error to occur with a probability of one-half if the power of the wanted signal fails to exceed the power of the strongest interfering signal; thus,

$$P_b(e) = \frac{1}{2} \Pr\left(p_0 < \max_{i(i=1,2,\ldots,n)} p_i \right) \tag{5.2}$$

The analysis closely resembles the analysis of outage probabilities.

2. Zhang, Pahlavan, and Ganesh [12] approximated cochannel interference as AWGN, resulting in the bit error probability

$$P_b(e|p_0,p_t) = \frac{1}{2} \operatorname{erfc} \sqrt{\frac{E_b}{N_0 + p_t T_b}} \tag{5.3}$$

Here, the variance of the Gaussian noise is increased with the *instantaneous* power of the joint interference signal.

3. Habbab, Kavehrad, and Sundberg [13] and later papers by Zhang and Pahlavan [14] and Linnartz, Goossen, and Hekmat [15] also assumed that cochannel interference behaves like AWGN, but now with error probability

$$P_b(e|p_0, \bar{p}_t) = \frac{1}{2} \text{erfc} \sqrt{\frac{E_b}{N_0 + \bar{p}_t T_b}} \tag{5.4}$$

In contrast to (5.3), the variance of the noise is increased with the *local-mean* power of the joint interference signal.

4. Wong and Steele [5] studied the BER for MSK modulation. The effect of co-channel interference on the decision variable, as extracted by a standard receiver, was taken into account for independent bit timing for all independent transmitters (6(a) and (b)). These results were subsequently averaged over the pdf of the received signal amplitudes caused by fading (2 and 6(c)).

Models 1 to 3 considerably simplify the analysis. In Section 5.2, it will be shown that model 3 can be recovered by simplifying model 4. The effect of one of the simplifications is studied in Section 5.5.

5.2 INSTANTANEOUS BER FOR BPSK SIGNALS WITH RAYLEIGH-FADING COCHANNEL INTERFERENCE[2]

First, precisely overlapping bit periods are assumed for all interfering signals; that is, none of the interfering carriers makes a phase reversal during the integration over the bit duration of the wanted signal. Interfering signals are assumed to have exactly the same carrier and bit clock frequency as the wanted signal: phase fluctuations (e.g., caused by frequency drifts or Doppler shifts) are considered negligible during each bit duration. In fact, at f_c = 900 MHz, random Doppler shifts are usually on the order of f_m = 50 Hz, whereas usual bit rates are at least 1200 b/s ($T_3 << T_6$). The momentary derivative of the phase $\dot{\theta}(t)$ may fall outside the Doppler spectrum, but the probability that $2\pi|\dot{\theta}(t)| >> f_m$ is small, and this event usually coincides with deep fades of the interference signal [10]. Hence, the effect of multipath reception on phase fluctuation during one bit interval is presumably negligible. The joint receiver input signal $v(t)$ is in the form

$$v(t) = \rho_0 \kappa_0 \cos(\omega_c t + \theta_0) + \sum_{i=1}^{n} \rho_i \kappa_i \cos(\omega_c t + \theta_i) + n(t) \tag{5.5}$$

[2]Section 5.2–5.6: Portions reprinted, with permission, from *Proceedings of 6th IEE Int. Conf. on Mobile Radio and Personal Communications*, Warwick, U.K., 9–11 Dec. 1991, pp. 241–247. © 1991 IEE.

where $\kappa_i(\kappa_i = \pm 1)$ represents the antipodal binary phase modulation of the ith carrier and $n(t)$ is an AWGN process. The received energy per desired bit is $E_b = p_0 T_b = (1/2)\rho_0^2 T_b$. In the detector, $v(t)$ is multiplied by a locally generated cosine ($2 \cos \omega_c t$) and integrated over the entire bit duration T_b. The decision variable v for synchronous bit extraction from the desired BPSK signal (having index 0) in the presence of n interferers with random (but constant) phase relative to the local oscillator is thus

$$v = \frac{2}{T_b} = \int_{kT_b}^{(k+1)T_b} v(t) \cos(\omega_c t) \, dt$$

$$= \rho_0 \kappa_0 \cos(\theta_0) + \sum_{i=1}^{n} \zeta_i \kappa_i + n_I$$

(5.6)

with n_I the inphase sample of the Gaussian noise, and $\kappa_i \rho_i$ is the sample of the ith interfering signal. As shown in Appendix A, the variance of this noise sample is

$$E[n_I^2] = \frac{N_0}{T_b}$$

(5.7)

In a Rayleigh-fading channel, the inphase components ζ_i ($\zeta_i = \rho_i \cos \theta_i$) of the n interferers are all independent Gaussian variables (see Section 2.1.3). Phase reversals caused by modulation of κ_i do not affect the Gaussian distribution of the n interference samples, provided that the interfering bits are exactly aligned to the bits of the wanted signal. The variance of the ith interference sample is

$$E[\zeta_i^2] = \bar{p}_i$$

(5.8)

Thus, AWGN as well as the Rayleigh-fading interference have a Gaussian pdf, but the correlation between the samples during successive bit intervals differs substantially. For AWGN, successive samples are uncorrelated, whereas for a slow-fading BPSK signal, successive samples have almost identical amplitude ζ_i. The Gaussian-distributed inphase samples $[\zeta_i]$ originating from the interference signals plus the Gaussian noise sample produce a new Gaussian variable. The corresponding conditional bit error probability for a receiver locked to the wanted signal with BPSK, for the event that $\kappa_0 = -1$, is

$$P_b(e|p_0, \bar{p}_t, \kappa_0 = -1) = \Pr\left(\sum_{i=1}^{n} \zeta_i \kappa_i + n_I > \rho_0\right)$$

$$= \int_{\rho_0}^{\infty} \frac{1}{\sqrt{2\pi(\bar{p}_t + (N_0/T_b))}} \exp\left(-\frac{\frac{1}{2}\rho^2 T_b}{N_0 + \bar{p}_t T_b}\right) d\rho$$

(5.9)

Because of symmetry, this BER also holds if the wanted signal carries a 1 (and, thus, $\kappa_0 = 1$). Therefore, writing the last integral in (5.9) as the complementary error function,

$$P_b(e|p_0,\bar{p}_t) = \frac{1}{2}\,\text{erfc}\sqrt{\frac{p_0 T_b}{K_i\bar{p}_t T_b + K_n N_0}} \tag{5.10}$$

with the constants $K_i = 1$ and $K_n = 1$, respectively. This expression confirms model 3 in Section 5.1.

If, analogously, one assumes interference with exact bit alignment in the cases of *quadrature phase shift keying* (QPSK) or *binary frequency shift keying* (BFSK) signals, one finds $K_i = 2$ and $K_n = 1$, or $K_i = 1$, and $K_n = 2$, respectively [16, 17]. It is well known that in classical time-invariant AWGN channels, antipodal modulation (such as BPSK) is 3 dB more effective than orthogonal modulation (such as FSK) to combat AWGN. This is confirmed by the above values of K_n. Interestingly, the results suggest that BPSK and BFSK are *equally* resistant to the cochannel interference in narrowband Rayleigh-fading channels and 3 dB *better* than QPSK when standard correlation receivers are employed. Of course, a more detailed model is required to verify whether this conclusion holds for the more realistic assumption of mutually independent transmitters (i.e., with random timing and clock offsets).

The expression (5.10) is now used to compute the average BER in cellular networks.

5.3 LOCAL-MEAN BER IN MICROCELLULAR AND MACROCELLULAR NETWORKS

For a Rician-fading wanted signal, the local-mean BER is found by integrating (5.10) over the Rician distribution (2.19) of the signal amplitude

$$\bar{P}_b \triangleq P_b(e|C_0,\bar{q}_0,\bar{p}_t) = \int_0^\infty \frac{\rho_0}{\bar{q}_0}\exp\left(-\frac{\rho_0^2 + C_0^2}{2\bar{q}_0}\right) I_0\left(\frac{\rho_0 C_0}{\bar{q}_0}\right)\frac{1}{2}\,\text{erfc}\sqrt{\frac{\frac{1}{2}\rho_0^2 T_b}{\bar{p}_t T_b + N_0}}\,d\rho_0 \tag{5.11}$$

Numerical results are given in Figure 5.2. The local-mean C/I ratio is defined as $\bar{p}_0/\bar{p}_t = (\bar{q}_{s0} + (1/2)C_0^2)/\bar{p}_t$. Various values of the Rician K-factor ($K = C_0^2/2\bar{q}_{s0}$) are given.

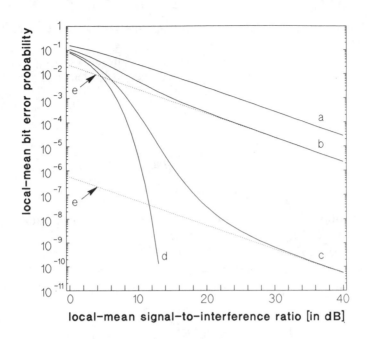

Figure 5.2 Local-mean BER versus the local-mean signal-to-joint-interference ratio. Rician parameter: (a)$K = 0$ (Rayleigh fading), (b)$K = 4$ (6 dB), (c)$K = 16$ (12 dB), and (d)$K \to \infty$ (no fading). Approximate expression for large C/I ratios ($\cdots$, e).

At high C/I ratios, bit errors mainly occur if the wanted signal happens to be in a deep fade, and, thus, for small ρ_0. The zeroth-order modified Bessel function $I_0(x)$ is always larger than one for real and positive arguments x, and $I_0(x) \approx 1 + (1/4)x^2 + \ldots \to 1$ for small x [9]. The behavior during fades (small ρ_0) mainly determines the BER at large C/I ratios. Thus, the integral (5.11) can be overbounded by

$$\bar{P}_b \geq e^{-K} \int_0^\infty e^{-x} \operatorname{erfc} \sqrt{\frac{\bar{p}_0 T_b \, x}{(1 + K)(\bar{p}_t T_b + N_0)}} \, dx$$

$$= \frac{1}{2} - \frac{1}{2} \sqrt{\frac{\bar{p}_0 T_b}{\bar{p}_0 T_b + (K + 1)(\bar{p}_t T_b + N_0)}} \qquad (5.12)$$

$$\to (K + 1)e^{-K} \frac{\bar{p}_t}{4\bar{p}_0} \quad \text{for} \quad \bar{p}_0 \gg \bar{p}_t \gg \frac{N_0}{T_b}$$

Figure 5.2 shows that (5.12) asymptotically approaches the exact expression (5.11) for large C/I ratios.

In the limiting case of Rayleigh fading, the instantaneous power p_0 of the test signal is exponentially distributed ($C_0 = 0$, $\bar{p}_0 = \bar{q}_{s0}$). This gives the local-mean BER

$$\bar{P}_b = \frac{1}{2} - \frac{1}{2} \sqrt{\frac{\bar{p}_0 T_b}{\bar{p}_0 T_b + \bar{p}_t T_b + N_0}} \tag{5.13}$$

for any C/I ratio. The special case of a noise-limited channel without cochannel interference (AWGN channel; $p_t = 0$) agrees with [1].

As seen from Figure 5.2, the Rician K-factor greatly affects the local-mean BER. Even if the scattered component is relatively weak ($K > 16$ (12 dB)), Rician fading substantially increases the BER above to the ideal (nonfading) case with $K \to \infty$.

5.4 AREA-MEAN BER IN MICROCELLULAR NETWORKS

In a microcellular network, the terminal transmitting the wanted signal is often within line of sight of the base station. The wanted signal is then not likely to experience shadowing, but interference signals may propagate over obstructed paths.

As discussed in Section 3.1.2.4, the pdf of the local-mean power sum $\bar{p}_t$ of a number of n log-normally distributed signals can be approximated by a log-normal distribution. The area-mean power and the standard deviation obtained with the method proposed by Schwartz and Yeh are given in Table 3.2. The conditional bit error probability (5.11) is averaged over the log-normal distribution (2.14) of the local-mean joint interference power $\bar{p}_t$.

In Figure 5.3, the area-mean C/I ratio is defined as $[(1/2)C_0^2 + \bar{q}_0]/\bar{p}_i$, which is in contrast to the use of *joint* interference powers in Figure 5.2. The definition adopted here allows the reader to use Figure 5.3 without first having to apply Schwartz and Yeh's method.

It can be shown (see [16] and the method in Section 5.5.1) that, for high C/I ratios, the BER tends toward

$$P_b(e|\bar{p}_0, \bar{p}_t) \to \frac{\bar{p}_t}{4\bar{p}_0} (K + 1) \exp\left(-K + \frac{1}{2} \sigma_t^2\right) \tag{5.14}$$

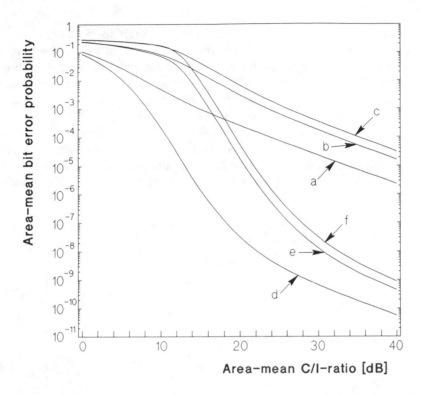

Figure 5.3 Area-mean BER in a microcellular network with six interferers versus area-mean signal-to-interference ratio $\bar{p}_0/\bar{p}_i$. Wanted signal: (a, b, c) $K = 4$ (6 dB); (d, e, f) $K = 16$ (12 dB). Rayleigh fading interfering signal: (a, d) without shadowing; (b, e) shadowing with $s_s = 6$ dB ($\bar{p}_t = 11.1\bar{p}_i$ and $s_t = 3.04$ dB); and (c, f) shadowing with $s_s = 12$ dB ($\bar{p}_t = 43.7\bar{p}_i$ and $s_t = 6.74$ dB).

5.5 AREA-MEAN BER IN MACROCELLULAR NETWORKS

In macrocellular networks, the wanted and interfering signals experience Rayleigh fading and shadowing. The local-mean BER is given by expression (5.13). To compute the area-mean BER, this expression must be averaged over the log-normal distribution of the local-mean power levels of the wanted and the joint interference signal. The integration variables $\bar{p}_0$ and $\bar{p}_t$ are substituted with logarithmic variables, defined analogously to (2.33). This gives

$$\bar{P}_b = \frac{1}{2} - \frac{1}{2\pi} \int_{-\infty}^{\infty}\int_{-\infty}^{\infty} dx dy \exp(-x^2 - y^2)$$

$$\cdot \sqrt{\frac{\bar{p}_0 T_b \exp(\sqrt{2}x\sigma_s)}{\bar{p}_0 T_b \exp(\sqrt{2}x\sigma_s) + \bar{p}_t T_b \exp(\sqrt{2}y\sigma_t) + N_0}}$$

(5.15)

No closed-form solution has been found for the integrals in (5.15). To obtain numerical results, the Hermite polynomial method [9] is used twice, resulting in the bit error probability being approximated by a double sum. Figure 5.4 presents the average bit error rate for channels without Gaussian noise ($N_0 = 0$). The area-mean C/I ratio is defined as $\bar{\bar{p}}_0/\bar{\bar{p}}_i$, which corresponds to r_u^β and $(3C)^{(1/2)\beta}$.

Shadowing of the desired signal is taken $\sigma_s = 1.36$ and 2.72 ($s_s = 6$dB and 12 dB). Six cochannel interfering signals with mutually independent shadowing with identical standard deviation ($\sigma_i = \sigma_s$ for $i = 1, 2, \ldots, 6$) are assumed. As seen from Table 3.1, the joint interference signal has a logarithmic standard deviation of $s_t = 3.04$ dB and 6.74 dB for $s_s = 6$ dB and 12 dB, respectively. For 12 dB of shadowing

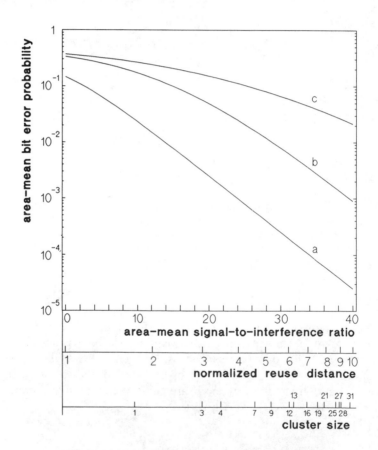

Figure 5.4 Area-mean BER for BPSK in macrocellular network with six interfering signals versus area-mean C/I ratio per interferer and corresponding reuse distance and cluster size L; (a) without shadowing; (b) shadowing with $s_s = 6$ dB ($\bar{\bar{p}}_t = 11.1\bar{\bar{p}}_i$, $s_t = 3.04$ dB); and (c) shadowing with $s_s = 12$ dB ($\bar{\bar{p}}_t = 43.7\bar{\bar{p}}_i$, $s_t = 6.74$ dB). UHF groundwave propagation with $\beta = 4$.

($\sigma_s = 2.72$), the area-mean BER remains unacceptably high, even for area-mean C/I ratios of 40 dB ($\bar{\bar{p}}_0/\bar{\bar{p}}_i = 10^4$) which corresponds to $\bar{\bar{p}}_0/\bar{\bar{p}}_t \approx 22.9$ (14 dB).

5.5.1 Approximate Expressions for Large C/I Ratios

In the event of negligible noise and high signal-to-interference ratios ($\bar{\bar{p}}_0 \gg \bar{\bar{p}}_t \gg N_0$), (5.15) can be approximated by

$$\bar{P}_b \approx \frac{1}{2\pi} \int_{-\infty}^{\infty}\!\!\int_{-\infty}^{\infty} dx\, dy\, \exp(-x^2 - y^2)\left[1 - \sqrt{1 - \frac{\bar{p}_t \exp(\sqrt{2}y\sigma_t)}{\bar{p}_0 \exp(\sqrt{2}x\sigma_s)}}\right] \quad (5.16)$$

After a first-order expansion of the square root, one finds

$$\bar{P}_b \approx \frac{\bar{\bar{p}}_t}{4\pi\bar{\bar{p}}_0} \int_{-\infty}^{\infty}\!\!\int_{-\infty}^{\infty} \exp(-x^2 - y^2 + \sqrt{2}y\sigma_t - \sqrt{2}x\sigma_s)\, dx\, dy$$

$$= \frac{\bar{\bar{p}}_t}{4\bar{\bar{p}}_0} \exp\left(\frac{\sigma_s^2 + \sigma_t^2}{2}\right) \quad (5.17)$$

where $\bar{p}_t$ depends on the standard deviation σ_i (see Table 3.2). It can be concluded that shadowing substantially affects the area-mean BER. For instance, with $s_i = 12$ dB for the wanted signal and each interfering signal ($s_t = 6.74$ dB and $\bar{p}_t = 43.7\bar{p}_i$, see Table 3.2), the area-mean BER is approximately $1.1 \cdot 10^3$ times larger than the BER in channels without shadowing ($s_s = 0$ dB, $s_t = 0$ dB, and $\bar{p}_t = 6\bar{p}_i$). If the shadowing of the wanted signal and of the joint interference signal are correlated, the BER is presumably less influenced.

5.6 EFFECT OF NONSYNCHRONOUS INTERFERENCE

In the previous sections, the receiver and interference models were greatly simplified to obtain mathematically tractable expressions for the BER. This required relatively strong assumptions and approximations, which were not qualified. It is the author's impression that the assumption of perfect bit alignment of all signals is particularly unrealistic in practical distributed communication networks: signals arriving from independent transmitters will certainly not contain exactly synchronized bit streams.

In this section, it is still assumed that the bit duration T_b is the same for all transmitters, but the starting time of the bits may now differ from terminal to terminal. The receiver is assumed to be in perfect bit synchronization with the wanted signal, and therefore no longer with the interfering signals. This is illustrated in Figure 5.5.

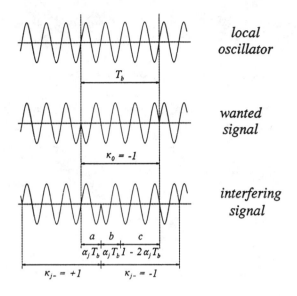

Figure 5.5 Wanted BPSK signal and the ith interfering BPSK signal with bit synchronization offset α_i. Interference introduced during period (a) cancels interference introduced during period (b).

The bit synchronization offset is denoted by α_i $(0 < \alpha_i < 1)$; that is, if a carrier reversal occurs, it happens at the instant $(k + \alpha_i)T_b$. We denote the bit value of the ith interfering signal before and after the phase reversal with $\kappa_{i\leftarrow}$ and $\kappa_{i\rightarrow}$, respectively. Thus, the interfering signal has the form $\kappa_{i\leftarrow}\rho_i \cos(\omega_c t + \theta_i)$ during the interval $\langle (k - 1 + \alpha_i)T_b, (k + \alpha_i)T_b \rangle$, and the form $\kappa_{i\rightarrow}\rho_i \cos(\omega_c t + \theta_i)$ during the interval $\langle (k + \alpha_i)T_b, (k + 1 + \alpha_i)T_b \rangle$.

The statistical decision variable becomes

$$v = \rho_0 \kappa_0 \cos(\theta_0) + \sum_{i=1}^{n} \zeta_i \{\kappa_{i\leftarrow}\, \alpha_i + \kappa_{i\rightarrow}\, (1 - \alpha_i)\} + n_I \qquad (5.18)$$

If the interfering transmitters are independent, the bit-timing offset α_i is likely to be uniformly distributed. Because of the symmetry of the case $0 < \alpha_i < 1/2$ with the case $1/2 < \alpha_i < 1$, it can be assumed that

$$f_{\alpha_i}(\alpha_i) = \begin{cases} 2 & 0 < \alpha_i < \frac{1}{2} \\ 0 & \text{elsewhere} \end{cases} \qquad (5.19)$$

If a phase reversal of the ith signal occurs at the instant $(k + \alpha_i)T_b$ (thus if $\kappa_{i\leftarrow} = -\kappa_{i\rightarrow}$), the signal in the period $\langle kT_b, (k + \alpha_i)T_b \rangle$ cancels the phase-reversed signal during the following interval $\langle (k + \alpha_i)T_b, (k + 2\alpha_i)T_b \rangle$. In this case, the variance of the interference sample is effectively reduced to $p_i(1 - 2\alpha_i)^2$. For a fixed value of α_i, the sample from an interfering signal with a phase reversal is Gaussian-distributed. However, if α_i is treated as a random variable, the interference sample in (5.9) becomes non-Gaussian.

5.6.1 Single Interfering Signal

If zeros and ones are equiprobable with independent probability of occurrence, the probability of a phase reversal is one-half. Thus, if the interference consists of only a single signal ($n = 1$) with a bit synchronization offset of $\alpha_i T_b$, the bit error probability is

$$\bar{P}_b(e|\bar{p}_0, \bar{p}_1, \alpha_1) = \frac{1}{2} - \frac{1}{4}\sqrt{\frac{\bar{p}_0 T_b}{\bar{p}_0 T_b + \bar{p}_1 T_b + N_0}}$$
$$- \frac{1}{4}\sqrt{\frac{\bar{p}_0 T_b}{\bar{p}_0 T_b + \bar{p}_1 T_b(1 - 2\alpha_1)^2 + N_0}}$$
(5.20)

For $\alpha_1 = 0$ (i.e., in the case of perfect bit synchronization), (5.20) reduces to (5.10). Integrating (5.20) over the uniform pdf (5.19) of α_1 gives, after some mathematical manipulations,

$$\bar{P}_b = \frac{1}{2} - \frac{1}{4}\sqrt{\frac{\bar{p}_0 T_b}{\bar{p}_0 T_b + \bar{p}_1 T_b + N_0}} + \frac{1}{4}\sqrt{\frac{\bar{p}_0}{\bar{p}_1}} \ln\left(\frac{\sqrt{\bar{p}_0 T_b + \bar{p}_1 T_b + N_0} - \sqrt{\bar{p}_1 T_b}}{\sqrt{\bar{p}_0 T_b + N_0}}\right)$$
(5.21)

Numerical results are shown as curve b of Figure 5.6.

5.6.2 Many Interfering Signals

If the interference is caused by the sum of many weak signals, the joint interference term in (5.18) becomes a Gaussian-distributed random variable. Since ζ_i and α_i are independent stochastic variables, the variance of a sample of an interfering signal that makes a phase reversal is

$$E[\zeta_i^2(1 - 2\alpha_i)^2] = E[\zeta_i^2]\,E[(1 - 2\alpha_i)^2] = \bar{p}_i \cdot \frac{1}{3}$$
(5.22)

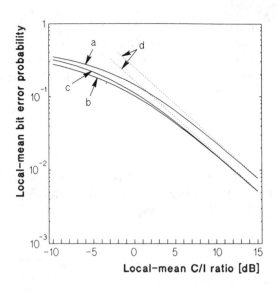

Figure 5.6 Local-mean BER of BPSK in a Rayleigh-fading channel versus local-mean signal-to-joint-interference ratio for $n = 1$ and ∞ interferers, with (a) perfect bit alignment and (b, c) random bit alignment for one (a, b) and many (a, c) interfering signals. Corresponding approximate expressions ($\cdots$, d).

Alternatively, if no phase reversal occurs, the variance is found from (5.8). If a phase reversal occurs with a probability of one-half, and if the bit alignment offset is random, the experienced total interference sample has the variance $1/2(1 + 1/3)\bar{p}_t$. This corresponds to a reduction of 1.8 dB of the effective interference power. Thus, for random bit offsets, and sufficiently many interfering signals, the average BER for coherent detection of a wanted BPSK signal becomes

$$\bar{P}_b(\epsilon | \bar{p}_0, \bar{p}_t, n \to \infty) = \frac{1}{2} - \frac{1}{2} \sqrt{\frac{\bar{p}_0 T_b}{\bar{p}_0 T_b + \frac{2}{3} \bar{p}_t T_b + N_0}} \qquad (5.23)$$

which is depicted as curve c in Figure 5.6.

As seen from Figure 5.6, the latter result for $n \to \infty$ closely approximates the BER for a single interferer ($n = 1$). A small effect is noticeable for local-mean carrier-to-interference ratios below 0 dB ($C/I < 1$). However, the behavior for $C/I > 1$ (0 dB) is of more practical interest: in this range, (5.21) and (5.23) produce almost identical results.

5.7 PROBABILITY OF SUCCESSFUL RECEPTION OF A BLOCK OF BITS[3]

In this section, we address a digital telephone network, with speech coded into blocks of L bits. Slow and fast Rayleigh fading of the wanted signal are considered. With slow fading, the amplitude and phase of each signal are assumed constant for the entire duration of the packet ($T_5 << T_6$, see Section 2.1.7). Fast fading, on the other hand, causes substantial fluctuations of the signal amplitudes and phases during packet reception. In an extreme case, the received signal fading is uncorrelated from bit to bit. Although a practical receiver is likely to lose carrier synchronization under such circumstances (if $T_3 \approx T_6$), the expressions proposed for modeling fast fading may be reasonable if $T_3 << T_6 << T_5$.

The model described in the previous sections is considered. We shall assume no bit synchronization offset between contending signals ($\alpha_i \equiv 0$ for $i = 1, 2, \ldots, n$). This leads to conservative results, because it overestimates the effect of interference. Moreover, in the case of random α_i, the worst-case interfering BPSK bit sequence $\kappa_{i,\rightarrow} = \kappa_{i,\leftarrow}$, (i.e., no phase reversal) is expected to occur during 50% of the bits in the test packet. This event gives the same error probability as the case $\alpha_i = 0$.

A block error correction code is assumed that can correct up to M errors in a block of L bits.

5.7.1 Fast Fading

With fast Rayleigh fading, the duration of a packet is substantially longer than the time constants of the multipath fading. Further, we assume that during one bit time the channel characteristics do not change ($T_3 << T_6 << T_5$). During the reception of a packet, each signal is expected to experience several fades. If it can be assumed that the received amplitude and phase of all signals are statistically independent from bit to bit even though the receiver remains perfectly locked to the wanted signal, the probability of successful reception for BPSK is obtained from

$$\Pr(s_0|\bar{p}_0,\bar{p}_t) = \sum_{m=0}^{M} \binom{L}{m}[1 - \bar{P}_b^{L-m}][\bar{P}_b^m] \qquad (5.24)$$

where the local-mean bit error probability is taken of the form of (5.11).

[3]Section 5.7: Portions reprinted, with permission, from *Proc. 1st IEEE Int. Conf. on Universal Personal Communications*, Dallas, TX, Sep. 29–Oct 2, 1992, pp. 308–313. © 1992 IEEE.

5.7.2 Slow Fading

For packets of sufficiently short duration, the received amplitude and carrier phase may be assumed to be constant throughout the duration of a packet. This condition is satisfied if the motion of the mobile terminal during a transmission time of the block is negligible compared to the wave length ($T_5 \ll T_6$).

The probability of not more than M bit errors in a block of L bits is found by averaging $\Pr(s_0|p_0, \bar{p}_t)$ over the Rician fading of the wanted signal, with

$$\Pr(s_0|\bar{p}_0, \bar{p}_t) = \int_0^\infty \frac{(1+K)e^{-K}}{\bar{p}_0} \exp\left(-(1+K)\frac{p_0}{\bar{p}_0}\right) I_0\left(\sqrt{4K(1+K)\frac{p_0}{\bar{p}_0}}\right)$$

$$\cdot \sum_{m=0}^{M} \binom{L}{m}\left[1 - \frac{1}{2}\operatorname{erfc}\sqrt{\frac{p_0 T_b}{N_0 + \bar{p}_t T_b}}\right]^{L-m}\left[\frac{1}{2}\operatorname{erfc}\sqrt{\frac{p_0 T_b}{N_0 + \bar{p}_t T_b}}\right]^m dp_0$$

$$(5.25)$$

Here we assumed that the interference sample is Gaussian-distributed with mean $\bar{p}_t$ and that interference samples during successive bits are statistically independent. This *Gaussian assumption* is reasonable if the number of interfering signals is large ($n \to \infty$) or if the interfering signal is fast fading. In Chapter 7, the accuracy of the Gaussian assumption is addressed for the special case of relatively low C/I ratios in random access-data networks.

If reception is not successful, which occurs with probability $1 - \Pr(s_0)$, either an *undetected block error* or a *block erasure* can occur. In the case of an undetected error, the subscriber experiences a burst of noise or crosstalk, whereas during an *erasure*, the subscriber experiences a short outage of the telephone link. If each block represents a voice segment of a duration of at least a few (tens) of milliseconds, it would be appropriate to call $1 - \Pr(s_0)$ the circuit outage probability. However, the RF outage probability is mostly defined as the probability that the instantaneous BER, or, in some papers, the instantaneous block erasure-or-error probability $1 - \Pr(s_0|p_0)$, exceeds a certain value P_z (see Section 3.1.1). For a further discussion of the probability of success, undetected error, and erasure (failure) of a block of bits, we refer to [18, 19].

5.7.3 Numerical Results for Microcellular Network

The principal difference between expression (5.24) for fast fading and (5.25) for slow fading is the sequence of the mathematical operation of averaging over the Rician fading of the envelope of the wanted signal and taking the Lth power of the bit error probability. Numerical results are shown in Figure 5.7 for slow and fast

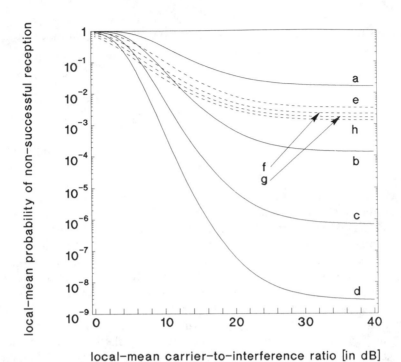

local-mean carrier-to-interference ratio [in dB]

Figure 5.7 Probability of successful reception $\Pr(s_0)$ of a block of $L = 63$ bits versus the local-mean carrier-to-interference ratio $\bar{p}_0/\bar{p}_r$. BPSK modulation. Fast (——, a–d) and slow (- - -, e–h) Rician fading of the wanted signal with $K = 6$ dB. Rayleigh-fading interference. Local-mean C/N ratio $C/N_M = 20$ dB ($\bar{p}_0 T_b/N_0 = 100$). Error correction distance $M = 0$ (a, e), $M = 1$ (b, f), $M = 2$ (c, g), $M = 3$ (d, h).

Rician fading with $K = 4$ (6 dB). Various error detection distances are considered ($M = 1$, 2, 3, and 4).

For slow fading, the error correction has a very limited effect on the probability of successful reception. This suggests that preferably very little error correction is to be used (small M), while the Hamming distance between code words is to be used mainly for effective error detection to avoid bursts of noise or crosstalk in the telephone circuit. This is in sharp contrast to the performance of idealized fast fading, which is greatly affected by the amount of error correction employed.

For substantial error correction coding (large M) at relatively high local-mean C/I ratios (and thus for relatively high reuse distances), a microcellular network with idealized fast fading channels appears to have a higher performance than one with slow fading channels. This is in contrast to results to be presented in Chapter 7 for random-access networks, where multiple competing signals arrive with small differences in received power. In such cases, fast fading gives a significantly poorer performance than slow fading. This is confirmed by the probabilities depicted in Figure 5.7 for small C/I ratios of, say, less than 5 dB.

5.8 CONCLUSION

The average bit error rate is a relevant measure of the performance of a digital mobile network with continuous wave transmission. The local- and area-mean BERs have been approximated for the case of a wanted signal in the presence of multiple fading interfering signals. The results were used to calculate the local- and area-mean BER in microcellular and macrocellular networks.

The analysis showed that the specular-to-scatter (Rician) ratio K of multipath fading and the standard deviation σ_s of any shadowing on the channel have a very substantial effect on the average BER. It appeared that the rapid decrease of the BER with the increasing power of the wanted signal, which occurs in time-invariant channels, is not experienced in mobile channels with fading. Even for high signal-to-interference-plus-noise ratios, the average BER remains relatively high. Comparing results for microcellular and macrocellular networks, it is seen that, within the usual parameter ranges of, say, less than 12 dB of shadowing ($0 < \sigma_s < 2.72$) and Rician fading with $0 < K < 1000$ (30 dB), shadowing affects the BER less than the Rician factor. It may be concluded that in microcellular networks with the terminal transmitting the wanted signal within line of sight, BERs are substantially smaller than in macrocellular networks. This is intuitively explained by the fact that for small K, as in Rayleigh-fading (macrocellular) channels, deep fades occur frequently, which largely contribute to the BER.

The effect of interference has been studied for bit intervals that are exactly overlapping and, more realistically, randomly aligned with the bit transitions of the wanted signal. It was seen that the assumption of exactly aligned bits overestimates the BER somewhat. Nonetheless, for conservative system and network design, this may be an acceptable approximation.

The probability of successful reception of a voice segment, coded as a block of bits, depends a great deal on the rate of channel fading. At high C/I ratios, only a few bits are in error. In this case, error correction coding and fast fading (and thus with bit errors randomly distributed over all voice segments) gives higher performance than slow fading. In contrast to this, bit errors may occur frequently because of a relatively small average C/I ratio in a cellular network with dense frequency reuse. In the latter case, the outage performance with fast fading is poor compared to slow fading. Further, we conclude that with slow fading, error correction coding only has a limited effect on the performance. This can be understood because with slow fading a block of bits is either received almost error-free or with many bit errors.

Perfect carrier synchronization has been assumed. Particularly in networks with burst transmissions, such as in TDMA inbound channels for mobile telephony or in packet radio networks, imperfect receiver synchronization may severely limit the performance. This aspect has not been addressed here and is recommended for further study.

REFERENCES

[1]   Proakis, J.G., "Digital Communications," 2nd ed. 1989, New York: McGraw-Hill, Inc., 1983.

[2]   Aghamohammadi, A., and H. Meyr, "On the Error Probability of Linearly Modulated Signals in Rayleigh Frequency-Flat Fading Channels," *IEEE Trans. on Comm.*, Vol. COM-38, No. 11, Nov. 1990, pp. 1966–1970.

[3]   Shimbo, O., and R. Feng, "Effects of Cochannel Interference and Gaussian Noise in M-ary PSK System," *COSMAT Tech. Rev.*, Vol. 3, No. 1, Spring 1973, pp. 183–207.

[4]   Steele, R., and V.K. Prabhu, "High-User-Desity Digital Cellular Mobile Radio Systems," *IEE Proceedings F*, Vol. 132, No. 5, August 1985, pp. 396–404.

[5]   Wong, K.H.H., and R. Steele, "Transmission of Digital Speech in Highway Microcells," *Journal of the Inst. of Electronic and Electrical Eng.*, Vol. 57, No. 6 (supplement), Nov./Dec. 1987, pp. s246–s254.

[6]   Kegel, A., H.J. Wesselman, and R. Prasad, "Bit Error Probability for Fading DPSK Signals in Microcellular Land Mobile Radio Systems," *Electron. Lett.*, Vol. 27, No. 18, 29 Aug. 1991, pp. 1647–1648.

[7]   Wozencraft, J.M., and I.M. Jacobs, *Principles of Communication Engineering*, New York: John Wiley and Sons, 1965.

[8]   Carlson, A.B., *Communication Systems*, 3rd ed., New York: McGraw-Hill, 1986.

[9]   Abramowitz, M., and I.A. Stegun, *Handbook of Mathematical Functions*, New York: Dover, 1965.

[10] Jakes, W.C., Jr., ed., *Microwave Mobile Communications*, New York: John Wiley and Sons, 1974.

[11] Shcikh, A.U.H., Y.D. Yao, and X. Wu, "The ALOHA Systems in Shadowed Mobile Radio Channels With Slow and Fast Fading," *IEEE Trans. on Veh. Tech.*, Vol. VT-39, No. 4, Nov. 1990, pp. 289–298.

[12] Zhang, K., K. Pahlavan, and R. Ganesh, "Slotted Aloha Radio Networks With PSK Modulation in Rayleigh Fading Channels," *Electron. Lett.*, Vol. 25, No. 6, 16 March 1989, pp. 413–414.

[13] Habbab, I.M.I., M. Kavehrad, and C-E.W. Sundberg, "ALOHA With Capture Over Slow and Fast Fading Radio Channels With Coding and Diversity," *IEEE Journal on Sel. Areas in Comm.*, Vol. SAC-7, No. 1, Jan. 1989, pp. 79–88.

[14] Zhang, K., and K. Pahlavan, "A New Approach for the Analysis of the Slotted ALOHA Local Packet Radio Networks," *Proc. Int. Conf. on Comm., ICC' 90*, Atlanta, 16–19 April 1990.

[15] Linnartz, J.P.M.G., H. Goossen, and R. Hekmat, "Comment on 'Slotted Aloha Radio Networks With PSK Modulation in Rayleigh Fading Channels,'" *Electron. Lett.*, Vol. 26, No. 9, 26 Apr. 1990, pp. 593–595.

[16] Linnartz, J.P.M.G., and A.J. 't Jong, "Average Bit Error Rate for Coherent Detection in Micro- and Macro-cellular Radio," *Proc. 6th IEE Int. Conf. on 'Mobile Radio and Personal Communications'*, Warwick, U.K., 9–12 Dec. 1991, pp. 241–247.

[17] Linnartz, J.P.M.G., "Effect of Multipath Fading and Shadowing on Error Performance in Narrowband Cellular Radio," *European Cooperation in the Field of Scientific and Technical Research*, COST 231, Document TD (91)-69, Leidschendam, 25–27 Aug. 1991.

[18]   Linnartz, J.P.M.G., A.J. 't Jong, and R. Prasad, "Performance Analysis of Interference and Noise Limited Digital Micro Cellular Personal Communication Systems," submitted for publication.

[19]   Linnartz, J.P.M.G., A.J. 't Jong, and R. Prasad, "Performance of Personal Communication Networks With Error Correction Coding in Microcellular Channels," *Proc. the 1st IEEE International Conf. on Universal Personal Communications*, Dallas, 29 Sept.–2 Oct. 1992, pp. 308–313.

Chapter 6
Random Multiple Access to Mobile Radio Channels

Chapters 3, 4, and 5 addressed the case of mobile communication from a known transmitter (sending the wanted signal) to the receiver in the presence of unwanted interference and AWGN. Terminals were assumed to transmit a continuous stream of information. In this chapter, data networks with a large number of participating terminals are considered. Each terminal is considered to generate messages in the form of data packets, which are to be transmitted to a common receiver. Transmissions are of a burst type: the average data rate being low, all communication is concentrated in occasional peaks of high data rates. As discussed in Chapter 1, burst traffic is commonly encountered in mobile information systems, such as paging and messaging networks, traffic information and route guidance systems, and fleet management systems. In such networks, the instantaneous flow of data from mobile terminals has to be organized and controlled dynamically. This is solved by random multiple-access schemes. The issue of random access of short messages is not restricted to networks for data communications. In complex systems for circuit-switched *voice* communication, a common inbound (mobile-to-fixed) random-access signaling channel is employed to handle call requests and specific signaling messages from mobile subscribers.

Scarcity of the radio spectrum requires efficient use of the available frequencies and has led to the development of a wide variety of multiple-access schemes (e.g., [1–4]): the same radio channel is used by a number of subscribers, each of which only uses the channel for a limited amount of time. In TDMA, the time interval reserved for each user is assigned according to a periodic scheme. A disadvantage of such fixed assignments is that a user must wait for his or her time slot before being allowed to transmit a packet. The delay may become unacceptably large if the number of subscribers is large, especially if slots assigned to nonactive users remain idle. In contrast to this, in dynamic access, resource reservation is flexible (demand assignment), or not made in advance at all (random access), so subscribers compete for access to the channel [4, 5]. The obvious disadvantage of random access is that two (or more) competing terminals may simultaneously attempt to transmit a data packet over the radio channel. Such a contention or access conflict is called a *collision*. To avoid collisions, various dynamic access control methods are possible at the expense of signaling capacity and control complexity.

This chapter reviews a number of simple random-access protocols, namely, slotted ALOHA [6, 7], *carrier sense multiple access* (CSMA) [8, 9], and ISMA [9, 10]. A method for the evaluation of their performance is presented. In contrast to most textbooks and scientific papers (e.g., by Kleinrock and Tobagi [8]), we address the performance experienced by individual terminals rather than the average performance of all participating terminals. Random-access techniques commonly used in wired *local areas networks* (LAN) were initially used in mobile radio networks. However, it soon appeared that the performance of many random-access schemes substantially differs for wired and radio channels (e.g., [11]), being highly dependent on the physical characteristics of the channel. Hence, the performance of the network is different for each participating terminal and depends on the location of the mobile terminal. The characteristics of the mobile channel are not included explicitly in this chapter: the physical radio link is modeled simply by stating the generic probability of successful reception of a data packet, given the presence of a certain number of contending packets. Detailed investigation of the effect of the channel will be deferred to Chapters 7, 8, and 9.

The organization of this chapter is as follows. First, in Section 6.1, some random-access schemes are briefly reviewed. A first step in our analytical evaluation is the formulation of the model for Poisson arrivals of data packets in Section 6.2. Section 6.3 discusses how the physical radio channel can be included in the consideration of random-access protocols. Performance measures from random-access channels are discussed in Section 6.4. Combining the various models and definitions from Sections 6.2 to 6.4, Section 6.5 goes on to determine the throughput and the near-far effect in slotted ALOHA. Similar results for ISMA and CSMA are determined in Section 6.6. Section 6.7 discusses the throughput of an ISMA channel when also taking account of the increase of the propagation delay with increasing separation between the mobile terminal and the base station. Section 6.8 contains the conclusions of this chapter.

6.1 DESCRIPTION OF SOME RANDOM-ACCESS PROTOCOLS

6.1.1 Slotted ALOHA

The earliest contention technique, known as ALOHA, was developed for a packet-radio network carrying data from remote terminals to a host computer at the University of Hawaii. The decision in 1969 to implement the ALOHA protocol was based on network simplicity [6, 12]: many users send their bursty packet traffic at high speed on a single (common) radio channel without mutual regulation, thus accepting the risk of access conflicts. Whenever a data packet is generated, this is transmitted on the random-access channel, without considering a priori whether or not it will interfere with other transmissions. If more than one packet is transmitted at the same moment, a collision occurs, so transmitted data packets may not correctly reach the destination. In the central receiver, the correctness of a received packet is determined. If the packet is deemed correct (i.e., if it contains a correctable number of bit errors), an acknowledgment is broadcast on a separate feedback channel. Since such outbound feedback transmissions do not experience competition from other transmitters, this feedback channel is likely to be substantially more reliable than the inbound collision channel. If a mobile terminal does not receive an acknowledgment of its transmission, it will therefore assume that its packet has been lost in a collision, and will retransmit the packet after waiting a random time.

The ALOHA access technique is illustrated in Figure 6.1 [13] using symbols proposed in the *Specification and Description Language* (SDL) of the CCITT [14].

An improvement can be made by dividing time into slots; terminals may only start a packet transmission at the beginning of a time slot. If the durations of all packets equal the slot time, contending packets overlap either completely or not at all. The event that a packet is lost due to *partially* overlapping interference is avoided. This has been shown to effectively double the throughput of the ALOHA channel [7].

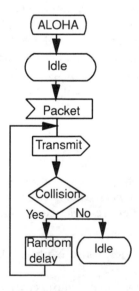

Figure 6.1 Flow diagram for the activities of each mobile terminal in ALOHA random access [13].

6.1.2 Protocols Exploiting Feedback Information

With increasing traffic loads, collisions occur more frequently in the ALOHA channel. This reduces the throughput and may eventually lead to instability if retransmitted packets continue to collide [15–19]. A number of other protocols have been proposed to mitigate this problem. *Collision resolution algorithms* (CRA) improve the stability of the channel and reduce the delay caused by retransmission waiting times by employing the feedback channel more effectively [15, 16]. For instance, the central station can report whether the inbound channel was idle, a collision occurred, or a successful packet reception happened. A more efficient retransmission strategy can be followed, exploiting knowledge of this feedback information. Examples of CRAs are the splitting (or tree) algorithm [14] and the stack algorithm [16]. Dynamic frame length ALOHA is another technique to ensure the stability of an ALOHA network [20].

    Busy-channel multiple access (BCMA) is the class of multiple-access schemes in which no new packet transmissions are allowed when the inbound channel is busy. Various strategies have been proposed to acquire this information about the channel state [8–10, 21–27].

6.1.3 Nonpersistent Carrier Sense Multiple Access

CSMA is one of the best-known examples of BCMA. CSMA employs feedback in the standard form of the positive acknowledgments of correct reception. Additionally, no new packet transmission is initiated by mobile terminals if they sense that the inbound channel is already busy [8]. Figure 6.2 gives a flow diagram of *nonpersistent* CSMA. Nonpersistent means that if the channel is sensed busy, the terminal attempts a new transmission only after waiting a random time. Thus, with nonpersistent CSMA, a transmission attempt does not necessarily lead to a transmission on the channel.

6.1.4 1-Persistent Carrier Sense Multiple Access

If, in 1-persistent CSMA, a mobile senses that the channel busy, it waits only until the channel becomes idle (thus, it *persists* on sensing) and then immediately transmits the packet. The terminal activities are depicted in Figure 6.3.

6.1.5 Slotted *p*-Persistent Carrier Sense Multiple Access

It is the impression of the author that a number of different interpretations of *p*-persistent CSMA are used in the analyses reported in technical literature (e.g., [8, 27]).

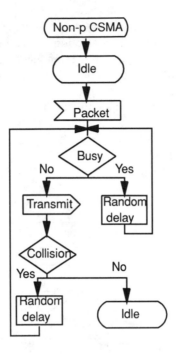

Figure 6.2 Flow diagram of the activities of each mobile terminal in nonpersistent CSMA [13].

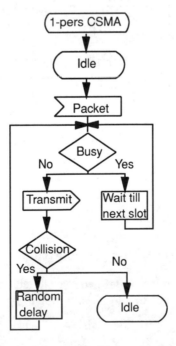

Figure 6.3 Flow diagram of the activities of each mobile terminal in 1-persistent CSMA [13].

This section describes slotted *p*-persistent CSMA as described in [8]. Time is divided into minislots with a duration that is a fraction of the packet duration. Any packet transmission occupies a fixed integer number of successive slots. A terminal with a packet to be transmitted starts transmission with probability *p* if the channel is sensed idle, and so waits with probability 1 − *p* until the next slot. If, on the other hand, the channel is busy, the terminal waits until the next slot and again senses the channel. This is repeated until either a destructive collision or a successful transmission of any packet occurs, in which case the packet is rescheduled. Figure 6.4 [13] illustrates the terminal activities.

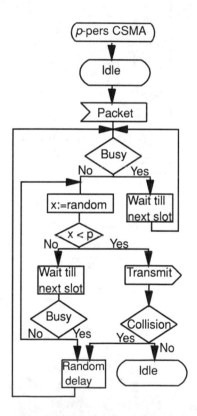

Figure 6.4 Flow diagram for the activities of each mobile terminal in slotted *p*-persistent CSMA [13].

6.1.6 Unslotted *p*-Persistent Carrier Sense Multiple Access

p-Persistent CSMA was originally defined for slotted channels [8]. Here, a modification of the protocol is made to allow evaluation of unslotted networks (see Figure 6.5).

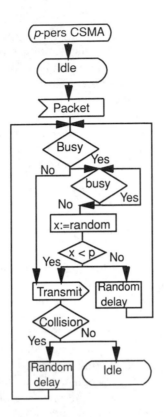

Figure 6.5 Flow diagram for the activities of each mobile terminal in unslotted *p*-persistent CSMA.

If a packet arrives at an instant when the channel is sensed idle, the packet is transmitted immediately. If, on the other hand, the channel is busy at the instant of arrival of the packet in the terminal buffer, the terminal performs a binary random experiment: with probability *p* the terminal transmits the packet as soon as the channel becomes idle. Such an attempt is considered successful unless the packet is destroyed in a collision. Alternatively, with probability $1 - p$, the packet is rescheduled. These unsuccessful attempts are considered to contribute to the *attempted packet traffic* G_t (see Section 6.2) even though no transmission occurred. Another attempt is performed after waiting a random time.

Intuitively, as compared to unslotted nonpersistent CSMA, unslotted *p*-persistent CSMA is expected to have lower delays because waiting terminals may start transmitting with some appropriately chosen probability *p* immediately after the channel becomes idle. Particularly in channels with capture, this may reduce the delay. Compared to the slotted *p*-persistent access scheme, unslotted *p*-persistent has the advantage that slot timing is not required, which is advantageous if substantial random propagation delays occur. Moreover, if packets arrive when the channel happens to be idle, the packet is transmitted immediately with unslotted *p*-persistent CSMA. Another major difference occurs if an initially busy channel becomes idle: in unslotted *p*-persistent CSMA, the terminal performs one binary experiment. If the outcome prohibits transmission, the packet is rescheduled, whereas in slotted *p*-persistent CSMA the terminal tries again immediately (in the next minislot).

According to the above description, unslotted nonpersistent CSMA becomes the limiting case of unslotted *p*-persistent CSMA for $p \to 0$. This observation is not valid for the slotted *p*-persistent CSMA. However, in the limit $p \to 1$, both protocols tend toward 1-persistent CSMA.

6.1.7 Inhibit Sense Multiple Access

Successful application of CSMA requires that all mobile terminals can receive each other's signals on the inbound frequency (e.g., as on a LAN). However, in radio nets, especially with fading channels, a mobile terminal might not be able to sense a transmission by all other terminals. This complication, known as the *hidden-terminal problem* [9], is avoided in ISMA, where the base station transmits a busy signal on an outbound channel (in order to forbid all other mobile terminals to transmit) as soon as an inbound packet is being received [10]. A disadvantage of ISMA is the necessity for a real-time (i.e., nonslotted, nonfading) feedback channel. A mobile terminal may also still not sense the presence of the busy tone, because of noise corruption or fading of the feedback signal [17]. This problem may be circumvented by broadcasting an active *idle* tone, rather than an active busy tone, or reduced by transmitting the busy reports with error-control coding.

Even if signaling messages on the feedback channel were always received correctly by all mobile terminals, collisions would nonetheless occur in ISMA for two reasons: (1) new packet transmissions may start during the transmission delay of the inhibit signal, and (2) packets from two or more persistent terminals, waiting for the channel to become idle, may collide immediately after the termination of the previous packet transmission.

6.2 PACKET TRAFFIC MODEL

All data packets are assumed of uniform duration, equal to the unit of time and, in the event of slotted ALOHA, equal to the duration of a time slot. The average number of attempted transmissions contributed by all users in the entire network, expressed in packets per unit of time (ppt), is denoted by G_t, known as the *attempted*, or *offered*, traffic. The number of participating terminals in the network is N. If the *i*th terminal ($i = 1, 2, \ldots N$) attempts to transmit a data packet in a time slot with probability $\Pr(i_{ON})$, the total attempted traffic is

$$G_t = \sum_{i=1}^{N} \Pr(i_{ON}) \qquad (6.1)$$

Most of the investigations reported in this and the following chapters address an infinite population ($N \to \infty$), albeit with bounded total offered traffic G_t. Accordingly, the probability of an attempt per terminal is assumed to approach zero ($\Pr(i_{ON}) \to 0$). The number of arrivals in a certain interval is assumed to be Poisson-distributed. This assumption may be reasonable for the packet traffic generated in some typical mobile networks, such as the call-request channel of a cellular telephone network.

The Poisson arrival rate λ is expressed by the average number of arrivals during a unit of time. So, for packets of unit duration, $G_t = \lambda$. The probability of n packet arrivals in a time interval T is [28, 29]

$$P_n(n) = \frac{(\lambda T)^n}{n!} \exp(-\lambda T) \qquad (6.2)$$

A generalized derivation of (6.2) is contained in Appendix B. The distribution of the number of arrivals in any time interval $\langle t, t + T \rangle$ is independent of starting time t. It can be shown that (6.2) also expresses the probability of n contenders to a given test packet already known to be present.

The probability of no arrivals during the period of duration τ_0 is

$$P_n(0) = \exp(-\lambda \tau_0) = \int_{\tau_0}^{\infty} f_\tau(\tau)d\tau \qquad (6.3)$$

where $f_\tau(\tau)$ is the pdf of the duration between two arrivals. Thus, interarrival times have the negative exponential distribution

$$f_\tau(\tau) = \lambda \exp(-\lambda \tau) \qquad (6.4)$$

with mean λ^{-1}.

Two extensions of this basic model are introduced below. A Poisson arrival rate of data packets depending on transmitting distance to the central base station and on time is considered.

Spatial Distribution: In general, packet offerings are not uniformly distributed over the service area of the network. For instance, remote terminals with weak signals may have to perform more retransmissions than nearby terminals. In the case of an infinite population of terminals, the attempted traffic may be considered to be continuously spread over the area. Abramson [11] defined $G(r)$ as the average number of attempted packet transmissions per unit of time and per unit of area offered by terminals at a distance r from the central receiver. Assuming no angular variations of traffic, the total traffic per unit of time is found from the polar integration

$$G_t = \int_0^\infty 2\pi r G(r)\, dr \qquad (6.5)$$

In a number of investigations, r is regarded as a stochastic variable, distributed with the probability density function $f_r(r)$. So, for a randomly selected packet, one may write

$$f_r(r) = \frac{1}{G_t} 2\pi r G(r) \qquad (6.6)$$

Nonstationarity: It is generally assumed that the arrival rate λ is constant with time. This assumption will be relaxed in Section 6.8, where the time-dependent arrival rate $\lambda(r, t)$ is introduced, resulting in a nonstationary Poisson process. As shown in Appendix B, this leads to

$$P_n(n) = \frac{(\bar{\lambda}T)^n}{n!} \exp(-\bar{\lambda}T) \qquad (6.7)$$

where $\bar{\lambda}$ is the average arrival rate during $\langle t_0, t_0 + T \rangle$; that is,

$$\bar{\lambda} \triangleq \frac{1}{T} \int_{space} \int_{t_0}^{t_0+T} \lambda(r,t)\ 2\pi r\, dt\, dr \qquad (6.8)$$

Appendix B focuses on integration over time, since the integration over space is considered in a number of research papers (e.g., [11, 30]).

6.3 IMPACT OF THE RADIO CHANNEL

Poisson packet arrivals are the foundation for a simple model for the probability of a collision with n interfering packets. Considering these collisions, classical papers such as [8] determine the throughput based on the two assumptions that, firstly, a data packet is always received correctly in the absence of collisions and, secondly, all packets involved in a collision are lost. In mobile radio nets, both assumptions should be reconsidered: channel imperfections may cause loss of a data packet even if no interference from other terminals occurs [31], and in the event of a collision, the strongest contending radio signal may capture the receiver, see e.g. [7, 30, 32]. Compared to remote terminals, nearby terminals therefore experience a higher probability of success in transmitting a packet; propagation fluctuations may also contribute to the capture effect (see Chapter 7).

In order to determine the probability of successful transmission in terms of the behavior of the physical layer, the following quantities are defined

$q_n(r)$ is the probability of successful reception of a test packet transmitted from a distance r from the central receiver, given the presence of interference from n other terminals with unknown positions.

q_n is the probability of successful reception, given that the arbitrary test packet is transmitted in the presence of n interfering signals, averaged over all expected positions of the mobile terminals, including the one transmitting the test packet. So, using (6.6),

$$q_n = \frac{1}{G_t} \int_0^\infty 2\pi r q_n(r) \, G(r) dr \tag{6.9}$$

C_{n+1} is the expectation value of the number of successful packets in collisions with $n + 1$ competing packets, with $C_{n+1} = (n + 1)q_n$. If receiver capture is mutually exclusive for each of the $n + 1$ packets (i.e., if successful reception of one packet ensures that a packet from another terminal cannot be correctly received), C_{n+1} also expresses the probability that one out of $n + 1$ simultaneously transmitted packets is received correctly.

These probabilities will be determined by taking account of the mobile channel characteristics from Chapter 7 onwards. In some equations, the more explicit notation $q_n(r_j)$ and $Q(r_j)$ is used, index j denoting the "test" packet.

6.4 MEASURES OF STEADY-STATE PERFORMANCE

Considering the arrival process of interfering packets offered to the channel and the probability of packet success, given a certain number of contenders, the efficiency of a certain random-access scheme will now be evaluated. A number of measures have been proposed to express the performance of the network. In this section, stationary behavior of the network is assumed.

The evaluation of the *total channel throughput* and the individual terminal *success probability* is performed here for steady-state behavior. Attempts to transmit a packet are considered a stationary Poisson process. If an attempt is unsuccessful (for instance, if the packet is lost in a collision), the packet is retransmitted after a long random waiting time. Similarly, if a nonpersistent terminal is inhibited from transmission, a new attempt is scheduled only after a long random waiting time. The assumption of Poisson arrivals implies that in the event of a collision, such random retransmission waiting times are required to be long enough to ensure that the retransmitted packets experience interference that is uncorrelated with the interference experienced during previous attempts. This steady-state analysis simplifies the dynamic behavior of the terminals in response to collisions. In practice, random waiting times should be kept as short as possible. However, if retransmissions are performed too soon, the steady-state assumption is inappropriate. In such cases, the random-access channel can become unstable. This will be addressed in Section 8.6.

6.4.1 Near-Far Effect

Because of propagation attenuation, the probability of successful access may depend on the distance r between the mobile terminal and the receiver. This leads to unfairness between terminals in the system. To evaluate this near-far effect, $Q(r)$ is defined as the unconditional probability of successful reception of a test packet generated for transmission from a terminal at a distance r, taking into account the probability of its permission to transmit and the statistics of the number of interfering packets n. So

$$Q(r) = \sum_{n=0}^{N} P_n(n) \, q_n(r) \qquad (6.10)$$

As will be described later, the probability $P_n(n)$ of n contending packets is in the form of (6.2) for ALOHA networks. For CSMA, ISMA, and CRA, $P_n(n)$ is not necessarily a Poisson distribution.

For an individual subscriber, the time-average success rate $Q(r)$ can differ widely from the ensemble average Q_t, where

$$Q_t = \frac{1}{G_t} \int_0^\infty 2\pi r G(r) \, Q(r) \, dr \qquad (6.11)$$

Analogous to the offered traffic per unit of area, the throughput per unit of area $S(r)$ may be considered, with $S(r) = Q(r)G(r)$.

6.4.2 Throughput

From the point of view of spectrum efficiency, the total throughput S_t is an important measure of the overall performance of a random-access network. It is defined as the average number of packets per unit of time received correctly at the base station:

$$S_t = \int_0^\infty 2\pi r Q(r) G(r) \, dr \qquad (6.12)$$

The ensemble-average probability Q_t of a successful attempt is found to be $Q_t = S_t/G_t$.

6.4.3 Queuing Delay

The packet delay can be considered to consist of a *terminal queuing delay* D_q, before the packet arrives at the first position in the terminal buffer, and an *access delay* D_a, caused by retransmission waiting times if the packet is lost in a collision. Models based on a Markov chain offer the possibility of evaluating the backlog, packet delay, and also the stability of the network. In this section, we address queuing delays within the terminal buffer, whereas a discussion of the access delay is deferred to Section 8.6.

A common approximation for packet queuing in a terminal buffer (e.g., [30]) assumes the probability of successful transmission $Q(r)$ to be stationary; thus, the network and the channel traffic is assumed in equilibrium. The Markov chain depicted in Figure 6.6 can be used to model the stochastic behavior of the number of packets in the buffer of a particular terminal.

P_0 is the probability that a new packet is generated and enters the queue. Here the probability of arrival of new packets is assumed to be independent of the number of packets in the terminal buffer. A packet in the first position of the queue leaves the system with probability $P_r Q(r)$, where P_r is the probability of permission to transmit the packet in the first position of the queue. Hence, for a queue with nonzero length, the total number of queued packets in the terminal increases with probability $P_1 = P_0(1 - P_r Q(r))$, and decreases with probability $P_2 = (1 - P_0)P_r Q(r)$. The delay experienced before this packet reaches the first position in the buffer is [33]

$$D_q(r) = \frac{P_0 + P_r^2 Q(r)^2}{P_r Q(r) - P_0} \tag{6.13}$$

Hence, the queuing delay D_q in the transmitter buffer can be expressed merely in

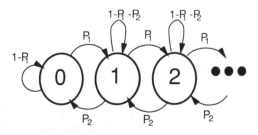

Figure 6.6 Markov chain model for packet queuing in a terminal buffer.

terms of P_0, P_r, and $Q(r)$. Therefore, we will focus on assessing the distant depen-dent probability of successful transmission Q_r); the queuing delay is not explicitly expressed during the analyses. It can be seen that the stability of the queue in the terminal buffer is ensured as long as $P_0 < Q(r) \cdot P_r$, and $Q(r)$ is stationary and independent from slot to slot.

6.5 NEAR-FAR EFFECT AND THROUGHPUT OF SLOTTED ALOHA

In slotted ALOHA, a mobile terminal transmits a packet arriving at a random instant t during the immediately following time slot, regardless of other transmissions in the same time slot (see Section 6.1). The number of simultaneous transmissions in an arbitrary time slot is thus found by studying the arrival process at all terminals during the previous time slot. For Poisson arrival processes, the probability that the test packet experiences interference from n other (contending) signals in the same time slot is also Poisson-distributed, according to (6.2). The probability $Q(r)$ of a suc-cessful transmission is

$$Q(r) = \sum_{n=0}^{\infty} \frac{G_t^n}{n!} e^{-G_t} q_n(r) \tag{6.14}$$

Using (6.12) and (6.2), the total throughput becomes

$$S_t = G_t e^{-G_t} \sum_{n=0}^{\infty} \frac{G_t^n}{n!} q_n \tag{6.15}$$

The classical collision channel without receiver capture [4, 8, 11, 12, 29] is re-covered by inserting

$$\begin{aligned} q_0(r) &= 1 \\ q_n(r) &= 0 \qquad \text{for } n = 1, 2, \ldots \end{aligned} \tag{6.16}$$

for any $r \geq 0$. Hence,

$$S_t = G_t \exp(-G_t) \tag{6.17}$$

For low traffic loads ($G_t \ll 1$ ppt), any packet is likely to be received successfully ($S_t \to G_t$). For high traffic loads ($G_t \gg 1$ ppt), however, collisions severely reduce the throughput ($S_t \ll G_t$). The maximum, $S_t = e^{-1} \approx 0.368$ ppt, is found for $G_t = 1$ ppt, though in a practical system this would lead to instability [16, 19, 20].

If significant propagation delays occur in the radio channel, the duration of each time slot has to be at least the duration of a packet transmission plus a guard time of the maximum round-trip delay t_p. In such cases, the duration of each time slot becomes $(1 + t_p)$, so the effective throughput becomes $S_t(1 + t_p)^{-1}$.

6.6 NEAR-FAR EFFECT AND THROUGHPUT OF INHIBIT SENSE MULTIPLE ACCESS

The ISMA technique was discussed in Section 6.1. Typical examples of packet transmissions on the common radio channel are shown in Figures 6.7 and 6.8 for unslotted nonpersistent and unslotted p-persistent ISMA, respectively. We now establish a terminology and a notation for mathematical analysis. The results described in the following sections also apply to CSMA, provided that each terminal can perfectly sense transmissions by any other terminal.

In ISMA, the radio system is supplemented by an outbound signal with the status of the channel: either busy or idle. An inbound packet arriving at the idle receiver will be called the *initiating packet*. When the base station receives an inbound initiating packet, a busy signal is immediately broadcast to all mobiles to inhibit them from transmission. In a real system, broadcasting the busy signal occurs after a short processing delay d_1. This delay is normalized to the unity duration of each data packet, and it is reasonable to consider only the case of $d_1 < 1$. After completion of all $n + 1$ contending transmissions, the base station starts transmitting

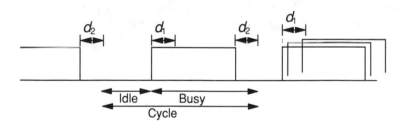

Figure 6.7 Typical event of packet transmissions in nonpersistent ISMA.

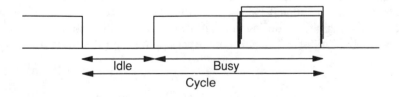

Figure 6.8 Typical event of packet transmissions in unslotted p-persistent ISMA.

an idle signal after a delay d_2. In ISMA, delays may occur, for instance, if a narrowband feedback channel prohibits rapid transitions.

In CSMA, these delays are mainly included to account for the time a mobile terminal takes to switch from reception to transmission mode (powerup) after sensing the radio channel for carriers from other active terminals [23].

A *busy period* is defined as a period during which the base station broadcasts a busy signal plus the preceding signaling delay d_1. This corresponds to the time interval between the first arrival of a packet until the moment that the channel becomes idle. Moreover, immediately after the channel becomes idle, persistent terminals that sense a busy signal may start to transmit. In such cases, the busy period is continued and lasts at least the transmission duration of two packets. Hence, the average duration of the busy period depends on the signaling delay d_1 and d_2, and on the persistency p in sensing the feedback channel [8]. The average duration of the busy period is denoted by B.

For memoryless Poisson arrivals, the duration of the *idle period* (i.e., the time interval between the release of the channel and the first packet arrival) is exponentially distributed with mean $I = G_t^{-1}$ (see (6.4)). An idle period plus the following busy period is called a cycle, with average duration $A = I + B$.

6.6.1 Nonpersistent ISMA

For nonpersistent CSMA and ISMA, rescheduling always occurs if the channel is busy at the instant of sensing. So, if a packet arrives at a nonpersistent terminal when the base station transmits a busy signal (denoted by event H_B), the attempt is considered to have failed. This occurs with probability

$$\Pr(H_B) = \frac{B - d_1}{I + B} \tag{6.18}$$

If the feedback channel is imperfect, a transmission may be started erroneously [21], but this case is not considered here. The packet is rescheduled for later transmission.

With probability $I/(I + B)$, a test packet, transmitted from a distance r, starts at an instant when the channel is idle; that is, the test packet becomes an initiating packet. This event is denoted by H_I. A collision can occur if one or more other terminals start transmitting during the time delay d_1 of the inhibiting signal. The conditional probability of n transmissions overlapping with the initiating test packet is

$$R_n(n|H_I) = \frac{(d_1 G_t)^n}{n!} \exp(-d_1 G_t) \tag{6.19}$$

Alternatively, the test packet itself starts during a period of duration d_1 when the channel is busy because of a transmission by another initiating terminal, but seems idle, since the inhibiting signal is not yet being broadcast. This event, denoted by H_d, occurs with probability $d_1/(B + I)$. The test packet thus experiences interference from the initiating packet, but possibly also from other arriving packets. The additional $n - 1$ contending signals occur with a Poisson arrival rate during the interval d_1. Hence, the conditional probability of n interferers is found from

$$R_n(n|H_d) = \frac{(d_1 G_t)^{n-1}}{(n - 1)!} \exp(-d_1 G_t) \qquad (6.20)$$

where $n = 1, 2, \ldots$. Taking into account the above three possible events H_B, H_I, and H_d, the unconditional probability of successful transmission $Q(r)$ is

$$Q(r) = \Pr(H_I) \, Q(r|H_I) + \Pr(H_d) \, Q(r|H_d)$$

$$= \frac{I}{B + I} \sum_{n=0}^{\infty} R_n(n|H_I) q_n(r|H_I) + \frac{d_1}{B + I} \sum_{n=1}^{\infty} R_n(n|H_d) q_n(r|H_d) \quad (6.21)$$

The probability of capture $q_n(r)$ depends, among other things, on the probability of acquiring receiver synchronization, which, in general, depends on the channel status (H_d or H_I) at the arrival of the packet. In some cases, only initiating packets (arriving at an idle receiver) can be handled. Because of necessary resynchronization of the receiver with respect to carrier, bit and packet timing in an unslotted network, and reinitialization of the communication protocol at the data-link level, packets arriving at a busy receiver may not be accepted. In this case, $q_n(r|H_d) = 0$.

The busy period is of average duration [8]

$$B = 1 + d_1 + d_2 - \frac{1}{G_t} [1 - \exp(-d_1 G_t)] \qquad (6.22)$$

If for a certain channel the conditional capture probabilities $q_n(r|H)$ are known, $Q(r)$ can be computed from (6.19) through (6.22) and $I = G_t^{-1}$.

From the above expressions it appears that $P_n(n)$ tends toward a Poisson distribution with mean $d_1 G_t$, provided that $I \gg B$ (and, thus, if G_t is relatively small). On the other hand, at high offered traffic loads (large G_t, $I \ll B$), the number of interferers to a test packet tends toward one plus a Poisson-distributed random number with mean $d_1 G_t$. The explanation is that for small G_t the test packet is usually an initiating packet, whereas for large G_t the test packet is only seldom an initiating packet. Thus, in contrast to the case for slotted ALOHA, the distribution of the number of interfering signals in ISMA is not Poisson, except in the limiting case $G_t \to 0$.

The total channel throughput S_t is expressed by the integration of $Q(r)$ into (6.21), namely,

$$S_t = \int_0^\infty \frac{2\pi r G(r)}{B+I} \exp(-d_1 G_t) \sum_{n=0}^\infty \left[\frac{d_1^n G_t^{n-1}}{n!} q_n(r|H_I) + \frac{n d_1^n G_t^{n-1}}{n!} q_n(r|H_d) \right] dr$$

$$= \frac{G_t e^{-d_1 G_t}}{B+I} \sum_{n=0}^N \frac{d_1^n G_t^n}{n!} \int_0^\infty \frac{2\pi r G(r)}{G_t} [q_n(r|H_I) + n\, q_n(r|H_d)]\, dr \qquad (6.23)$$

In the above analysis, we considered a particular test packet. Alternatively, the total throughput S_t can be derived considering Poisson arrivals of packets at the receiver in the base station without addressing any particular test packet [8, 24, 25]. This approach is based on the fact that each cycle contains an initiating packet plus a Poisson-distributed number of interfering packets. Hence,

$$S_t = \frac{1}{B+I} \exp(-d_1 G_t) \sum_{n=0}^\infty \frac{d_1^n G_t^n}{n!} C_{n+1} \qquad (6.24)$$

For channels without capture ($C_1 = 1$ and zero otherwise), results in [8] are recovered from (6.24). Also, for instantaneous inhibiting signaling (d_1, $d_2 \to 0$), collisions will not occur in nonpersistent ISMA, and (6.24) reduces to $S_t \to G_t(1 + G_t)^{-1}$ [8].

The expressions in (6.23) and (6.24) are identical, provided that $C_{n+1} = q_{n|H_I} + n q_{n|H_d}$. This appears correct, since, in any cycle, the initiating packet has capture probability $q_n(r|H_I)$; each of the n other packets has capture probability $q_n(r|H_d)$.

6.6.2 Unslotted p-Persistent ISMA

We now consider unslotted p-persistent ISMA without signaling delay ($d_1 = d_2 = 0$), though in practice only nonzero (positive) values can be realized. A busy period can consist of a number of packet transmissions in succession because any terminal may start transmitting as soon as the previous transmission by another terminal is finished. If a packet arrives during an idle period (event H_I), the probability of correct reception of this first packet is $q_0(r)$. During the transmission of this packet, a random number of k terminals sense that the channel is busy with probability $P_n(k)$.

When the channel goes idle, each of the k terminals starts transmitting with probability p. For a test packet arriving during a busy period (event H_B), the probability of n interfering packets is thus

$$P_n(n|H_b) = \sum_{k=n}^{\infty} \binom{k}{n} P_n(k)(1-p)^{k-n}p^n = \frac{(pG_t)^n}{n!}\exp(-pG_t) \qquad (6.25)$$

In particular, the probability that the busy period is terminated (i.e., none of the k terminals starts transmitting) is

$$P_n(0|H_b) = \exp(-pG_t) \qquad (6.26)$$

The probability $P_m(m)$ of transmissions by other persistent terminals during m units of time, concatenated to the first packet transmission, is

$$P_m(m) = \exp(-pG_t)[1 - \exp(-pG_t)]^m \qquad (6.28)$$

On the average, a busy period thus has the total duration

$$B \triangleq E_m[1+m] = 1 + \exp(-pG_t)\sum_{m=0}^{\infty} m[1 - \exp(-pG_t)]^m = \exp(+pG_t) \qquad (6.28)$$

where E_m denotes the expectation over m. The probability of a successful transmission $Q(r)$ is

$$Q(r) = \Pr(H_I)\,Q(r|H_I) + \Pr(H_B)\,Q(r|H_B) \qquad (6.29)$$

$$= \frac{I}{B+I}q_0(r) + \frac{B}{B+I}\sum_{n=0}^{\infty} pP_n(n|H_b)\,q_n(r)$$

Using $I = G_t^{-1}$ (6.25), (6.28), and (6.29) become

$$Q(r) = \frac{q_0(r) + pG_t\displaystyle\sum_{n=0}^{\infty}\frac{(pG_t)^n}{n!}q_n(r)}{1 + G_t\exp(pG_t)} \qquad (6.30)$$

After integration, the total channel throughput S_t is obtained from

$$S_t = G_t\frac{q_0 + \displaystyle\sum_{i=1}^{\infty}\frac{(pG_t)^i}{i!}C_i}{1 + G_t\exp(pG_t)} \qquad (6.31)$$

In the special case $p = 0$, we recover the result for unslotted nonpersistent ISMA without signaling delay ($d = 0$). The classical case of unslotted 1-persistent ISMA on wired channels is recovered by inserting $p = 1$, $q_0 = 1$, and $q_n = 0$ for $n = 1$, 2, ... into (6.31), yielding [8]

$$S_t = \frac{G_t + G_t^2}{1 + G_t \exp(G_t)} \qquad (6.32)$$

6.7 NEAR-FAR EFFECT AND THROUGHPUT OF NONPERSISTENT ISMA IN CHANNELS WITH PROPAGATION DELAYS[1]

In nonpersistent ISMA, the central base station continuously broadcasts the channel status on the outbound channel. Because of the finite speed of radio waves, this feedback signal is received by a mobile terminal only after a certain delay. Hence, remote terminals have a retarded view of the channel status, whereas nearby terminals have more up-to-date information. The previous sections dealt with a propagation delay that is identical for all participating terminals. The impact of propagation delays increasing with propagation distance was investigated in [34, 35] for wired LAN networks. This section deals with radio data networks, assuming delays that consist of a fixed part, called the *processing delay*, and a distant-dependent part, called the *propagation delay*. Again, the processing delay is caused by the necessary operations performed in the central station (for instance, to recognize an incoming packet signal) or by transmitter powerup times in the mobile terminals. Propagation delays may be negligible compared to the processing delays in a typical cellular mobile data network employing a low bit rate, such as 1200 b/s. At present, however, new mobile information systems are designed for increasingly high data rates. In a high-capacity mobile data net with a service area of 30 km, packets of 200 bits, and a bit rate of 200 kb/s, thus with a packet duration of 1 ms, the maximum round-trip delay is 20% of the duration of a packet, so propagation delays may seriously affect the system performance.

Propagation delays, unlike processing delays, lead to unfairness in the probability of performing a successful packet transmission, since remote terminals have more retarded information of the receiver status than nearby terminals. It will be shown (in Section 6.7.6) that with reasonably low packet traffic loads, propagation delays cause the probability of successful access to be reduced approximately linearly with the distance to the common receiver. This further increases the near-far effect that occurs if weak radio signals from remote terminals have an increased probability of being lost in excessive noise or interference, as addressed in the previous sections.

---

[1]Portions of Section 6.7 are reprinted, with permission, from IEEE Transactions on Communications, paper 91–239, © 1993, IEEE.

The analysis of channels with propagation delays is complicated by the fact that packet arrivals at the receiver do not comply with an on-off switched Poisson process, as was considered so far: as the leading edge of the busy signal propagates outwards, more and more terminals will be inhibited from potential transmissions, so the packet arrival rate at the base station will decrease gradually rather than step-wise. Similarly, when the channel is released by the transition from a busy signal to an idle signal, the resulting arrival rate of packets will increase only gradually. A perfect feedback channel is considered, so each mobile terminal is assumed to receive the signaling messages without errors. Fading, noise, or interference corrupting the signaling are neglected. Also, the time resolution is assumed to be perfect. In practice, this would require a feedback channel of infinite bandwidth. For feedback channels with finite bandwidth, the transitions between busy and idle states contribute to the fixed processing delay. Imperfections of the feedback channel are addressed in [21].

6.7.1 The Wake—ISMA Channel Traffic Model

Extending the "switch" model considered in Section 6.6, the following additional time intervals are defined for a terminal at a distance r from the central station:

- The *idle period*, with duration $i(r)$, is the interval of time during which the following two conditions are satisfied simultaneously: the terminal receives an idle signal and a packet transmitted by the terminal will indeed arrive at an idle receiver. Because of propagation delays, the duration $i(r)$ of the idle period is a function of the location of the terminal.
- The *vulnerable period*, with duration $v(r)$, is the interval of time during which the terminal receives an idle signal, while a packet transmitted in this interval is bound to arrive at an already busy receiver.
- The *inhibited period*, with duration h, is the period during which a busy signal is received.

A subsequent vulnerable and inhibited period was simply called a *busy period* in Section 6.6. The total duration of a *cycle* (i.e., the period from the beginning of an inhibited period until the beginning of the next inhibited period) is $a = h + i(r) + v(r)$. Evidently, a and h cannot depend on the distance r. Since packet arrivals are random, a, h, $i(r)$, and $v(r)$ are stochastic variables. Corresponding expectation values are denoted by capitals: A, H, $I(r)$, and $V(r)$, respectively.

In the analysis of channels without propagation delays, these periods can be drawn on a one-dimensional time axis (see [8] or Figures 6.7 and 6.8). To consider the effect of propagation delays and the resulting signaling wake, the distance to the base station is taken as the second dimension in the time-space diagram given in Figure 6.9.

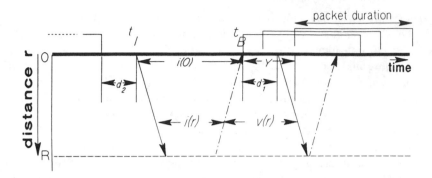

Figure 6.9 Time-space relation for inbound (— · —) and outbound (——) signals traveling over the radio channel.

A packet transmitted in the idle period arrives at an idle receiver in the central station. During a certain cycle, the first (and only) packet arriving at the idle receiver is called the *initiating* packet. After a processing delay of normalized duration d_1, the central station starts transmitting a busy signal. This busy status is maintained for the entire duration of the packet reception. If a number of overlapping packets is received, the busy signal is also broadcast. After the reception of the last of the colliding signals terminates, the idle signal is transmitted with a delay d_2. The processing delays d_1 and d_2 are parameters of the central station and are considered to be fixed.

We define the instant t_I as the instant when the base station starts broadcasting an idle signal. This signal arrives at a distance r from the base station at the instant $t_I + r/u$, with u the speed of light, normalized to the packet duration and the radius of the service area. A terminal only has permission to transmit if it receives an idle signal. Thus, the first instant when a packet from a distance r can arrive at the base station is $t_I + 2r/u$. The instant of arrival of the initiating packet at the central station is denoted by t_B, where $t_B = t_I + i(0)$. This event implies that, for a terminal at distance r, the idle period ended at $t_B - r/u$, and the vulnerable period was entered. Taking account of the possibility that the initiating packet arrives before the idle signal has reached all terminals ($i(0) < t_p$), the idle period at a distance r is found as the interval

$$\left\langle t_I + \frac{r}{u}, \ \max\left(t_I + \frac{r}{u}, \ t_I + i(0) - \frac{r}{u} \right) \right\rangle$$

Thus, $i(r) = \max(0, \ i(0) - 2r/u)$.

After receiving the initiating packet, the central station broadcasts the busy signal with a normalized delay d_1 (thus, at the instant $t_I + i(0) + d_1$). The vulnerable period at a distance r is found as the time interval

$$\left\langle \max\left(t_I + \frac{r}{u}, t_I + i(0) - \frac{r}{u} \right), t_I + i(0) + d_1 + \frac{r}{u} \right\rangle$$

with duration $v(r) = d_1 + \min(2r/u, i(0))$. Colliding packets may arrive until $t_B + d_1 + 2r/u$. Termination of the inhibited period occurs $1 + d_2$ units of time after the arrival of the last packet in the cycle considered. If only the initiating packet is present, this occurs at $t_B + 1 + d_2 + r/u$. However, in general, the duration of the inhibited period depends on the stochastic arrival times of interfering packets. The inhibited period nonetheless always ends before $t_B + d_1 + t_p + 1 + d_2 + r/u$.

6.7.2 Packet Arrival Rate in Channels With Propagation Delays

A terminal has permission to transmit only if it receives an idle signal. Only after a duration of $t_p = 2R/u$, with R the normalized cell radius ($R = 1$) and t_p the normalized maximum round-trip delay, packets can arrive from any location in the circular area. We assume $d_1 + t_p < 1$. Between t_I and $t_I + t_p$, the arrival rate steadily grows from 0 to G_t. Taking into account the time-space volume from which data packets can arrive, the arrival rate is found as $\lambda(t) = \lambda_t(t - t_I)$, with

$$\lambda_t(t) = \begin{cases} 0 & t < 0 \\ \int_0^{1/2ut} 2\pi r G(r) dr = G_t \dfrac{t^2}{t_p^2} & 0 < t < t_p \\ G_t & t > t_p \end{cases} \qquad (6.33)$$

Here it has been assumed that the central station does not start transmitting a busy signal again during the interval $\langle t_I, t_I + t \rangle$. The arrival rate $\lambda(t)$ is depicted in Figure 6.10. Using (6.7), the probability that no packet arrives at the central receiver during the period $\langle t_I, t_I + x \rangle$ is

$$P_{N_{\langle t_I, t_I + x \rangle}}(n = 0) = \exp\left(-\int_0^x \lambda_t(t)\, dt \right) \qquad (6.34)$$

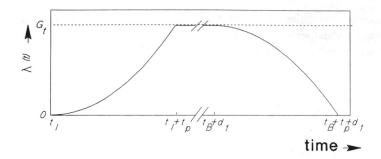

Figure 6.10 Rate of arrival $\lambda(t)$ versus time after the start of an idle period (t_I). $t_B + d_1$: the instant when base station starts transmitting the busy signal, with $t_I + t_p < t_B + d_1$.

The *cumulative distribution function* (cdf) $F_{i0}(x)$ of the lapse of idle time $i(0)$ at the central station is

$$F_{i(0)}(x) \triangleq \Pr\{i(0) < x\} = 1 - P_{N_{(t_I, t_I + x)}}(n = 0) = 1 - \exp\left(-\int_0^x \lambda_I(t) \, dt\right) \quad (6.35)$$

It can be seen that, in contrast to the conventional case with *stationary* Poisson arrivals ($\lambda(t)$ constant), the interarrival time is not exponentially distributed.

$$\lambda(t) = \begin{cases} \lambda_I(t - t_I), & t_I < t < t_I + t_p \\ \lambda_B(t - t_B - d_1), & t_B + d_1 < t < t_B + t_p + d_1 \end{cases}$$

6.7.3 Duration of the Idle Period

The average duration $I(r)$ of an idle period is obtained from

$$I(r) \triangleq E\left[\max\left(0, i(0) - \frac{2r}{u}\right)\right] = \int_{2r/u}^{\infty} \left(x - \frac{2r}{u}\right) \frac{dF_{i(0)}(x)}{dx} \, dx$$

$$= \int_{2r/u}^{t_p} \left(x - \frac{2r}{u}\right) \lambda_I(x) \exp\left\{-\int_0^x \lambda_I(t) \, dt\right\} dx$$

$$+ \int_{t_p}^{\infty} \left(x - \frac{2r}{u}\right) \lambda_I(x) \exp\left\{-\int_0^{t_p} \lambda_I(t) dt - \int_{t_p}^x \lambda_I(t) dt\right\} dx \quad (6.36)$$

After some mathematical operations, one finds

$$I(r) = \int_{2r/u}^{t_p} \exp\left(-G_t \frac{x^3}{3t_p^2}\right) dx + \frac{1}{G_t} \exp\left(-\frac{1}{3} G_t t_p\right)$$

$$= \frac{1}{3}\sqrt[3]{\frac{3t_p^2}{G_t}} \left[\gamma\left(\frac{1}{3}, \frac{1}{3} G_t t_p\right) - \gamma\left(\frac{1}{3}, \frac{2}{3} G_t \frac{r^3}{R^2 u}\right)\right] + \frac{1}{G_t} \exp\left(-\frac{1}{3} G_t t_p\right)$$

(6.37)

where the incomplete gamma function is defined as [36]

$$\gamma(\alpha, x) \triangleq \int_0^x e^{-t} t^{\alpha-1} \, dt$$

(6.38)

As a verification of the analysis, it is confirmed that for channels without propagation delay ($t_p = 0$), the average duration of the idle period (6.37) correctly tends toward $I(r) = G_t^{-1}$. Figure 6.11 gives $I(r)$ as a function of r, and the approximation max(0, $I(0) - 2r/u$) for small G_t.

Figure 6.12 presents $I(r)$ versus G_t for various r and the bounds $G_t^{-1} \leq I(0) \leq G_t^{-1} + t_p$ and $G_t^{-1} - t_p \leq I(1) \leq G_t^{-1}$. It can be seen that for a terminal at a certain distance, the average duration of the idle period rapidly reduces to zero if the attempted traffic load G_t exceeds a certain intensity. If an interfering packet always

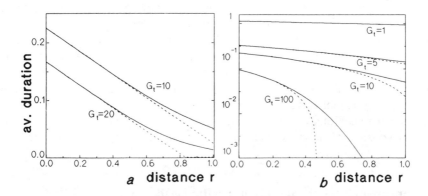

Figure 6.11 Average duration of the idle period $I(r)$ (——) in packet times on a linear axis (a, left) and on a logarithmic axis (b, right), versus distance r, and linear approximation (- -) by max(0, $I(0) - 2r/u$). Round-trip delay $t_p = 0.2$.

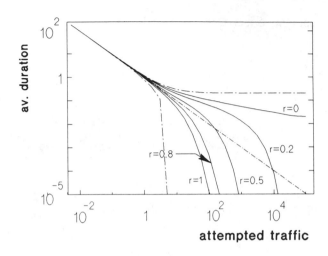

Figure 6.12 Average duration $I(r)$ of the idle period (——) as a function of attempted traffic load G_t for various distances. Maximum round-trip delay $t_p = 0.2$. ($\cdot$——$\cdot$) upperbound $I(r) < G_t^{-1} + t_p$ and lowerbound $I(r) > G_t^{-1} - t_p$ and ($\cdot$—$\cdot$) average duration in channels without propagation delays (G_t^{-1}).

destroys the initiating packet (no capture), the probability $Q(r_0)$ of a successful transmission of a packet from a distance r_0 is found from

$$Q(r_0) = \frac{I(r_0)}{A} \text{ Pr(no interfering packets)} \quad (6.39)$$

where A is the average duration of a complete cycle. In (6.39), $I(r_0)$ is the only factor depending on r_0. Apart from a multiplicative factor independent of r_0, the near-far unfairness is thus determined by the distance dependence of the idle period.

Compared to the investigation of $I(r)$, the determination of A, which includes assessment of the average duration of the vulnerable and inhibited periods, is relatively complicated. This is caused by the necessity of taking into account the presence of the busy signal corresponding to the foregoing cycle if $i(0)$ happens to be shorter than the maximum round-trip delay t_p. Therefore, we initially simplify the analysis by considering an approximate model, particularly appropriate for light traffic loads.

6.7.4 Approximate Model for Light Traffic Loads

Initially, we assume that $i(0)$ is always larger than the round-trip delay t_p, which is reasonable if $F_{i(0)}(t_p) = 1 - \exp[(-1/3)G_t t_p] \approx (1/3)G_t t_p$ is small. If this condition is satisfied, each idle period is likely to have nonzero duration $i(r) = i(0) - 2r/u$ > 0 for any location in the service area. In other words, this assumption ensures that a propagating busy signal corresponding to a previous cycle does not affect the durations of the next vulnerable and inhibited periods. In this case, the duration of the vulnerable period loses its stochastic nature and becomes identical to $v(r) = d_1 + 2r/u$ in each cycle.

We now address the duration of the inhibited period by initially considering the arrival rate of interfering packets. After arrival of the initiating packet at the instant t_B, the central station starts broadcasting a busy signal at the instant $t_B + d_1$. This signal reaches a terminal at distance r at the instant $t_B + d_1 + r/u$. Taking account of the time interval in which transmissions can occur, packets from the area $0 < r < R_0$ and destined to collide continue to arrive at the central station with the rate $\lambda(t) = \lambda_B(t - t_B - d_1)$, where

$$
\lambda_B(t) = \begin{cases}
G_t \dfrac{R_0^2}{R^2} & t < 0 \\[2ex]
\displaystyle\int_{1/2ut}^{R_0} 2\pi r G(r)\, dr = G_t\left[\dfrac{R_0^2}{R^2} - \dfrac{t^2}{t_p^2}\right] & 0 < t < 2R_0/u \qquad (6.40) \\[2ex]
0 & t > 2R_0/u
\end{cases}
$$

We insert $R_0 = R = 1$ to take account of all arriving packets, including those from the most remote terminals. R_0 is introduced here because later we will consider systems with receiver capture, in which weak interfering packets ($r > R_0$) do not cause any harmful interference.

The arrival rate of interfering packets decreases smoothly (see Figure 6.10) and becomes zero only at the instant $t_B + d_1 + t_p$. The time lapse y between the arrival of the initiating packet and the last interfering packet in the vulnerable period has the cdf

$$
F_Y(y) = \begin{cases}
0 & y < 0 \\
\Pr(\text{no arrivals in } \langle t_B + y, t_B + d_1 + t_p\rangle) & y > 0
\end{cases} \qquad (6.41)
$$

$$
F_Y(y) = \begin{cases}
0 & y < 0 \\[1ex]
\exp\left(-\displaystyle\int_{-d_1+y}^{t_p} \lambda_B(\tau)\,d\tau\right) & 0 < y < d_1 + t_p \\[2ex]
1 & y > d_1 + t_p
\end{cases} \qquad (6.42)
$$

Inserting (6.39) gives

$$
F_Y(y) =
\begin{cases}
0 & y < 0 \\[2mm]
\exp\left\{-\dfrac{2}{3}t_p G_t - d_1 G_t + y G_t\right\} & 0 < y < d_1 \\[3mm]
\exp\left\{-G_t[t_p + d_1 - y] + \dfrac{1}{3}G_t\left[t_p - \dfrac{(y - d_1)^3}{t_p^2}\right]\right\} & d_1 < y < d_1 + t_p \\[3mm]
1 & y > d_1 + t_p
\end{cases}
$$

(6.43)

At $y = 0$, this cdf exhibits a step of a size equal to the probability that no interfering packet is present. The average duration of the inhibited period is $H = E[y] + 1 - d_1 + d_2$, or

$$
H = yF_y(y)\Big|_0^{d_1+t_p} - \int_0^{d_1+t_p} F_Y(y)\,dy + 1 - d_1 + d_2
$$

$$
= t_p + e^{-2/3t_p G_t}\left[-\frac{1}{G_t}(1 - e^{-d_1 G_t}) - \int_0^{t_p} \exp\left(+ xG_t - \frac{x^3}{t_p^2}G_t\right)dx\right] + 1 + d_2
$$

(6.44)

For light attempted traffic loads ($G_t \to 0$), the average duration of the inhibited period tends toward $H \to 1 - d_1 + d_2$. This result does not depend on the round-trip delay t_p: in this case, collisions are rare, so the inhibited period nearly always corresponds to the transmission time of one packet, adjusted for the processing delays.

For large offered traffic loads ($G_t \to \infty$), $H \to 1 + t_p + d_2$: many collisions are likely to occur, and interfering packets may arrive until $t_B + d_1 + t_p$. It should be noted, however, that (6.39) to (6.44) may become inaccurate for large traffic loads because of the assumptions discussed at the beginning of the section. For channels without propagation delays ($t_p \to 0$), $H \to [1 - \exp(-d_1 G_t)]G_t^{-1} + 1 + d_2$ [8]. Figure 6.13 portrays (6.44) and an exact solution [26], which will be discussed later.

The overall probability of a successful access $Q(r)$ is found from the probability $I(r)/A$ that the attempt occurs in the idle period and the probability that no harmful interfering packet is transmitted in the vulnerable period. In a realistic radio channel, an initiating packet may be received correctly despite the presence of weak interfering packets. The capture effect will be studied in more detail for slotted ALOHA

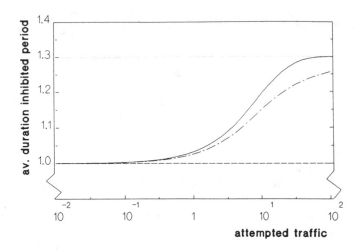

Figure 6.13 Average duration H of the inhibited period in packet times versus attempted traffic load G_t, according to (——) approximate and (———·) exact method. $t_p = 0.2$, $d_1 = d_2 = 0.1$. Bounds: (- -) $H > 1 - d_1 + d_2 = 1$ and (···) $H < 1 + d_2 + t_p = 1.3$.

and ISMA with fixed (distance-independent) propagation delays in Chapter 7. None-theless, to gain some initial insight into the combined (near-far) effects of delay and attenuation, we consider a basic capture model: the vulnerability-circle model, pre-sented in 1977 by Abramson [11] for ALOHA channels, can conveniently be intro-duced here. An initiating packet from distance r_0 is assumed to be received correctly if no interfering packets arrive from a distance closer than $c_z r_0$, with c_z a system constant. Furthermore, we assume that no noninitiating packets can be received cor-rectly because the receiver is already busy receiving and synchronized to another (i.e., to the initiating) packet at the time of arrival. The probability of no packets harmfully interfering with the initiating packet is found from integrating the arrival rate in (6.39) and taking $R_0 = \min(R, c_z r_0)$. So,

$$\Pr(\text{capture} \mid r_0) = \exp\left[-\left(\frac{\min(R, c_z r_0)}{R} \right)^2 \left(d_1 + \frac{2}{3} t_p \right) G_t \right] \qquad (6.45)$$

The probability of a successful transmission (6.39) now becomes

$$Q(r_0) = \frac{I(r_0)}{A} \Pr(\text{capture} \mid r_0)$$

$$= I(r_0) \frac{\exp\left[-G_t \left(\frac{\min(R, c_z r_0)}{R} \right)^2 \left(d_1 + \frac{2}{3} t_p \right) \right]}{I(0) + d_1 + H} \qquad (6.46)$$

Here, (6.37) and (6.44) are to be inserted. Figure 6.14 presents numerical results of $Q(r)$ for $G_t = 2$ ppt, respectively.

The total channel throughput S_t expressed in the average number of successfully received packets per unit of time is obtained as

$$S_t = \int_0^R G(r)Q(r)2\pi r\,dr \qquad (6.47)$$

and illustrated in Figure 6.15.

For CSMA, it is often suggested (e.g., [8]) that the effect of variable propagation delays be approximated by considering all terminals to be at an identical (worst-case) distance to the base station. This corresponds to inserting the modified delays d_1' and d_2', with $d_1' = d_1 + R/u$ and $d_2' = d_2 + R/u$, in formulas such as (6.22) and (6.23). For ISMA, as addressed throughout this section, three approximate methods are compared in Figure 6.15 for the special case $d_1 = 0.1$, $d_2 = 0.1$, and $t_p = 0.2$ (so $R/u = t_p/2 = 0.1$).

Method (a) adds the *one-way* propagation delay to d_1 and d_2 ($d_1' = d_1 + R/u$ = 0.2, $d_2' = d_2 + R/u = 0.2$);

Method (b) adds the *round-trip* delay to d_1 and d_2 ($d_1' = d_1 + t_p = 0.3$, $d_2' = d_2 + t_p = 0.3$).

Method (c) adds the round-trip delay to d_1 but does not account for propagation delays in d_2 ($d_1' = d_1 + t_p = 0.3$, $d_2' = d_2 = 0.1$).

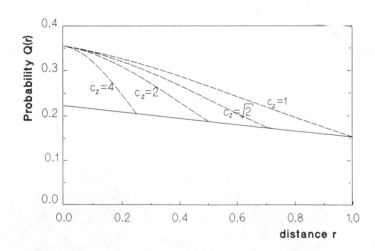

Figure 6.14 Probability of successful transmission $Q(r)$ versus distance r for $t_p = 0.2$ and $d_1 = d_2 = 0.1$, (– –) for various vulnerability parameters c_z, and (——) without capture. The total channel traffic load $G_t = 2$ ppt.

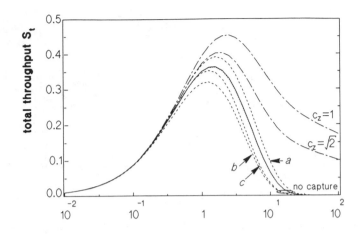

Figure 6.15 Throughput S_t versus attempted traffic G_t in ppt. Normalized delays: $d_1 = 0.1$, $d_2 = 0.1$, $t_p = 0.2$.
No capture: (——) exact method and (– –) approximate method proposed in Section 6.7.6; (– –, a,b,c) approximation by constant, distance-independent delays (see text); (—·) channels with variable delay and with capture.

Although method (a) might offer worst-case results for CSMA [8], for ISMA it is optimistic; that is, it overestimates the throughput. On the other hand, method (b) yields relatively conservative estimates, but may serve as a worst-case performance bound. It can be seen that, in the practical range, $G_t \approx 1$ ppt or less, approximation (c) gives about 5% lower results than the exact analysis, and may offer a sufficiently tight bound for practical system design.

This conclusion is explained intuitively by the fact that method (c) takes into account the effect that round-trip delays substantially widen the vulnerable period, and thus increase the expected number of interfering signals. Method (a) overestimates throughput mainly because it tends to underestimate the number of interfering signals. As can be seen from (6.45), for a uniform distribution of the attempted traffic, the probability of no packets interfering with an initiating packet is $\exp(-d_1 - 2/3t_p)$, thus d_1 is effectively prolonged by 133% of the one-way propagation delay.

The effect of receiver capture ($c_z = 1$ and $\sqrt{2}$) can be seen to be significant, particularly for high offered traffic loads ($G_t > 1$ ppt). This shall be addressed in more detail in Chapters 7 to 9.

6.7.5 Comparison With Exact Analysis

The wake diagram in Figure 6.16 illustrates the effect of the event $i(0) < t_p$ on the vulnerable period. This event also affects the average duration of the inhibited period

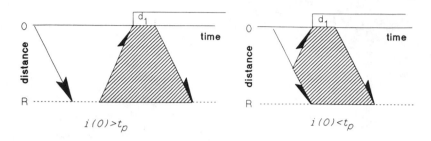

Figure 6.16 Time-space (wake) diagram and vulnerable period ($///$) for the events $i(0) < t_p$ and $i(0) > t_p$.

and the probability of no harmful interference. Given the duration of the idle period $i(0)$, the duration of the vulnerable period is determined by $v(r) = d_1 + \min(2r/u, i(0))$. The probability of no harmfully interfering packets can be computed by integrating over the appropriate time-space volume, which is a tedious exercise. Moreover, the duration of the inhibited period is a stochastic variable, dependent, among other things, on the duration of the foregoing vulnerable period, and thus indirectly on the duration of the foregoing idle period.

The unconditional average duration of the inhibited period is obtained from

$$H = \int_0^\infty H(x) f_{i(0)}(x)\, dx \tag{6.48}$$

where $H(x)$ is the average duration of the inhibited period for a given $i(0) = x$. Evaluation of (6.48) also becomes tiresome, because an analytic expression of the integrand only exists for subintervals of the domain of the integration variable, and various segments of the time-space volume are to be treated separately in the integration. This was Venema's endeavor in his graduation work [26].

Numerical results are given by (——) in Figure 6.13 and by (— · —) in Figure 6.15 to discuss the accuracy of the approximate model presented here. It can be seen in Figure 6.13 that the duration of H is overestimated by (6.44). This leads to an underestimation of $Q(r)$ and S_t. Moreover, if $i(0) < t_p$, the time-space volume from which interfering packets may arrive is smaller than assumed in (6.45). Hence, the number of interfering signals is overestimated. This also leads to an underestimation of the channel performance. Nonetheless, even the combined effect of these approximations turns out to be very small, and is barely distinguishable in Figure 6.15 (compare ($\cdot\cdot$) dotted and (——) solid line).

6.8 CONCLUSION

This chapter discussed protocols for random multiple access and analytical models for studying their performance in mobile channels. The channel performance experienced by individual mobile terminals may differ widely from the mean performance averaged over all terminals. Two types of near-far effects occur in a mobile random access network:

1. *Effects because weaker signals are received from remote terminals*. Calculation of the effect of path losses required extension of the known techniques to compute the total throughput of ALOHA, CSMA, and ISMA in wired channels. Here, the probability of capture and total channel throughput have only been expressed in terms of the probability that a test signal survives a collision with n contenders. As will be shown in Chapter 7, the latter probability can be obtained from the models for channel fading and receiver capture developed in Chapters 3 to 5.

2. *Effects caused by propagation delays*. The performance of nonpersistent ISMA was also studied for radio networks with substantial propagation delays. The principal contribution of variable propagation delays to the near-far unfairness is caused by the wake-effect of the idle period decreasing almost linearly with the distance between terminal and the central station. The unfairness caused by propagation delays increases the unfairness caused by the fact that weak signals from remote terminals experience a higher probability of being lost in a collision. It was seen that for reasonable traffic loads, say $G_t < 5$ ppt, propagation delays cause less unfairness than propagation attenuation. At high traffic loads in nonpersistent ISMA networks, delays are expected to become the major cause of unfair access probabilities for remote terminals: beyond a certain offered traffic load, terminals at a certain distance experience a highly limited probability of successfully transmitting a data packet.

Exact analysis of the ISMA with propagation delays appeared to be a tedious task if cycles can overlap. Fortunately, an approximate technique, taking exact account of the duration of the idle period, but ignoring the effect overlapping cycles have on the duration of the vulnerable and busy periods, proved to give relatively accurate results. It was seen that propagation delays principally affect the delay in the vulnerable period. A further simplification and approximation appeared reasonable: the worst-case *round-trip* propagation delay may be included in the duration of the vulnerable period. This approximation, based on fixed (distant-independent) propagation delays, was seen to lead to somewhat pessimistic results for the throughput.

The present development of public networks for packet switched mobile data communication and wireless office automation systems tends to focus on wireless communication links over shorter and shorter ranges. This seems to divert interest from the propagation delays in the network. Note, however, that these propagation delays are to be considered in relation to the packet duration, which decreases inversely proportional to the chosen bit rate.

REFERENCES

[1]   Verhulst, D., M. Mouly, and J. Szpirglas, "Slow Frequency Hopping Multiple Access for Digital Cellular Radiotelephone," *IEEE J. Sel. Areas Comm.*, Vol. SAC-2, No. 4, July 1984, pp. 563–574.

[2]   Kegel, A., "Mobile Networks" (in Dutch), in Lecture Notes L78, "Communication Systems; Selected Areas," Delft University of Technology, Dept. of E.E., 1990.

[3]   Miya, K., ed., *Satellite Communications Networks*, KDD Engineering & Consulting, Inc., Tokyo, 1985.

[4]   Tanenbaum, A.S., *Computer Networks*, 2nd ed., London: Prentice-Hall International Editions, 1989.

[5]   Stallings, W., *Data and Computer Communication*," 2nd ed., New York: Macmillan Publishing Company, 1988.

[6]   Abramson, N., "The ALOHA System-Another Alternative for Computer Communications" *Proc. 1970 Fall Joint Computer Conf.*, AFIPS Press, Vol. 37, 1970, pp. 281–285.

[7]   Roberts, L.G., "ALOHA Packet System With and Without Slots and Capture," *Comput. Comm. Rev.*, Vol. 5, Apr. 1975, pp. 28–42.

[8]   Kleinrock, L., and F.A., Tobagi, "Packet Switching in Radio Channels: Part 1—Carrier Sense Multiple Access Modes and Their Throughput-Delay Characteristics," *IEEE Trans. on Comm.*, Vol. COM-23, No. 12, Dec. 1975, pp. 1400–1416.

[9]   Tobagi, F., and L. Kleinrock, "Packet Switching in Radio Channels Part II—The Hidden Terminal Problem in Carrier Sense Multiple Access and the Busy Tone Solution," *IEEE Trans. on Comm.*, Vol. COM-23, No. 12 Dec. 1975, pp. 1417–1433.

[10]   Krebs, J., and T. Freeburg, "Method and Apparatus for Communicating Variable Length Messages Between a Primary Station and Remote Stations at a Data Communications System," U.S. patent No. 4,519,068, 1985.

[11]   Abramson, N., "The Throughput of Packet Broadcasting Channels," *IEEE Trans. on Comm.*, Vol. COM-25, No. 1, Jan. 1977, pp. 117–128.

[12]   Abramson, N., "Development of the ALOHAnet," *IEEE Trans. on Inf. Theory*, Vol. IT-31, No. 2, March 1985, pp. 119–123.

[13]   Nijhof, J.A.M., "Datacommunication L102," hand-out during lectures, Delft University of Technology, Dept. of E.E., 1990.

[14]   Rec. Z.100 and annexes A, B, C and E, Rec. Z.110, "Functional Specification and Description Language (SDL), Criteria for Using Formal Description Techniques (FDTs)," *CCITT Blue Book*, Vol. X, Fascicle X.1, IXth Plenary Assembly, Melbourne, 14–25 Nov. 1988.

[15]   Bertsekas, D., and R. Gallager, *Data Networks*, London: Prentice-Hall International, Inc., 1987, ISBN 0–13-196981-1.

[16]   Tsybakov, B., "Survey of USSR Contributions to Random-Access Communications," *IEEE Trans. on Inf. Theory*, Vol. IT-31, No. 2, March 1985, pp. 142–165.

[17]   Carleial, A.B., and M.E. Hellman, "Bistable Behavior of ALOHA-Type Systems," *IEEE Trans. on Comm.*, Vol. COM-23, No. 4, April 1975, pp. 401–410.

[18]   Kleinrock, L., and S.S. Lam, "Packet Switching in a Multiaccess Broadcast Channel-Performance Evaluation," *IEEE Trans. on Comm.*, Vol. COM-23, No. 4, April 1975, pp. 410–423.

[19]   Schoute, F.C., "Determination of Stability of a Markov Process by Eigenvalue Analysis, Applied to the Access Capacity Enhancement ALOHA Protocol" PTI report SR2200–82–3790, Hilversum, 12 Dec. 1982. Parekh, S., F. C. Schoute, and J. Walrand, "Instability and Geometric Transience of the ALOHA Protocol," Memorandum No. UCB/ERL M86/73, Berkeley, 1986.

[20]   Schoute, F.C., "Dynamic Frame Length ALOHA," *IEEE Trans. on Comm.*, Vol. COM-31, No. 4, April 1983, pp. 565–568. (see also F. C. Schoute, A.W. Doorduin, and L. M. Hooijman, "Method and Apparatus for Preventing Overloading of the Central Controller of a Telecommunication System," U.S. Patent No. 4,497,978, 1985.)

[21]   Andrisano, O., G. Grandi, and C. Raffaelli, "Analytical Model for Busy Channel Multiple Access (BCMA) for Packet Radio Networks in a Local Environment," *IEEE Trans. On Veh. Tech.*, Vol VT-39, No. 4, Nov. 1990, pp. 299–307.

[22]   Murase, A., and K. Imamura, "Idle-Signal Casting Multiple Access with Collision Detection (ICMA-CD) for Land Mobile Radio," *IEEE Trans. on Veh. Tech.*, Vol. VT-36, No. 2, May 1987, pp. 45–50.

[23]   "Technical Characteristics and Test Conditions for Non-speech and Combined Analogue Speech/Non-speech Radio Equipment With Internal or External Antenna Connector, Intended for the Transmission of Data, for Use in the Land-Mobile Service" (DRAFT), European Telecommunications Standards Institute, I-ETS [A] version 3.3.2, Valbonne, France, 1990.

[24]   Zdunek, K.J., D.R. Ucci, and J.L. Locicero, "Throughput of Nonpersistent Inhibit Sense Multiple Access With Capture," *Electron. Lett.*, Vol. 25, No. 1, 5 Jan. 1989, pp. 30–32.

[25]   Prasad, R., and J.C. Arnbak, "Capacity Analysis of Non-persistent Inhibit Sense Multiple Access in Channels With Multipath Fading and Shadowing," *Proc. 1989 Workshop on Mobile and Cordless Telephone Communications*, IEE, London, Sept. 1989, pp. 129–134.

[26]   Venema, R.J., "Inhibit Sense Multiple Access in Systems With Propagation Delays," M.Sc.E.E. graduation thesis A333, Delft University of Technology, The Netherlands, 18 Sept. 1990.

[27]   Tsiligirides, T., and D.G. Smith, "Analysis of a *p*-Persistent CSMA Packetized Cellular Network With Capture Phenomena," *Computer Communications*, Vol. 14, No. 2, March 1991, pp. 94–104.

[28]   Gross, D., and C.M. Harris, "Fundamentals of Queueing Theory," New York: John Wiley and Sons, 1974.

[29]   Kleinrock, L., *Queueing Systems, Vol. 1: Theory, Vol. 2: Computer Applications*, New York: John Wiley and Sons, 1975, 1976.

[30]   Arnbak, J.C., and W. van Blitterswijk, "Capacity of slotted-ALOHA in a Rayleigh Fading Channel," *IEEE J. Sel. Areas Comm.*, Vol. SAC-5, No. 2, Feb. 1987, pp. 261–269.

[31]   Roberts, J.A., and T.J. Healy, "Packet Radio Performance Over Slow Rayleigh Fading Channels," *IEEE Trans. on Comm.*, Vol. COM-28, No. 2, Feb. 1980, pp. 279–286.

[32]   Metzner, J.J., "On Improving Utilization in ALOHA Channels," *IEEE Trans. on Comm.*, Vol. COM-24, No. 4, April 1976, pp. 447–448.

[33]   Goodman, D.J., and A.A.M. Saleh, "The Near/Far Effect in Local ALOHA Radio Communications," *IEEE Trans. on Veh. Tech.*, Vol VT-36, No. 1, Feb. 1987, pp. 19–27.

[34]   Molle, M.L., K. Sohraby, and A.N. Venetsanopoulos, "Space-Time Model of Asynchronous CSMA Protocols for Local Area Networks" *IEEE J. Sel.* Areas Comm., Vol. SAC-5, No. 6, July 1987, pp. 956–968.

[35]   Kamal, A.E., "A Discrete-Time Approach to the Modeling of Carrier-Sense Multiple-Access on the Bus Topology," *IEEE Trans. on Comm.*, Vol. 40, No. 3, March 1992, pp. 533–540.

[36]   Abramowitz, M., and I.A. Stegun, eds., *Handbook of Mathematical Functions*, New York: Dover, 1965.

[37]   Pronios, N.B., and A. Polydoros, "Utilization Optimization of CDMA Systems," *Bilcon 90 Int. Conf. on New Trends in Comm., Control and Signal Processing*, Ankara, Turkey, 2–5 July 1990, pp. 560–566.

Chapter 7
Models for Receiver Capture in Mobile Random-Access Networks

As we saw in Chapter 6, the performance of random-access protocols strongly depends on the probability of a successful transmission of an individual data packet over the common radio channel. The probability that a selected test packet experiences a collision with n contending packets was presented for various protocols. The probability that the test packet is received successfully despite its n contenders depends, among other things, on the characteristics of the radio channel, such as the rate of fading, the type of modulation, the receiver design, and the distribution of the terminals over the service area. To find the probability of successful reception of a packet, this chapter summarizes and compares a number of models for receiver capture.

The organization of this chapter is as follows. Section 7.1 discusses the capture effect. It also compares capture with the event of *receiver success* as studied in coding theory. Section 7.2 presents the quasi-uniform spatial distribution of the attempted packet traffic over the service area. This special case allows relatively convenient analysis of the network behavior for most of the capture models considered. Each of the Sections 7.3 to 7.7 discusses a particular capture model in more detail and presents numerical results for the probability of capture and the channel throughput. Section 7.8 presents the probability of successful access according to the models discussed in Sections 7.3 to 7.7. The suitability of the capture ratio model is addressed in more detail in Section 7.9.

7.1 RECEIVER CAPTURE

The term *capture effect* was originally used to describe the effect of the threshold in analog nonlinear modulation (such as FM) [1, 2]. In studies of collision-type random-access networks for data communication, *capture effect* has been adopted to describe the fact that any practical radio receiver is resistant against noise and interference, at least to some extent. If packets compete for successful reception at the receiver, one signal may *capture* the receiver [3–5]. In this context, however, the word *effect* has lost its meaning of indicating a remarkably rapid impairment of the quality of reception with only a slight decrease of the C/I ratio.

An ad hoc definition of receiver capture in now proposed. It may not immediately provide a rigid base for analytical evaluation, but it represents an extension of the FM-capture effect in the case of discrete messages, such as data packets:

> Receiver capture is the phenomenon in which the decisions required from the receiver to estimate or reconstruct a potential message transmitted on the channe, are dominated by a single signal, despite the presence of interfering (contending) signals and noise.

Hence, receiver capture is essentially an event (denoted by c_j) of the physical layer of the communication system. Conditions relevant to the probability of capture are, among other things, received power levels, the type of modulation, robustness of receiver synchronization, characteristics of the interference, and channel fading. According to this definition, the type of channel coding does not directly affect the probability of capture.

This means that capture probabilities do not uniquely determine the throughput of the channel. Taking also the effects of channel coding into consideration, *transmitter success* (event s_j) and *receiver success* (event $s_1 \bigvee s_2 \bigvee \cdots \bigvee s_N$) can be defined as follows [6]:

- Transmitter success is the event in which the bit sequence of the packet from transmitter j is identical to the bit sequence decoded by the receiver; that is, the packet from terminal j is received error-free or contains only a correctable number of bit errors.
- Receiver success is the event in which the receiver decodes a bit sequence that has been transmitted by at least one of the competing transmitters.

Evidently, receiver success and receiver capture are highly correlated, but the two events do not necessarily coincide: receiver capture does not necessarily imply correct decoding of the test packet. Also, in some fortuitous cases, receiver success may occur while receiver capture does not occur: the receiver may, by chance, make a right decision, even if all signals are buried in noise or mutual interference [6, 7].

Because of the discrete nature of digital messages, events of receiver or transmitter success can be sharply defined. In contrast to this, the distinction between capture and noncapture is more arbitrary. However, exact computation of probabilities of success is always a complicated task. This is illustrated by the large number of different approximate models used to compute these probabilities. In theoretical analyses, basically two classes of models exist:

A   Models focusing on the physical layer: It is assumed that successful reception occurs if (and only if) a limited number of conditions are fulfilled, under which *capture* is likely to occur. Mostly, received signal power levels are considered as the criterion for capture. In these models, the probability of bit errors and the effect of coding are not addressed explicitly. These models can be acceptable if $\Pr(s_j|c_j) \to 1$ and $\Pr(s_j|\text{NOT}c_j) \to 0$ (e.g., if a clear receiver threshold is present).

B   Models focusing at the data-link layer: These models allow investigation of the probability of success, erasure, or undetected error during a particular time slot, from the point of view of the common receiver or a particular mobile terminal. To confine the complexity of the calculations, such models often simplify receiver behavior (e.g., by assuming perfect carrier and bit synchronization). Such a model is acceptable if the receiver manages to capture a signal even if the C/I ratio is small. Practical coherent narrowband receivers, however, often fail to acquire synchronization to a short (burst-type) transmission if the interference is strong. In the latter case, theoretical block success rates for perfectly synchronized detectors tend to overestimate the probability of successful reception.

Examples of models that belong to category A are:

A1 Classification of power level: Metzner [3] proposed to intentionally divide the participating terminals into two groups: one group employs substantially higher transmitting power than the other group. He assumed that receiver capture occurs either if a collision involves exactly one packet from a high-power terminal and an arbitrary number of packets from low-power terminals, or if only a packet from the low-power group is present.

A2 In 1977 Abramson [4] proposed the *vulnerability circle* to evaluate capture in a radio network: a packet transmitted from distance r is received correctly if and only if no other packet is generated within a vulnerability circle of radius $c_z r$, with c_z a system constant [4, 8].

A3 Kuperus and Arnbak [9] considered a *capture ratio*: a packet is received correctly if and only if its received power exceeds the joint interference power by at least a threshold factor (or capture ratio) z. During the capture period, the received power is considered constant [5, 9–24].

A4 Sinha and Gupta [25] assumed correct packet reception if the entire packet is received during a *nonfade interval*. In [25–27], the model was considered for noise limitations of the network performance. Linnartz [8] studied this model for a contention-limited network.

A5 If colliding data packets do not arrive exactly simultaneously, the receiver may lock to the *first arriving* signal. This initiating packet is only received correctly if the joint interference power that accumulates during the following vulnerable period is sufficiently low. Packets arriving at a busy receiver are considered to be lost. For instance, a spread-spectrum receiver may lock to an arriving packet and effectively reject interference from later packets [28, 29].

Models of the second category, based on bit error probabilities, are:

B1 Receiver success for a receiver that always achieves perfect carrier and bit synchronization to any packet. If the number of bit errors is sufficiently small (i.e., smaller than or equal to the error correction (Hamming) distance M), successful reception is assumed [7, 30]. To obtain the probability of receiver success, the probabilities of transmitter success of all transmitters in the network are added. This model is reasonable only if one may assume that different transmitters never transmit the same bit sequence.

B2 Transmitter success. Zhang, Pahlavan, and Ganesh [31] considered a receiver that randomly selects a favorite test packet and acquires perfect bit and carrier synchronization to this packet. If the test packet happens to be selected by the receiver and detection occurs without error or with a correctable number of bit errors, it is assumed to be received successfully.

Hybrid models are:

D1 Habbab, Kavehrad, and Sundberg [32] assumed a receiver to lock to the signal from the *nearest* transmitter. If this packet is received with M or less bit errors, successful reception is assumed.

D2 Sheikh, Yao, and Wu [35] considered a bit error to occur (with probability one-half) if and only if the received signal power fails to exceed the power contained in the bit of the strongest interfering packet. Transmitter success occurs if not more than M out of L bits are in error.

In the following sections, we study the probability $q_n(r)$ that a test packet from a terminal located at a distance r is received successfully, conditional on the presence of n competing transmissions from terminals with unknown locations. Moreover, results from Chapter 6 are used to express the unconditional probability $Q(r)$ of successful transmission, taking into account the influence of the access protocol.

7.2 QUASI-UNIFORM SPATIAL DISTRIBUTION

In this chapter, the near-far effect and Rayleigh fading are considered, while shadowing is ignored ($\sigma_s = 0$, $\bar{\bar{p}}_j = \bar{p}_j$). The normalized local-mean power $\bar{p}_j$ received from the jth mobile terminal at a normalized distance r_j from the central receiver is assumed to be of the form $\bar{p}_j = r_j^{-4}$; thus, the attenuation exponent in (2.13) is taken as $\beta = 4$. If the position of the terminal is unknown, the pdf of the mean power is found from

$$|f_{\bar{p}_j}(\bar{p}_j)d\bar{p}_j| = \left| \frac{2\pi r_j G(r_j)}{G_t} dr_j \right| \qquad (7.1)$$

with $G(r)$ the offered traffic per unit area and G_t the total offered traffic (see Chapter 6). Throughout this chapter, we assume the quasi-uniform spatial distribution of the offered channel traffic suggested in [5], namely,

$$G(r) = \frac{G_t}{\pi} \exp\left(-\frac{\pi}{4} r^4 \right) \qquad (7.2)$$

The median of this distribution is $r_M \approx 0.734$, since

$$\frac{1}{G_t} \int_0^{r_M} 2\pi r G(r) dr = \text{erf}\left(\frac{\sqrt{\pi}}{2} r_M^2 \right) = \frac{1}{2} \qquad (7.3)$$

where the error function erf is defined as [15]

$$\text{erf}(x) \triangleq \frac{2}{\sqrt{\pi}} \int_0^x \exp\{-\lambda^2\}d\lambda \qquad (7.4)$$

For $\beta = 4$, the median received power is $\bar{p}_M = r_M^{-4} \approx 3.545$. Figure 7.1 shows that

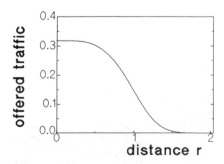

Figure 7.1 Quasi-uniform distribution of the offered traffic $G(r)$ versus distance r normalized for $G_t = 1$ ppt.

(7.2) is an approximation of the exact uniform distribution

$$G(r) = \begin{cases} \dfrac{1}{\pi} G_t, & 0 < r < 1 \\ 0 & r > 1 \end{cases} \tag{7.5}$$

by a smooth analytical function. The reason for adopting this quasi-uniform spatial distribution is the convenient analytical expression found for the pdf of the joint power of n uncorrelated signals after incoherent addition [5]: n-fold convolution of (7.1) with (7.2) gives

$$f_{\bar{p}_t}(\bar{p}_t|n) = \begin{cases} \dfrac{n}{2} \bar{p}_t^{-3/2} \exp\left(-\dfrac{\pi n^2}{4\bar{p}_t}\right) & n = 1, 2, \cdots \\ \delta(\bar{p}_t) & n = 0 \end{cases} \tag{7.6}$$

where $\bar{p}_t$ ($\bar{p}_t = \Sigma \bar{p}_j$) is the local-mean power of the joint interference signal.

Further, Rayleigh fading in a narrowband channel is assumed for each signal. The instantaneous amplitude ρ_j of the jth carrier is Rayleigh-distributed with mean $\bar{p}_j$, so the corresponding inphase and quadrature carrier components ζ_j and ξ_j are independently Gaussian-distributed, with mean $\bar{p}_j$. Combining the statistical fluctuations caused by the spatial distribution of the mobile terminals (7.2) and those caused by Rayleigh fading, the unconditional pdfs of the amplitude ρ_j and the inphase component ζ_j are [31, 34]

$$f_{\rho_j}(\rho_j) = \frac{\sqrt{2\pi}\, \rho_j}{[(\pi/2) + \rho_j^2]^{3/2}} \tag{7.7}$$

and

$$f_{\zeta_j}(\zeta_j) = \frac{1}{\sqrt{2\pi}\, [(\pi/2) + \zeta_j^2]} \tag{7.8}$$

respectively. Any retransmitted or rescheduled packet is assumed to experience un-correlated fading and path loss.

7.3 MODEL BASED ON VULNERABILITY CIRCLE

7.3.1 Review

In 1977, Abramson [4] suggested the vulnerability-circle as a model for receiver capture. A test packet transmitted from a terminal at a distance r_j is received correctly if no other packet is transmitted within a circle of radius $c_z r_j$, with c_z a system constant (model A2). This is equivalent to assuming that capture occurs if and only if the area-mean power $\bar{\bar{p}}_j$ of the test packet exceeds the largest area-mean power $\bar{\bar{p}}_k$ among the interfering signals by at least a factor of $z = c_z^\beta$, (thus, if $\bar{\bar{p}}_j \geq c_z^\beta \bar{\bar{p}}_i$ for all $i = 1, 2, \ldots, n$). For slotted ALOHA, the probability of successful transmission is,

$$Q(r_j) = \Pr\left(\begin{array}{c} \text{no interfering arrivals from} \\ 0 < r < c_z r_j \end{array}\right)$$

$$= \exp\left(-\int_0^{c_z r_j} 2\pi r G(r) dr\right) \tag{7.9}$$

The ALOHA-net at the University of Hawaii [4] employed fixed transmitters at islands located at ranges of several tens of kilometres. Rayleigh fading and shadowing, which are ignored in (7.9), may not have played a mayor role in the signal attenuation, so the propagation model $p_j = r_j^{-\beta}$ might have been appropriate in this case. The vulnerability-circle has also been used in many other studies, particularly in the evaluation of multi-hop packet radio systems.

7.3.2 Analysis for Quasi-Uniform Offered Traffic

Abramson [4] considered exactly-uniform offered traffic with infinite extension of the service area, thus $G(r) = G_0$ for all $r > 0$. With our choice of a quasi-uniform spatial distribution (7.2), one finds for slotted ALOHA

$$Q(r_j) = \exp\left\{-\int_0^{c_z r_j} 2r G_t \exp\left(-\frac{\pi}{4} r^4\right) dr\right\}$$

$$= \exp\left\{-G_t \operatorname{erf}\left(\frac{\sqrt{\pi}}{2} c_z^2 r_j^2\right)\right\} \tag{7.10}$$

To compute the throughput for ISMA, we initially consider a test packet that arrives during the vulnerable period of an initiating packet (event H_d). The probability that the initiating signal (with index 1) arrives from outside the vulnerability circle is

$$q_1(r_j|H_d) = \Pr(r_1 < c_z r_j) = \int_{c_z r_j}^{\infty} 2r \exp\left(-\frac{\pi}{4} r^4\right) dr = \text{erfc}\left(\frac{\sqrt{\pi}}{2} c_z^2 r_j^2\right) \quad (7.11)$$

with $\text{erfc}(\cdot)$ the complementary error function (i.e., $\text{erfc}(x) = 1 - \text{erf}(x)$). The probability $H(r_j)$ that no further interfering transmission starts during the vulnerable period from within the vulnerable circle is (analogous to (7.10))

$$H(r_j) = \exp\left\{-d_1 G_t \, \text{erf}\left(\frac{\sqrt{\pi}}{2} c_z^2 r_j^2\right)\right\} \quad (7.12)$$

Using (6.23), the probability of a successful transmission becomes for nonpersistent ISMA

$$Q(r_j) = H(r_j) \frac{1 + d_1 G_t \, \text{erfc}\left(\frac{\sqrt{\pi}}{2} c_z^2 r_j^2\right)}{G_t(1 + d_1 + d_2) + \exp(-d_1 G_t)} \quad (7.13)$$

For unslotted p-persistent ISMA without signaling delays, one finds

$$Q(r_j) = \frac{1 + pG_t \exp(pG_t)\Pr\left(\begin{array}{c}\text{no interferers from} \\ 0 < r < c_z r_j\end{array}\right)}{1 + pG_t \exp(pG_t)}$$

$$= \frac{1 + pG_t \exp\left\{+ pG_t \, \text{erfc}\left(\frac{\sqrt{\pi}}{2} c_z^2 r_j^2\right)\right\}}{1 + G_t \exp(pG_t)} \quad (7.14)$$

Thus, for the vulnerability circle model, $Q(r)$ can be expressed without explicit derivation of $q_n(r)$. Numerical results for $Q(r)$ are included later, in Figure 7.14.

7.4 MODEL BASED ON CAPTURE RATIO[1]

Model A3 assumes a test packet to capture the receiver in the base station if, and only if, its instantaneous power p_j exceeds the instantaneous *joint* interference power

---

[1]Portions of Sections 7.4, 7.6, and 7.8 reprinted with permission from IEEE Transactions on Vehicular Technology, Vol. 41, No. 1, Feb. 1992, pp. 77–90. © IEEE 1992.

p_t by at least a threshold factor (or *capture ratio*) z. The received power is assumed to be constant during the reception of a packet ($T_5 \ll T_6$, see Section 2.1.7).

7.4.1 Review

The effect of cumulation of interference power if more than one interfering signal is present was considered in a number of papers. Kuperus and Arnbak [9] and Fronczak [24] derived receiver capture probabilities in a Rayleigh-fading channel for the particular spatial distribution that gives an exponential distribution of the signal power at the receiver. Namislo [10] reported the probability of capture for a slotted ALOHA network with near-far effect. The results were obtained by means of Monte-Carlo simulation. Verhulst et al. [11] studied slow frequency hopping in cellular telephone networks. Their results included an analytical expression for the throughput of slotted ALOHA in a channel with slow Rayleigh fading, but without shadowing and near-far effect. Arnbak and Van Blitterswijk [5] elaborated the method in [9] for a channel with combined Rayleigh fading and near-far effect caused by the quasi-uniform spatial distribution (7.2). Coherent and incoherent addition of interference signals were compared. Linnartz [12] showed that the use of integral transforms of spatial distributions can greatly facilitate the analysis of the throughput of mobile ALOHA networks with Rayleigh fading and near-far effect. This technique was used to quantify the effect in which remote terminals have to perform more retransmissions than nearby terminals [13]. An investigation of various spatial distributions and a further comparison of the vulnerability circle (A2) with the capture-ratio model (A3) has been presented by Lau and Leung in [14], focusing on channels without Rayleigh fading.

The influence of shadowing was discussed by Prasad and Arnbak in [15]. The technique by Schwartz and Ych (see Chapter 3) was employed to compute the joint interference power cumulated from multiple log-normal signals. In [16] and [17], this technique was extended for combined Rayleigh fading and shadowing with coherently and incoherently adding signals, respectively. The influence of combined near-far effect and shadowing on the throughput of the ALOHA channel was studied by Linnartz and Prasad in [18]. Capture probabilities for channels with combined Rayleigh fading, shadowing, and near-far effect, using integral transforms are developed in [20].

The total throughput of nonpersistent ISMA networks with capture in Rayleigh-fading channels was reported by Zdunek et al. [21]. Prasad further extended the analysis to include the effects of shadowing [35] and combined shadowing and near-far effects [19]. Linnartz [22] reported the probability of successful transmission of an a priori selected test packet. This allowed computation of the probability of successful transmission as a function of the distance between the terminal and the receiving base station. Near-far effects caused by an exactly uniform spatial distribution and Rayleigh fading were considered. Analytical expressions in closed-form were derived for $Q(r)$ in the case of slotted ALOHA, unslotted nonpersistent ISMA, and unslotted 1-persistent ISMA. Analogous to the results reported by the author for an exactly uniform spatial distribution in [22], the next section gives the probability of capture for the *quasi*-uniform spatial distribution (7.2).

7.4.2 Analysis for Quasi-Uniform Offered Traffic

Similar to the evaluation of the outage probability in CW communication in Section 3.2, the probability of capture (event c_j) conditional on the local-mean power of the test packet is expressed as a Laplace image, namely,

$$\Pr(c_j|\bar{p}_j) \triangleq \Pr\left(\frac{p_j}{p_t} > z \middle| \bar{p}_j\right) = \mathcal{L}\left\{f_{p_t}, \frac{z}{\bar{p}_j}\right\} \tag{7.15}$$

where $f_{p_t}(\cdot)$ is the pdf of the joint interference power p_t caused by cumulation of n contending packets. The lower limit of the integral over p_t (with dummy variable x) is taken "$0-$" to ensure that (7.15) and (3.16) are also valid if, with some nonzero probability, interference is absent in a noiseless channel ($n = 0$, $N_A = 0$). In the special case

$$f_{p_t}(p_t) = \delta(p_t) \tag{7.16}$$

Equation (7.15) becomes

$$\Pr(c_j) = \mathcal{L}\left\{\delta(\cdot), \frac{z}{\bar{p}_j}\right\} = 1 \tag{7.17}$$

That is, capture always occurs if no interference and noise are present, which agrees with the assumptions of model A3.

For the case of multiple interfering signals ($n = 2, 3, \ldots$), we will distinguish between coherent and incoherent cumulation of interference signals in Chapter 8. Here we address the case that the interference power p_t is caused by *incoherent* cumulation of n independently fading signals. The pdf of joint interference power is the n-fold convolution of the pdf of the individual signal power levels. Laplace transformation results in the multiplication of n factors (see Chapter 3), so, if all interfering packets arrive from random positions in the service area and if $N_A = 0$,

$$q_n(r_j) = \mathcal{L}\left\{f_{p_\pm}, \frac{z}{\bar{p}_0}\right\} = \mathcal{L}^n\left\{f_{p_i}, \frac{z}{\bar{p}_0}\right\} = \{q_1(r_j)\}^n \tag{7.18}$$

(For the quasi-uniform spatial distribution (7.2), one finds after applying [35]),

$$q_n(r) = \left[1 - \frac{\pi}{2} \sqrt{z} r^2 \, \text{erfc}\left(\frac{\sqrt{z\pi}}{2} r^2 \right) \exp\left(\frac{z r^4 \pi}{4} \right) \right]^n \tag{7.19}$$

The effect of noise was also considered in Chapter 3. If additive noise with power $p_N = N_A = N_0 B_T$ is present, the probability of successful transmission (7.19) is to be multiplied by a factor of $\exp\{-z N_A r^\beta\}$. Figure 7.2 portrays $q_n(r)$ for $n = 0, 1, \ldots, 6$ and a receiver threshold of 6 dB ($z = 4$). The median signal-to-noise ratio is 20 dB; that is, the noise power is $N_A \approx 0.035$ (see Section 7.2).

Using the expressions presented in Chapter 6 and the series expansion of the exponential function, the probability of a successful transmission for slotted ALOHA is found in the form of

$$Q(r) = \exp\{-z N_A r^\beta\} e^{-G_t} \sum_{n=0}^{\infty} \frac{G_t^n}{n!} q_n(r) \tag{7.20}$$

$$= \exp\{-G_t(1 - q_1(r)) - z N_A r^\beta\}$$

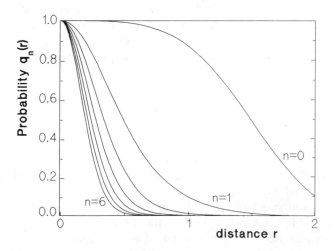

Figure 7.2 Probability of capture $q_n(r)$ versus distance r for $n = 0, 1, \ldots, 6$, according to capture-ratio model. Receiver threshold $z = 4$ (6 dB), Median $C/N = 20$ dB ($N_A \approx 0.035$).

with $q_1(r)$ in the form of (7.19) with $n = 1$. Analogously, the probability of correct reception for nonpersistent ISMA becomes

$$Q(r) = \exp\{-d_1G_t(1 - q_1(r)) - zN_Ar^\beta\} \frac{1 + q_1(r)d_1G_t}{G_t(1 + d_1 + d_2) + e^{-d_1G_t}} \quad (7.21)$$

For unslotted p-persistent ISMA with zero signaling delay, $Q(r)$ is

$$Q(r) = \exp\{-zN_Ar^\beta\} \frac{1 + pG_t \exp\{pG_tq_1(r)\}}{1 + G_t \exp(pG_t)} \quad (7.22)$$

Numerical results for $Q(r)$ in these cases are shown in Figure 7.15.

7.5 MODEL BASED ON NONFADE INTERVAL

In Section 7.4, the duration of a packet was assumed to be short with respect to the rate of channel fading. This is reasonable if mobile terminals move less than a fraction of the wavelength during the transmission of a packet ($T_5 \ll T_6$). For usual vehicle speeds, this requires that data packets are shorter than a few milliseconds [36]. In practical narrowband UHF networks, this condition is not always satisfied. During the transmission of a packet, the received signal power may fade. To investigate this effect, approximations suggested in Chapter 4 will now be considered.

For a wanted Rayleigh-fading signal with local mean power $\bar{p}_j$ in the presence of Nakagami-fading interference with mean power $\bar{p}_t$, the average nonfade duration (expressed in seconds) was found in the form of (4.31). Analogous to Section 4.10, the probability that a test packet is located in a nonfade interval can be found if a number of additional assumptions are made:

1. Threshold crossings are memoryless, so that nonfade durations are exponentially distributed. It may be argued that nonfade durations corresponding to the maximum Doppler shift prevail in mobile reception (Section 4.8). However, if the speed of the vehicle in an urban environment is also seen as a stochastic variable, nonfade interval lengths tend to be spread over a wider range. Lacking further details, an exponential distribution yields convenient mathematical expressions and appears plausible.
2. The joint interference-plus-noise signal behaves as a nonfading signal. This is correct in the limiting case $n \to \infty$, but underestimates the duration of nonfade intervals when the interference is dominated by a single fading signal ($m \downarrow 1$) by about 13% (see (4.32)).

Similar to the analysis in [25–27] for noise-limited channels, the probability of successful reception is found from the requirements that C/I is larger than z at the start of the packet and that the packet duration must be shorter than the time to the next fade. Rewriting (4.56) gives

$$\Pr(c_j|\bar{p}_t, \bar{p}_j) = \exp\left\{ -z\frac{\bar{p}_t + N_A}{\bar{p}_j} - \sqrt{2\pi}\, T_s f_m \sqrt{z\frac{\bar{p}_t + N_A}{\bar{p}_j}} \right\} \quad (7.23)$$

The probability of capture $q_n(r_j)$, after averaging (7.23) over the joint interference power (7.6) and substituting $\bar{p}_t = s^{-4}$, becomes

$$q_n(r_j) = \int_0^\infty \Pr(c_j|\bar{p}_j, \bar{p}_t) f_{\bar{p}_t}(\bar{p}_t|n) d\bar{p}_t$$

$$= 2\int_0^\infty ns \exp\left\{ -z(s^{-4} + N_A)r_j^4 - \frac{\pi}{4}n^2 s^4 - r_j^2 \sqrt{2\pi z(s^{-4} + N_A)}\, f_m T_s \right\} ds \quad (7.24)$$

for $n = 1, 2, \ldots$. For $n = 0$, $q_0(r_j)$ is found from (7.23) with $\bar{p}_t = 0$.

Figure 7.3 confirms that the effect of the packet duration T_s is significant. This suggests that during a collision, the fade margin can be relatively small. In such

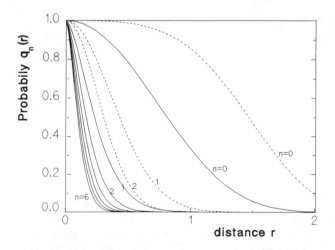

Figure 7.3 Probability of successful reception $q_n(r_j)$ versus distance r_j for packets of normalized duration (- -) $f_m T_s = 0.0166$ and (———) $f_m T_s = 1$. Median $C/N = 20$ dB ($N_A \approx 0.035$). Number of interfering signals $n = 0, 1, \ldots, 6$. Receiver threshold $z = 4$ (6 dB).

cases, the assumption of constant received power (as in Section 7.4) appears optimistic and will lead to overestimating channel performance if packet durations are larger than or on the same order of magnitude as the time constants of the fading.

The case $f_m T_s = 0.0166$ corresponds to, for example, packets of $L = 16$ bits transmitted at a bit rate of $r_b = 16$ kb/s in a mobile radio channel at $f_c = 900$ MHz by a terminal with velocity $v = 5.55$ m/s (20 km/h), thus with a maximum Doppler shift of $f_m = 16.6$ Hz. Computer simulation of this case will be reported in Section 7.7 for noncoherent detection of DPSK signals.

The probability of successful access, $Q(r_j)$, in a slotted ALOHA network is found by considering a Poisson-distributed number of interferers with unknown positions. So,

$$Q(r_j) = e^{-G_t} \Pr(c_j | \bar{p}_j = r_j^{-\beta}, \bar{p}_t = 0) + \sum_{n=1}^{\infty} \frac{G_t^n}{n!} e^{-G_t} \int_0^{\infty} 2ns$$

$$\cdot \exp\left\{ -z(s^{-4} + N_A)r_j^4 - \frac{\pi}{4} n^2 s^4 - r_j^2 \sqrt{2\pi z(s^{-4} + N_A)} f_m T_s \right\} ds \quad (7.25)$$

where we substituted $\bar{p}_t = s^{-4}$.

Figure 7.4 shows that (7.25) with $f_m T_s = 0$ is pessimistic compared to the exact result (7.20) for packets of infinitely short duration. A possible explanation of this

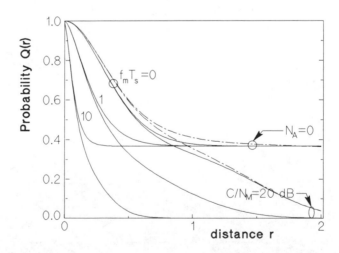

Figure 7.4 Probability of successful access $Q(r)$ versus distance r for slotted ALOHA with packets of normalized duration (——) $f_m T_s = 0$, 1, and 10. Noise-free channel ($N_A = 0$) and noise power $N_A \approx 0.035$ ($C/N_M = 20$ dB). Receiver threshold $z = 4$ (6 dB) (—·—). Exact solution (7.20) for packets of zero duration (- -). Quasi-uniform offered traffic with $G_t = 1$ ppt.

discrepancy is that in the event of very short packet durations ($T_s \rightarrow 0$) in slots with one (or few) interfering signal(s), fades of the test packet may coincide with fades of the interference. In such events, our assumption (number 2 at beginning of this section) of nonfading interference gives relatively pessimistic estimates of the capture probabilities. However, considering the arguments raised in Section 4.10, it may be concluded that the assumption of packets of zero duration ($T_s \rightarrow 0$) is always optimistic, even if ($T_5 \ll T_6$).

For $f_m T_s > 1$ ($T_5 > T_6$), the performance is substantially degraded by fades during packet reception. Figure 7.4 indicates that for relatively long packets ($T_s > 10 f_m^{-1}$) in a channel with a median C/N-ratio of 20 dB, the majority of terminals experience a failure of almost every transmission attempt ($Q(r) \approx 0$ if $r > 0.5$). This is in sharp contrast to the case for shorter packets, when all terminals may have an acceptable probability of success ($Q(r) > 0.35$ if $T_s \rightarrow 0$ and $0 < r < 1$).

Figure 7.16 gives numerical results for (7.25) and similar results for $Q(r)$ in the case of unslotted ISMA.

7.6 MODELS BASED ON PROBABILITY OF TRANSMITTER SUCCESS

We now address the models of the second category (category B). A test packet of L bits is assumed to be received successfully if and only if the bit sequence detected by the receiver entirely matches the bit sequence in the test packet. Error detection and correction coding [6] are not considered explicitly in this section.

7.6.1 Review

Habbab, Kavehrad, and Sundberg presented the throughput of slotted ALOHA nets with BPSK modulation in [37]. The analysis was later extended in [32]. The model for successful reception was based on criteria of terminal position and packet success rates. Slow ($T_5 \ll T_6$), and fast ($T_3 \ll T_6 \ll T_5$) Rayleigh fading of the test packet were considered. In [6, 7, 31, 32, 37], the joint interference signal was approximated as Gaussian noise, similarly to our analysis in Section 5.7. Zhang and Pahlavan [34] addressed slow fading channels with interference samples of constant envelope during successive bits of a packet. In [31, 34, 38], a receiver was assumed to lock perfectly onto a randomly chosen signal out of the $n + 1$ colliding packets, and all other packets were assumed to have zero probability of capturing the receiver, even though they might arrive with substantially higher power. Other synchronization models will be addressed in the following sections.

7.6.2 Formulation of Receiver Model

The model described in Chapter 5 is reconsidered. We shall assume no bit synchronization offset between contending signals ($\alpha_i \equiv 0$). This slightly overestimates the

effect of interference, so it may yield slightly pessimistic results. The received joint
signal $v_t(t)$ has the form

$$v_t(t) = \rho_j \kappa_j \cos(\omega_c t + \theta_j) + \sum_{i=1}^{n} \rho_i \kappa_i \cos(\omega_c t + \theta_i) + n(t) \qquad (7.26)$$

where κ_i ($\kappa_i = \pm 1$) represents the phase reversals arising from BPSK modulation of
the ith carrier, and $n(t)$ is the AWGN process. The phasor diagrams in Figures 7.5
through 7.7 present the inphase and quadrature signal components after fictitious
correlation with $\cos \omega_c t$ and $\sin \omega_c t$; thus, $2/T_b \cdot \int_{T_b} v_t(t) \cos(\omega_c t)dt$, and $2/T_b \cdot \int_{T_b} v_t(t)$
$\sin(\omega_c t)dt$, respectively. AWGN, although not strictly band-limited at the receiver
front-end, can also be depicted in the inphase/quadrature (I/Q) diagram because of
the bandpass nature after the integration over the bit duration T_b.

The decision variable for synchronous bit extraction is obtained by multiplying
the received (composite) signal $v(t)$ by a locally generated cosine $2 \cos(\omega_c t + \theta_r)$,
where θ_r denoted the phase of the local oscillator in the receiver, and integrating
over the entire bit duration T_b. In Figures 7.5 to 7.7, this corresponds to determining

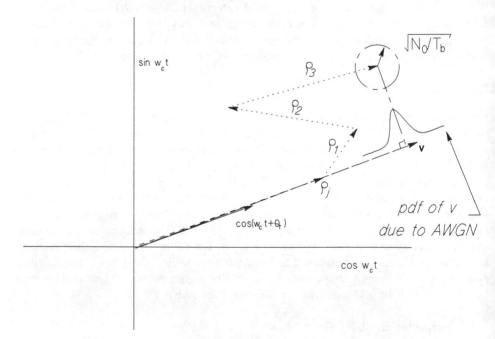

Figure 7.5 Phasor diagram for coherent detection for a test signal in the presence of three interferers
and AWGN: (– –) test signal; (···) interfering signal. Receiver locked to the test packet.

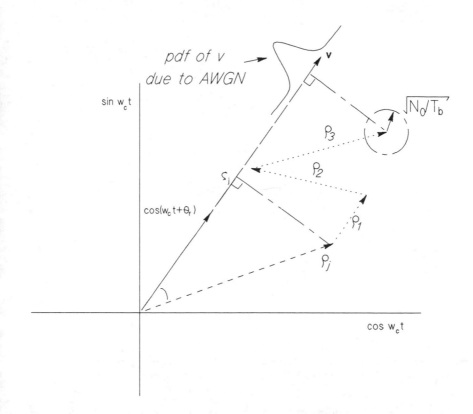

Figure 7.6 Phasor diagram for coherent detection for a test signal in the presence of three interferers and AWGN: (– –) test signal; (· · ·) interfering signal. Receiver locked to an interferer.

the projection of the received resultant vector on a line with direction θ_r. For a test packet j in the presence of n interferers, the decision variable is

$$v = \frac{1}{T_b} \int_0^{T_b} 2v(t) \cos(\omega_c t + \theta_r)dt = \rho_j \kappa_j \cos(\theta_j - \theta_r) + \sum_{i=1}^{n} \rho_i \kappa_i \cos(\theta_i - \theta_r) + n_I$$

(7.27)

Three idealized cases for the phase of the local oscillator in the coherent detector are compared; namely, a receiver locked to the test signal ($\theta_r = \theta_j$, Figure 7.5), a receiver locked to one of the interferers ($\theta_r = \theta_k$, Figure 7.6), and a receiver with arbitrary, but constant, phase ($\theta_r - \theta_i = $ constant, Figure 7.7). The latter two events correspond to extreme cases of carrier phase errors in the receiver caused by signals competing with the test packet.

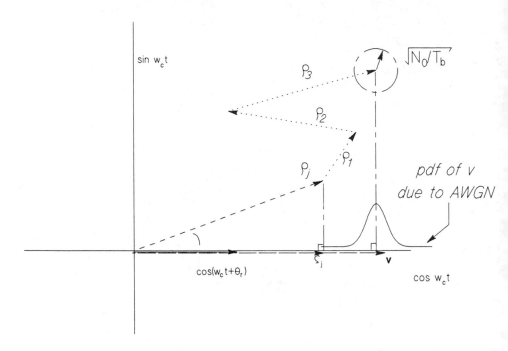

Figure 7.7 Phasor diagram for coherent detection for a test signal in the presence of three interferers and AWGN: (– –) test signal; (· · ·) interfering signal. Arbitrary but constant receiver phase.

In a Rayleigh-fading channel, the inphase components $\zeta_i (\zeta_i = \rho_i \cos(\theta_i - \theta_r))$ of the n interferers (with random phase relative to the local oscillator in the receiver), and the noise sample n_t are all Gaussian variables without mutual correlation. The rate of fading determines the correlation of amplitude and phase in successive bits. We will now distinguish between fast and slow fading.

7.6.3 Analysis for Quasi-Uniform Offered Traffic and Fast Fading

With fast Rayleigh fading, the duration of a packet is substantially longer than the time constants of the multipath fading. On the other hand, we assume that during one bit time the channel characteristics do not change ($T_3 << T_6 << T_5$) [6, 7, 31, 32]. If it can be assumed that the received amplitude and phase of all signals are statistically independent from bit to bit even though the receiver remains perfectly locked to the test packet $j(\theta_r = \theta_j)$, the transmitter success probability for BPSK without error correction coding is obtained from

$$q_n(r_j) = \int_0^\infty f_{\bar{p}_t}(\bar{p}_t|n)[1 - \bar{P}_b(e|\bar{p}_j = r_j^{-4}, \bar{p}_t)]^L d\bar{p}_t \qquad (7.28)$$

where the local-mean bit error probability is taken in the form of (5.13). Ignoring shadowing ($\bar{p}_j = \bar{\bar{p}}_j = r_j^{-4}$) and considering a quasi-uniform spatial distribution of offered traffic (7.6), with $\bar{p}_t = \bar{\bar{p}}_t = s^{-4}$, the probability of error-free reception is found to be

$$q_n(r_j|\theta_r = \theta_j) = \int_0^\infty 2ns \exp\left(-\frac{\pi}{4} n^2 s^4\right) \left[\frac{1}{2} + \frac{1}{2}\sqrt{\frac{T_b r_j^{-4}}{T_b r_j^{-4} + T_b s^{-4} + N_0}}\right]^L ds \quad (7.29)$$

for $n = 1, 2, \ldots$. For $n = 0$, $q_0(r)$ is found from (5.13) with $\bar{p}_t = 0$. This probability is depicted in Figure 7.8 for $n = 0, 1, \ldots, 6$. The median C/N ratio is taken 20 dB: the corresponding spectral power density of the noise is $N_0 = 0.01\bar{p}_M T_b \approx 0.035 T_b$. Data packets of $L = 16$ bits are considered.

Two capture models are now defined:

MODEL BF1: This model assumes perfect synchronization for the test packet (Figure 7.5):

$$q_n(r_j) = q_n(r_j|\theta_r = \theta_j) \quad (7.30)$$

for any j, so

$$C_{n+1} = (n + 1)q_{n|\theta_r=\theta_j} = (n + 1) \int_0^\infty 2\pi r q_n(r_j|\theta_r = \theta_j)G(r_j)dr_j \quad (7.31)$$

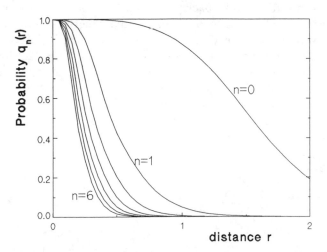

Figure 7.8 Probability of success $q_n(r)$ versus the distance r for fast fading and $n = 0, 1, \ldots$ interfering signals. Perfect carrier synchronization $\theta_r = \theta_j$, $C/N_M = 20$ dB: $N_A \approx 0.035$. Packet length is $L = 16$ bits.

This model is based on the two assumptions that (1) in the case of error-free reception of the (rapidly fading) test signal, the signal is also sufficiently strong to maintain perfect receiver synchronization during the entire packet, while (2) for a weak signal, the probability of error-free reception is negligible irrespective of the carrier phase synchronization, provided that L is sufficiently large.

MODEL BF2: The receiver is assumed to lock onto a randomly selected signal k out of the $n + 1$ competing packets, where k does not necessarily represent the test packet or the strongest contending packet [31, 34, 38]. Correct reception of the test packet occurs if and only if the receiver locks to the test packet; that is, if the receiver selects $k = j$ (Figure 7.5), and no bit error occurs [27]. This produces the pessimistic relations

$$q_n(r_j) = \Pr(k = j)q_n(r_j|\theta_r = \theta_j) = \frac{1}{n+1} q_n(r_j|\theta_r = \theta_j) \quad \text{and} \quad C_{n+1} = q_{n|\theta_r=\theta_j} \quad (7.32)$$

For a numerical comparison of models BF1 and BF2, we refer to [7]. The results for the two models appeared essentially different for large offered traffic loads. Numerical results for $Q(r)$ for model BF1 are included in Figure 7.17.

7.6.4 Analysis for Quasi-Uniform Offered Traffic and Slow Fading With Gaussian Interference

In a contention-limited network, packets should be short with respect to the time constants of the channel fading to fully exploit the benefits from receiver capture. For packets of sufficiently short duration, the received amplitude and carrier phase may be assumed to be constant throughout the duration of a packet for each of the $n + 1$ signals. Inserting (5.13) into (7.28) and ($T_5 << T_6$). Packet error rates for a slow fading (test) signal were reported in [39] for channels with AWGN.

The probability of error-free reception of a data packet of L bits from terminal j with a known local-mean received signal power is found by averaging $\Pr(s_j)$ over the Rayleigh fading of the wanted signal and over the local-mean interference power, with

$$q_n(r_j|\theta_r = \theta_j) = \int_0^\infty \int_0^\infty 2nsr_j^4 \exp\left\{ -\frac{\pi}{4} n^2 s^4 - p_j r_j^4 \right\}$$

$$\cdot \left[1 - \frac{1}{2} \text{erfc} \sqrt{\frac{p_j T_b}{N_0 + s^{-4}T_b}} \right]^L dsdp_j \quad (7.33)$$

Here we assumed that the receiver is locked to the test packet ($\theta_r = \theta_j$, Figure 7.5), that the interference sample is Gaussian-distributed with mean $\bar{p}_t$ $(s^{-4} = \bar{p}_t)$, and that interference samples during successive bits are statistically independent. This *Gaussian assumption*, suggested in [32], is reasonable if the number of interfering signals is large $(n \to \infty)$, or if the interfering signal is fast-fading. However, the case of a slow-fading test signal with fast-fading interference is at odds with the fact that *all* signals are competing for receiver capture and a priori have identical characteristics. Numerical results from (7.33) are in Figure 7.9 for $n = 0, 1, \ldots, 5$.

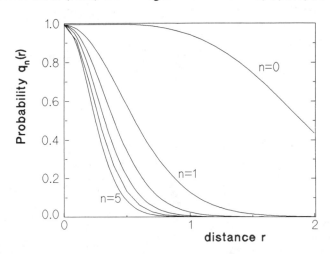

Figure 7.9 Probability of transmitter success $q_n(r)$ versus the distance r for slow-fading test packet and $n = 0, 1, \ldots, 5$. Gaussian interfering signals. Perfect carrier synchronization $\theta_r = \theta_j$, $C/N_M = 20$ dB ($N_A \approx 0.035$). Packet length $L = 16$ bits.

Various models for receiver synchronization can be defined:
MODEL BSG1: Perfect synchronization is assumed: $q_n(r_j) = q_n(r_j | \theta_r = \theta_j)$.
MODEL BSG2: The receiver randomly locks to a favorite packet:

$$q_n(r_j) = (n + 1)^{-1} q_n(r_j | \theta_r = \theta_j).$$

MODEL BSG3: The receiver locks to the signal from the nearest transmitter, while more remote packets are always lost in a collision [32].

Here, we confine ourselves to model BSG1. Corresponding results for $Q(r)$, taking account of the behavior of the random access protocol, are in Figure 7.14.

7.6.5 Analysis for Quasi-Uniform Offered Traffic and Slow Fading With Constant-Envelope Interference

For angle-modulated signals with constant envelope, a Gaussian approximation of the pdf of the interference signal (as in (7.33)) may become inaccurate: successive interference samples in the decision variable occur with constant inphase amplitude $\zeta_1, \zeta_2, \ldots, \zeta_n$ throughout the entire packet. Zhang and Pahlavan [31] studied correlated interference samples in successive bits. Their model is extended here. In particular, the distinction between the a priori selected test packet and the packet that acquires receiver synchronization is further developed.

As illustrated by Figure 7.5, for known amplitudes ρ_i, phases θ_i and bit contents κ_i of the test signal $(i = j)$ and all interference signals $(i = 1, 2, \ldots, n)$, the decision variable v has the Gaussian pdf

$$f_v(v|\{\rho_i, \theta_i, \kappa_i\}_{i=j,i=0}^n) = N\left[\mu = \sum_{\substack{i=0 \\ i=j}}^{n} \rho_i\kappa_i \cos(\theta_r - \theta_i), \sigma^2 = \frac{N_0}{T_b}\right] \quad (7.34)$$

If the test packet contains a 1 $(\kappa_j = 1)$, the detector makes a bit error if the decision variable happens to be less than zero. For reasons of symmetry of the events $\kappa_j = \pm 1$, the bit error probability, conditional on the set of bits $\{\kappa_i\}$ in the interfering packets and on the envelope and phase of each signal, is

$$P_b(e|\{\rho_i, \theta_i, \kappa_i\}) = \frac{1}{2} \text{erfc}\left(\frac{\rho_j \cos(\theta_r - \theta_j) + \sum_{i=1}^{n} \kappa_i\rho_i \cos(\theta_r - \theta_i)}{\sqrt{2N_0/T_b}}\right) \quad (7.35)$$

The integrated noise n_I is uncorrelated during each successive bit. Further, the mean value of (7.33) may change because of (possible) phase reversals. With n interfering signals, there are 2^n different sets $\{\kappa_i\}$, and all are assumed to be equally likely, so

$$P_b(e|\{\rho_i, \theta_i\}_{i=j,i=0}^n)$$

$$= 2^{-n-1} \sum_{\kappa_1=\pm 1} \cdots \sum_{\kappa_n=\pm 1} \text{erfc}\left(\frac{\rho_j \cos(\theta_r - \theta_j) + \sum_{j=1}^{n} \kappa_i\rho_i \cos(\theta_r - \theta_i)}{\sqrt{2N_0|T_b}}\right)$$

$$(7.36)$$

Each inphase component $\zeta_i(\zeta_i = \rho_i \cos(\theta_r - \theta_i))$ either has a Gaussian probability density (if θ_i is random) or has a Rayleigh probability density (if the receiver is locked to the ith signal).

For BPSK, the capture probability conditional on the carrier phasors $\{(\rho_i, \theta_i)\}$ received from the interference signals is

$$\Pr(s_j|\{\rho_i, \theta_i\}_{i=j,i=0}^n) = [1 - P_b(e|\{\rho_i, \theta_i\}_{i=j,i=0}^n)]^L \qquad (7.37)$$

with (7.36). For differentially encoded BPSK, phase reversal of the carrier of the test packet (and thus inverting all L bits) does not affect the probability of a bit error. Consequently, the conditional packet error probability for coherent detection of differentially encoded BPSK follows from

$$\Pr(s_j|\{\rho_i, \theta_i\}_{i=0}^n) = [1 - P_b(e|\{\rho_i, \theta_i\}_{i=0}^n)]^L + [P_b(e|\{\rho_i, \theta_i\}_{i=0}^n)]^L \qquad (7.38)$$

The effect that one extra bit is required for differential encoding is ignored here; in fact, with coherent detection of BPSK, a reference carrier phase is also required, which may necessitate a preamble of a least one bit. The three cases of carrier synchronization of the receiver in Figures 7.5 to 7.7 are considered: (1) a receiver locked to the test packet, (2) a receiver locked to another packet, and (3) a receiver with a constant but arbitrary phase with respect to each signal.

(a) The capture probability $q_n(r|\theta_r = \theta_j)$ for a receiver locked to the test packet (Figure 7.5) is obtained by integrating over the inphase amplitudes $\{\rho_j, \zeta_1, \zeta_2, \ldots, \zeta_n\}$. So,

$$q_n(r_j|\theta_r = \theta_j) = \int_0^\infty d\rho_j \int_{-\infty}^\infty d\zeta_1 \ldots \int_{-\infty}^\infty d\zeta_n \, \Pr(s_j|\{\rho_i, \theta_i\}, \theta_r = \theta_j) f_{\rho_j|r_j} f_{\zeta_1} \ldots f_{\zeta_n} \qquad (7.39)$$

with ρ_j Rayleigh-distributed and $\zeta_1, \zeta_2, \ldots, \zeta_n$ distributed according to (7.7).

(b) If the carrier recovery circuit in the receiver happens to be locked to the kth interference signal ($k = 1, 2, \ldots, n; k \neq j$, Figure 7.6), the decision variable, evaluated for a test packet with a random (but constant) phase relative to the receiver ($\theta_r = \theta_j$ = constant), goes into

$$v = \kappa_j \zeta_j + \kappa_k \rho_k + \sum_{i=1,i\neq k}^n \kappa_i \zeta_i + n_I \qquad (7.40)$$

The capture probability is found by integration:

$$q_n(r|\theta_r = \theta_k) = \int_{-\infty}^\infty d\zeta_j \int_{-\infty}^\infty d\zeta_1 \ldots \left(\int_0^\infty d\rho_k \right) \ldots \int_{-\infty}^\infty d\zeta_n$$
$$\cdot \Pr(s_j|\{\rho_i, \theta_i\}_{i=j,i=0}^n, \theta_r = \theta_k) f_{\zeta_j|r_j} f_{\zeta_1} \ldots (f_{\rho_k}) \ldots f_{\zeta_n} \qquad (7.41)$$

with ζ_j Gaussian-distributed, $\zeta_1, \zeta_2, \ldots, \zeta_n$ distributed according to (7.7), and ρ_k distributed according to (7.6).

(c) The phase reference θ_r of the coherent detector is assumed fixed but arbitrary; that is, it is not related to the phase of any of the $n + 1$ signals (Figure 7.7). Each signal produces a Gaussian inphase component. The capture probability of error-free detection turns into

$$q_n(r_j|\text{no lock}) = \int_{-\infty}^{\infty} d\zeta_j \int_{-\infty}^{\infty} d\zeta_1 \ldots \int_{-\infty}^{\infty} d\zeta_n \, \text{Pr}(s_j|\{\rho_i, \theta_i\}_{i=j,i=0}^{n}) f_{\zeta_j|r_j} f_{\zeta_1} \ldots f_{\zeta_n}$$

(7.42)

The probability $q_n(r)$ is depicted in Figure 7.10 for a single interfering signal ($n = 1$) for the various cases of receiver synchronization. Figure 7.11 reports the effect of the number of interferers for a receiver perfectly locked to the test packet, computed from (7.39) with (7.37). These results have been obtained by numerical computation of the above $n + 1$ dimensional integrals. Due to the wide dynamic range of the signals, numerical results obtained from the computations become less accurate for $r < 0.1$. Computation time exponentially increases with n, so results for $n = 4, 5, \ldots$ are tiresome to obtain.

For a coherent receiver locked to the test packet, BPSK and DPSK are seen to give almost equal capture probabilities. Figure 7.10 suggests that if large synchronization errors can occur (e.g., if the receiver locks to an interfering signal or

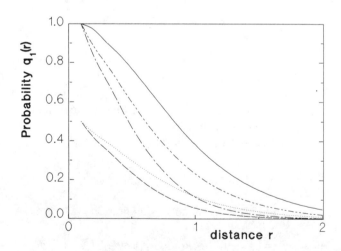

Figure 7.10 Probability success $q_1(r)$ for a test packet in the presence of $n = 1$ interfering signal versus distance r for slow-fading BPSK signals: (——) receiver locked to test packet; receiver locked to interfering packet, (—·—) with and (– –) without differential encoding; random but constant phase of the local oscillator, (—··—) with and (···) without differential encoding. $C/N_M = 20$ dB ($N_A \approx 0.035$). Packet length $L = 16$ bits.

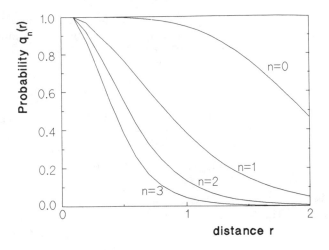

Figure 7.11 Model BS1: probability success $q_n(r)$ for a receiver perfectly locked to the test packet, in the presence of $n = 0$, 1, 2, and 3 interfering signals, versus distance r for slow-fading BPSK signals. $C/N_M = 20$ dB ($N_A \approx 0.035$). Packet length $L = 16$ bits.

if the receiver has an arbitrary phase), differential encoding increases the probability of correct reception by almost a factor of 2.

We now postulate five capture models based on the three idealized cases of carrier synchronization in Figures 7.5 through 7.7.

MODEL BS1: Analogous to models BF1 and BSG1, perfect carrier synchronization is always assumed for the test packet. We apply (7.30) for each of the signals in the collision. Particularly if two signals with approximately equal power are competing, this assumption is optimistic. Results for $q_n(r)$ ($n = 1$, 2, and 3) in Figure 7.11 are substantially more optimistic than the results for Gaussian interference in Figure 7.9.

MODEL BS2: The receiver locks to any one of the $n + 1$ contending signals. Capture occurs if and only if the receiver locks to the test packet *and* the detected bit sequence is identical to the bit sequence of the test packet. The capture probability is found from (7.32), (7.37), and (7.39); numerical results for $q_n(r)$ are a factor ($n + 1$) smaller than results from model BS1.

The models BS3 to BS5 address the effect of inaccuracies in receiver synchronization with respect to the phase of the test packet, caused by interfering signals. Extreme cases are considered: the receiver is assumed to lock perfectly to the carrier of one of the $n + 1$ signals, but this does not exclude other packets from being received correctly. The probability of correct detection of the test packet is evaluated for events of reception with and without carrier phase offset. To this end we use (7.39) to (7.42).

Carrier synchronization to an interferer may be reasonable in some unslotted networks: the receiver may maintain carrier synchronization to the initiating packet. The capture probability for this packet is found from

$$q_n(r_j|H_I) = q_n(r_j|\theta_r = \theta_j) \qquad (7.43)$$

with (7.39), whereas for each successive test packet (j) in the vulnerable period, the capture probability is taken to have the form

$$q_n(r_j|H_d) = q_n(r_j|\theta_r = \theta_1) \qquad (7.44)$$

with (7.41), where θ_1 denotes the phase of the initiating packet.

MODEL BS3: The receiver locks to one of the $n + 1$ BPSK signals. The probability that one of the $n + 1$ packets is received correctly is $C_{n+1} = q_n(r_j|\theta_r = \theta_j) + nq_n(r_j|\theta_r = \theta_k, k \neq j)$.

MODEL BS4: This model is identical to BS3 except that we assume coherent detection of differentially encoded BPSK.

MODEL BS5: The carrier phase reference of the coherent detector is constant but arbitrary; that is, it is not related to the phase of any of the $n + 1$ signals. With DPSK, the probability of correct packet reception follows from (7.42) and $C_{n+1} = (n + 1)q_n$.

The models BS3 to BS5 require constant carrier phases, unaffected by mobile fading or by differences in carrier frequencies during a packet time ($T_5 \ll T_6$). Consequently, these models become inappropriate with fast fading. A comparison of these five models is given in Figure 7.20 for $Q(r)$ in the event of nonpersistent ISMA.

7.7 COMPUTER SIMULATION

Rayleigh fading in a narrowband mobile radio channel has been simulated by splitting the signal into an inphase component, in the form $\kappa_i \cdot \cos(\omega_c t)$, and a quadrature component, in the form $\kappa_i \cdot \sin(\omega_c t)$, and multiplying these components by independent Gaussian variables ζ_i and ξ_i, respectively. The factors ζ_i and ξ_i are low-pass filtered according to the Doppler spectrum in a typical mobile channel [36]. The block diagram of the fading generator as implemented by Pluymers [30] in computer software is given in Figure 7.12.

Packets of $L = 16$ bits with DPSK modulation with a bit rate of $r_b = 16$ k/ps on a carrier frequency of $f_c = 900$ MHz are considered. The speed of all vehicles is 20 km/h ($v = 5.55$ m/s). Hence, the maximum Doppler shift is $f_m = vf_c/c = 16.6$ Hz, so the Rayleigh fading is relatively slow compared to a packet duration of $T_s = 1$ ms ($f_m T_s = 0.0166$). For the quasi-uniform distribution of the

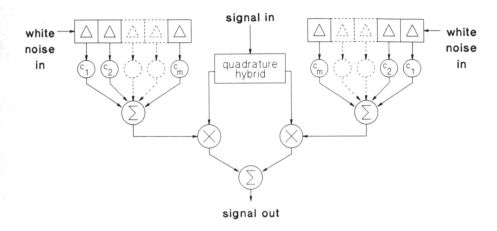

Figure 7.12 Channel simulator for frequency-nonselective Rayleigh fading, based on tapped-delay line filters.

interfering traffic, the probability $q_n(r_j)$ of successful transmission of a test packet from r_j in the presence of $n = 1$ and 2 interfering packets is depicted in Figure 7.13 (marked by $\triangle$ and $\square$, respectively).

The capture probability found from the simulation (see Figure 7.13) is lower than the mathematical results from the model for transmitter success for coherent detection of BPSK (see Figure 7.11). Possible explanations can be that (1) noncoherent detection is simulated, whereas coherent detection is considered in (7.39); and (2) despite the slow terminal speeds, signal levels still fluctuate during a packet.

In Figure 7.13, simulation results are compared with models A3 and A4 with receiver threshold $z = 1$ (perfect capture). Both models appear relatively pessimistic for $r < 0.7$, but model A3 is relatively optimistic for $r > 0.8$. A possible explanation of the high simulated probabilities of capture for $r < 0.7$ might be the assumption that the receiver is in perfect bit-synchronization with the test packet, with non-aligned bit timing for interfering signals.

7.8 COMPARISON OF MODELS

In the previous sections, various models to compute $q_n(r)$ have been proposed. The probabilities $q_n(r)$ have been inserted in expressions for the probability $Q(r)$ of a successful transmission for various protocols, as derived in Chapter 6. Next, the models A2, A3, A4, BF1, BSG1, and BS1 are compared in Figures 7.14 to 7.19. The local-mean signal-to-noise ratio for a signal from r_M is taken 20 dB ($C/N_M =$ 20 dB); thus $N_A \approx 0.035$ ($\beta = 4$). The offered traffic is quasi-uniformly distributed with $G_t = 1$ ppt. Results are indicated by solid lines (——) for slotted ALOHA

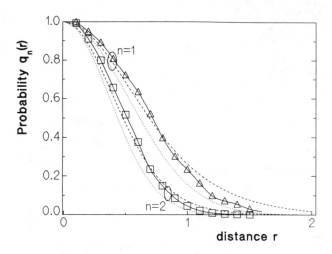

Figure 7.13 Probability of capture $q_n(r)$ versus distance r for noncoherent detection of DPSK. Packet length $L = 16$ bits at a bit rate of $r_b = 16$ kbps. $\triangle$: one interferer; $\square$: two interferers. Vehicle speed 20 km/h ($v = 5.55$ m/s). Carrier frequency $f_c = 900$ MHz. Median C/N is 20 dB ($N_A \approx 0.035$). Corresponding results for $(- -)$ capture-ratio model and for $(\cdots)$ the nonfade model, assuming $z = 1$ (0 dB).

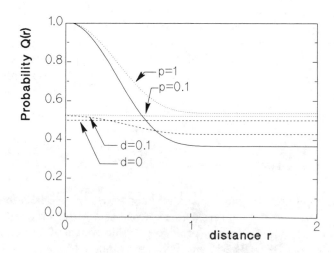

Figure 7.14 Model A2: vulnerability circle. Probability $Q(r)$ of successful transmission versus distance r for (——) slotted ALOHA, $(- -)$ unslotted nonpersistent ISMA, and $(\cdots)$ unslotted p-persistent ISMA. Vulnerability parameter $c_z = \sqrt{2}$ ($c_z^4 = z = 4$; 6 dB).

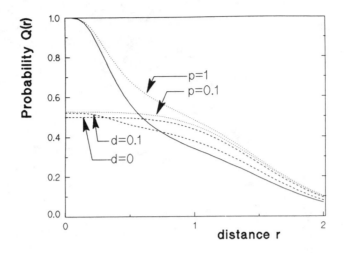

Figure 7.15 Model A3: capture ratio for slow fading. Probability $Q(r)$ of successful transmission versus distance r. Receiver threshold $z = 4$ (6 dB).

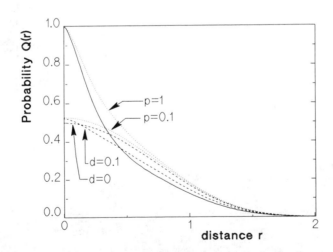

Figure 7.16 Model A4: nonfade duration. Probability $Q(r)$ of successful transmission versus distance r. Receiver threshold $z = 4$ (6 dB). Normalized packet duration $f_m T_s = 1$.

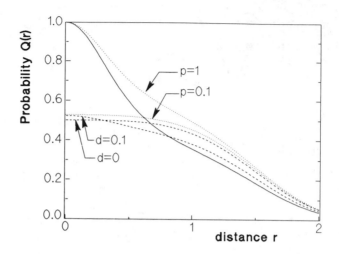

Figure 7.17 Model BF1: transmitter success for fast fading. Probability $Q(r)$ of successful transmission versus distance r. Packet length $L = 16$ bits.

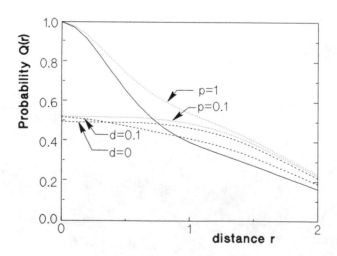

Figure 7.18 Model BSG1: transmitter success for slow fading with Gaussian interference. Probability $Q(r)$ of successful transmission versus distance r. Packet length $L = 16$ bits.

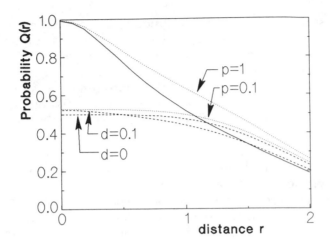

Figure 7.19 Model BS1: transmitter success for slow fading. Probability $Q(r)$ of successful transmission versus distance r. Packet length $L = 16$ bits.

without guard times, by broken lines (– –) for unslotted nonpersistent ISMA, and by dots ($\cdots$) for unslotted p-persistent ISMA.

The effect of packet length appears significant (compare Figures 7.15 and 7.16); appropriate system design may require careful selection of the number of bits per packet and the bit rate. Longer packets experience a higher probability of being lost because of channel fades. On the other hand, shortening packet duration by increasing r_b may lead to intersymbol interference, may require larger guard times to ensure "clean" slotted arrival of packets from various locations, and may also increase the relative duration of propagation delays of feedback signals in ISMA.

If a capture model assumes constant received signal amplitudes, the event of a deep fade of the interference is accounted for as a contribution to $q_1(r)$: model A3 appeared more optimistic than model A4 with $T_s \rightarrow 0$. However, capture of a test packet during fades of the interference is possible only with extremely short packet lengths (see Section 4.10).

For a slow-fading channel, the Gaussian approximation of the interference signal is inappropriate for $n = 1$: $q_1(r)$ is underestimated by more than 30% for $r > r_M$. For $r > 1$, the models BS1 and BSG1 differ by more than a factor of 2. Based on observations to be discussed in Section 7.9, it is the impression of the author that if error correction coding is applied, the differences between the exact model BS1 and the Gaussian approximation BSG1 may become less significant.

For any n, the probability of transmitter success in a fast fading channel is seen to decrease rapidly beyond a certain distance. Such a "knee" is less apparent for slow fading: distant terminals then experience a relatively high probability of success. With slow Rayleigh fading, a packet from a distant terminal might, under certain circumstances, even survive a collision with a packet from a more nearby terminal. However, if the slow-fading channel is evaluated at a receiver threshold of $z = 4$, such high probabilities of capture are not confirmed. This indicates that in (7.39) a substantial portion of the successful packets captures the receiver at remarkably low C/I ratios, whereas a practical narrowband receiver may loose synchronization in such cases. Results for a capture ratio of few decibels, to account for a synchronization margin, may be more realistic. This suggests that considering packet error rates for a perfectly synchronized receiver are presumably optimistic.

Slotted ALOHA is seen to result in the most significant unfairness because of prevailing packets from nearby users. In contrast to this, nonpersistent ISMA without delay ($d = 0$) provides a uniform chance of access for all terminals, although $Q(r)$ degrades for terminals beyond $r = 1$ because of noise. A signaling delay ($d > 0$) is known to degrade the average performance of ISMA networks for receivers with capture [16, 19, 21] and without capture [40]. Nonetheless, it appears that nearby users benefit from a signaling delay. This is explained by a high probability of capture for strong packets arriving during the inhibit signaling delay.

For nonpersistent ISMA in slow-fading channels, the probability of a successful transmission according to the models BS1 to BS5 are compared in Figure 7.20.

For terminals near the receiver, ignoring capture in the event H_d, as occurs in model BS2, gives pessimistic estimates for the probability of successful transmission. Even if a packet that arrives during the vulnerable period fails to seize full

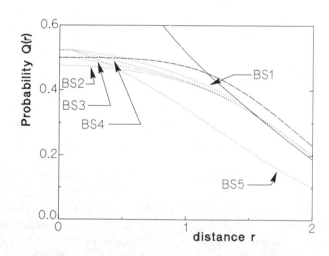

Figure 7.20 Probability of transmitter success $Q(r)$ versus distance r for nonpersistent ISMA ($\cdots$) in a slow fading channel. Comparison of model BS1 to BS5. Inhibit signaling delay $d = 0.1$. (——) Slotted ALOHA according to model BS1, and (– –) nonpersistent ISMA without delay ($d = 0$). Quasi-uniform offered traffic with $G_t = 1$ ppt. Median C/N is 20 dB. Slow fading. Packet length $L = 16$ bits.

carrier synchronization, as, for example, is assumed in the models BS3 to BS5, it nonetheless has a significant probability of being received correctly. Results are compared with slotted ALOHA and with ISMA without inhibit signaling delay (thus, without collisions).

Results for the various capture models considered here mainly differ for small r. For weak signals arriving from remote terminals, capture is generally limited to cases with no contenders at all, so

$$Q(r) \rightarrow \frac{I}{I + B} \exp(-d_1 G_t) \, q_0(r) \quad \text{for } r \rightarrow \infty \tag{7.45}$$

Hence, probabilities for models BS1 to BS4 converge for large r. (Gardiner and Jabbar [27] considered CSMA in a channel without capture ($q_n(r) = 0$ for $n = 1$, 2, …, though with limitations by multipath fading if $n = 0$. Their results correspond to (7.45) with (7.25) and $\bar{p}_t = 0$.)

Figure 7.21 summarizes the probability C_{n+1} that one out of $n + 1$ packets captures the receiver for the models A3, BF1, BF2, and BS1 to BS5. If a data packet arrives without interference ($n = 0$), a median C/N ratio of 20 dB causes an average outage probability of less than 4% ($q_0 \approx 0.96$) in a slow-fading channel and 8% ($q_0 \approx 0.92$) in a fast-fading channel. If interference is present ($n = 1, 2, …$), the models BS1 to BS5 produce widely different estimates of C_{n+1}.

It can be seen that model BS1 becomes unacceptable for determining C_{n+1}: it would appear far too optimistic to ignore carrier synchronization errors during the

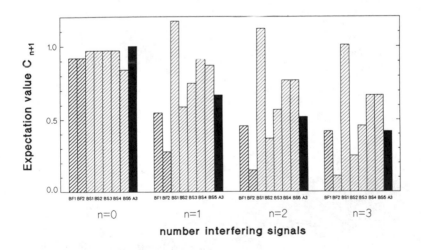

Figure 7.21 Expectation value C_{n+1} for models A3, BF1, BF2, BS1 to BS5. Quasi-uniform offered traffic. Median C/N ratio 20 dB. Packet length $L = 16$ bits. Receiver threshold (model A3) $z = 4$.

detection of the test packet. Receiver synchronization impairments in a practical receiver are likely to be a more significant cause of packet loss than excessive interference samples in the coherent detector. The other extreme, model BS2, is believed to underestimate capture performance, since it ignores the fact that any narrowband receiver is likely to lock to the strongest signal rather than to a random signal. The threshold model A3 with $z = 4$ (6 dB) tends to produce results that represent a good compromise of estimates for C_{n+1} according to all the models BS1 to BS5. For $n > 1$, the estimates of C_{n+1} by model BS3, particularly, closely agree with a capture ratio of $z = 4$ (6 dB).

The steady-state throughput S_t of the network in its entirety is studied in Figure 7.22 for model A3. For low offered traffic loads ($G_t < 1$ ppt), slotted ALOHA and ISMA have almost equal total throughput. 1-persistent ISMA yields higher throughput than nonpersistent ISMA. For reasonably high traffic loads ($3 < G_t < 10$ ppt), nonpersistent ISMA with a small signaling delay, say $d < 0.1$, outperforms slotted ALOHA and 1-persistent ISMA. Anticipating a further discussion of model A3 in Chapter 8, the theoretical limits for $G_t \to \infty$ are

$$\lim_{G_t \to \infty} S_t = \lim_{i \to \infty} iq_{i-1} = \frac{2}{\pi \sqrt{z}} \tag{7.46}$$

for slotted ALOHA without guard time and unslotted p-persistent ISMA with $0 <$

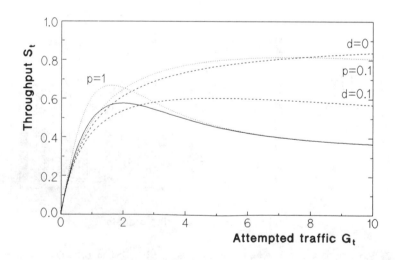

Figure 7.22 Total throughput S_t versus offered traffic G_t for the capture ratio model A3. Receiver threshold $z = 4$ (6 dB). Quasi-uniform offered traffic. Median C/N ratio 20 dB.

$p \leq 1$ and without signaling delays ($d_1 = d_2 = 0$, $t_p = 0$). For nonpersistent ISMA with signaling delays,

$$S_t \rightarrow \frac{2}{\pi\sqrt{z}} \frac{1}{1 + d_1 + d_2} \qquad (7.47)$$

Thus, in the limit $G_t \rightarrow \infty$, the use of feedback signaling decreases the throughput if delays occur. However, shows that for reasonably high offered traffic, say $G_t <$ 10 ppt, nonpersistent ISMA with $d_1 = d_2 = 0.1$ has a significantly higher (steady-state) throughput than slotted ALOHA. At these traffic loads, collisions occur frequently, particularly in slotted ALOHA. Their effect is, however, significantly less disastrous than in channels without capture.

Similar curves have been computed for the models based on transmitter or receiver success (e.g., [7, 31]. However, if packets of finite length ($L < \infty$) are considered with $G_t \rightarrow \infty$, the event that multiple terminals transmit exactly the same bit sequence becomes significant. In such cases, one will obtain the unrealistic result of multiple transmitter successes per time slot. This suggests $S_t \rightarrow \infty$, indicating that in each time slot the receiver decodes a bit sequence that corresponds to the message transmitted simultaneously by an infinite number of terminals. However, with a finite packet length L, the number of terminals that can uniquely identify themselves cannot exceed 2^L [6]. So, assuming $G_t \rightarrow \infty$ or $G_t > 2^L$ is unrealistic.

7.9 DISCUSSION

Several models for the probability of successful reception have been proposed. The capture-ratio model A3 proved a convenient technique to compute the performance of mobile random-access networks. The fact that the probability of capture could be expressed in closed form suggests that refinement of the analysis (e.g., to include the effect of shadowing) might also be feasible. For the other models studied in this chapter, analysis for multiple coexisting types of fading will presumably become substantially more complicated. Nonetheless, much attention is paid in the technical literature to the effects of the detection process for particular modulation and coding techniques, whereas the effect of different received power levels is often highly simplified. It is the author's strong impression that this attention, though appropriate for wideband spread-spectrum systems with adaptive power control, may not result in a realistic assessment of the performance of systems with narrowband channel with multipath fading, shadowing, and near-far effects without power control. It was argued previously [5, 41] that adaptive power control degrades the performance of random-access channels because it reduces the probability of capture. Moreover, power control may be difficult to implement in random access channels because each mobile terminal only occasionally transmits a data packet, so no continuous feedback information on the attenuation in the inbound channel is available.

We now propose a technique to evaluate the suitability of the capture ratio model. The probability of successful reception is in the general form of

$$q_n = \int_0^\infty \int_0^\infty \cdots \int_0^\infty \Pr(s_j|p_j, p_1 \ldots p_n) f_{p_j}(p_j) f_{p_1}(p_1) \ldots f_{p_n}(p_n) \, dp_j dp_1 \ldots dp_n \quad (7.48)$$

The capture-ratio model A3 simply assumes

$$\Pr(s_j|p_j, p_0, \ldots, p_n) = U\left(p_j - z \sum_{i=1}^n p_i\right) \quad (7.49)$$

with $U(\cdot)$ the unit step function ($U(x) = 0$ if $x < 0$, and 1 otherwise). In particular, the immunity of a perfectly locking receiver to a single (constant-envelope) interfering signal leads to a very abrupt transition (see (7.37) and (7.36) with $N_0 \to 0$), so a step function near a C/I ratio of 0 dB ($z = 1$ in 7.5.1) may be considered. For a realistic radio receiver, a smoother transition than this step function may exist, depending on the type of modulation, the synchronization capabilities of the receiver, and the character of the interference.

For Gaussian-distributed interference, the error probability (5.10) (see also Section 7.6.4) leads to a smoother transition. If an error correction code is used, which can correct up to t errors, the probability of correct reception of a data packet of L bits from terminal j (denoted by event s_j), given the instantaneous received signal power, is

$$\Pr(s_j|p_j, \bar{p}_t) = \sum_{m=0}^t \binom{L}{m} (1 - P_b)^{L-m} P_b^m \quad (7.50)$$

where we assumed that bit error are statistically independent from bit to bit because of Gaussian interference. Figure 7.23(b) illustrates the packet success probability (7.50) versus the C/I ratio for packets of $L = 16$ and 128 bits for various t.

To find the inaccuracy arising from the capture-ratio model A3, the probability of transmitter success $\Pr(s_j)$ is considered to satisfy the bounds

$$\begin{cases} 0 \quad \leq \Pr(s_j|C/I) < \epsilon_1 & C/I < z_1 \\ 0 \quad < \Pr(s_j|C/I) < 1 & z_1 \leq C/I \leq z_2 \\ 1 - \epsilon_2 < \Pr(s_j|C/I) \leq 1 & C/I > z_2 \end{cases} \quad (7.51)$$

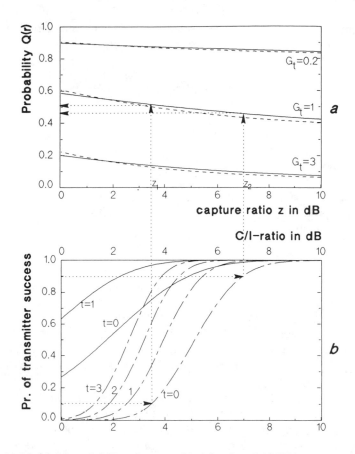

Figure 7.23 (a) Model A3: probability of capture $Q(r_M)$ for slotted ALOHA versus receiver threshold z. (———) Quasi-uniform offered traffic (test packet transmitted at $r = r_M \approx 0.734$), and (– –) Rayleigh fading only, with $G_t = 0.2$, 1, 3 ppt. Noise-free channel $N_A = 0$. (b) Probability of correct reception of a packet of $L = 16$ bits (———) and $L = 128$ bits (—·) versus the C/I-ratio. BPSK modulation with Gaussian interference for various error-correction distances t.

where ϵ_1 and ϵ_2 are small (ϵ_1, $\epsilon_2 \ll 1$) and depend on the choice of z_1 and z_2, respectively. In fact, these inequalities bound the inaccuracy of (7.49). Such bounds can be computed from (7.50) for Gaussian or constant-envelope interference using the bit error probabilities (5.10) and (7.35), respectively. For a mixed-type of interference, worst-case bounds can be defined similar to (7.51).

For slotted ALOHA, model A3 computes the performance measures $Q(r)$ and S_t from $\Pr(C/I > z)$, while the probability of transmitter success $\Pr(s_j)$ might be more appropriate. The service area-mean probability $\Pr(s_j)$ can be found by averaging $\Pr(s_j|C/I)$ over the pdf of the signal-to-interference ratio:

$$\Pr(s_j) = \int_0^\infty \Pr(s_j|C/I = z_1) \frac{d\Pr(C/I < z)}{dz}\bigg|_{z=z_1} dz_1 \qquad 7.52$$

In most cases, $\Pr(s_j|C/I)$ is not exactly known because the character of the interference is a stochastic variable, depending, among other things, on n. Nonetheless, the absolute error e in $\Pr(s_j)$ can be bounded by

$$\begin{aligned}
|e| &\leq \epsilon_1 \Pr(C/I < z_1) + \epsilon_2 \Pr(C/I > z_2) + \Pr(z_1 \leq C/I \leq z_2) \\
&= \epsilon_1(1 - \Pr(C/I > z_1)) + \epsilon_2 \Pr(C/I > z_2) + \Pr(C/I > z_2) - \Pr(C/I > z_1) \\
&\leq \max(\epsilon_1, \epsilon_2) + \Pr(C/I > z_1) - \Pr(C/I > z_2) \quad \forall z_1 < z_2
\end{aligned}$$

$$(7.53)$$

Thus, the capture-ratio model may be considered to yield accurate results if the effect of the choice of z on $Q(r)$ is limited. Figure 7.23(a) compares the probability of successful access $Q(r)$ for a terminal at the median distance $r = r_M = 0.734$ versus the capture ratio. Channels with Rayleigh fading and quasi-uniform offered traffic (7.6) (——) and channels with Rayleigh fading only (– –) are considered. In the latter case, $Q(r) = \exp\{z(z + 1)^{-1}G_t\}$ [5,11]. The total offered traffic is $G_t = 0.2$, 1 and 3 ppt.

Results in Figure 7.23 indicate that the capture-ratio model becomes increasingly appropriate with increasing packet length L and error-correction distance t. Particularly if the offered traffic is modest ($G_t << 1$), the capture ratio models give relatively accurate results. In channels with large fluctuations of the received power, the model is more accurate than in channels with Rayleigh fading only. This can be understood intuitively, since $\Pr(z_1 < C/I < z_2)$ is small if large fluctuations of the received signal power levels occur. In channels with Rayleigh fading only, instantaneous differences in received signal power are smaller. Nonetheless, the effect of z appeared small even for channels without near-far effect and shadowing: in narrowband communication, capture is unlikely to occur if multiple signals arrive with almost identical power, irrespective of z ($z > 1$), so $Q(r)$ is then largely determined by the probability that no contending packet is present ($Q(r) \rightarrow \Pr(n = 0)$).

In most analyses in the next two chapters, numerical results given for $z = 4$ (6 dB), which is more pessimistic than the values suggested by Figure 7.23(b). This is believed to be a more realistic value for practical receivers for BPSK signals in narrowband channels, because imperfect synchronization may impair the capture performance, compared to idealized model as BS1. For analog FM modulation by an FFSK subcarrier, the threshold model is believed reasonable even if fluctuations in received power are relatively small, because of the classical FM capture effect. In this case, practical values for z are about 10 dB, corresponding to the FM threshold. Then results for $z = 4$ may tend to be optimistic.

7.10 CONCLUSION

A large number of channel aspects influence the probability of successful packet reception. Complete investigation of the combined effects of all relevant aspects appears prohibitively complex. However, a number of simplified models have been investigated and compared.

Mobile slotted ALOHA and 1-persistent ISMA networks exhibit a significant near-far effect: packets from nearby terminals have a substantially higher probability of being received successfully than from remote terminals. This effect is much reduced in networks employing nonpersistent ISMA. However, at light offered traffic loads, slotted ALOHA and 1-persistent ISMA offer higher throughput.

The relative packet duration, compared to the rate of multipath fading, has a substantial effect on the probability of successful access in a mobile network. The performance rapidly deteriorates if the packet duration becomes larger than the time constant of the multipath fading. The throughput for fast fading channels (i.e., with independent amplitudes for successive bits in a packet) is substantially smaller than for slow fading with constant signal amplitudes during packet reception. This is in contrast to the observation in Section 5.7 that a network for cellular CW communication with high mean C/I ratios performs better in a fast-fading environment.

The approximation of one or multiple contending (constant-envelope) signals by a Gaussian-distributed joint interference signal appeared pessimistic in a random-access network with narrowband channels. This seems to be in contrast to the case in spread-spectrum networks where this approximation is frequently applied.

The capture-ratio model produced convenient mathematical expressions for the probability of successful reception. However, the accuracy of simplifying the detection process was also disputed. The discussion in this chapter showed that in *narrowband* mobile channels, this simplification is acceptable, since relatively large fluctuations of the received signal power occur. Some approximation of the detection process in a small range of the C/I ratio near the threshold appeared acceptable. This is a major distinction between narrowband and spread-spectrum communication; in the latter case, the capture ratio model may be unrealistic. These observations give a strong justification and motivation for studying the capture-ratio model for narrowband channels in more detail in the next chapter. Comparison with models for transmitter success probabilities showed that $z = 4$ (6 dB) may be a reasonable choice for the receiver threshold.

In slow Rayleigh-fading channels, receiver synchronization may become a more critical issue than bit errors caused by excessively strong interference samples. Synchronization impairments have been studied from a number of simplified models; further investigation is recommended but is thought outside the scope of this book.

REFERENCES

[1]   Wozencraft, J.M., and I.M. Jacobs, *Principles of Communication Engineering*, New York: John Wiley and Sons, 1965.

[2]   Carlson, A.B., *Communication Systems*, McGraw-Hill, 1986.

[3]   Metzner, J.J., "On Improving Utilization in ALOHA Channels," *IEEE Trans. on Comm.*, Vol. COM-24, No. 4, April 1976, pp. 447–448.

[4]   Abramson, N., "The Throughput of Packet Broadcasting Channels," *IEEE Trans. Commun.*, Vol. COM-25, No. 1, Jan. 1977, pp. 117–128.

[5]   Arnbak, J.C., and W. van Blitterswijk, "Capacity of Slotted-ALOHA in a Rayleigh Fading Channel," *IEEE J. Sel. Areas Commun.*, Vol. SAC-5, No. 2, Feb. 1987, pp. 261–269.

[6]   Linnartz, J.P.M.G., and J.J.P. Werry, "Error Correction and Error Detection Coding in a Fast Fading Narrowband Slotted ALOHA Network With BPSK Modulation," *Proc. International Symposium on Communication Theory & Applications*, 9–13 September 1991, Crieff, U.K., paper 37.

[7]   Linnartz, J.P.M.G., H. Goossen, and R. Hekmat, "Comment on 'Slotted Aloha Radio Networks With PSK Modulation in Rayleigh Fading Channels,'" *Electron. Lett.*, Vol. 26, No. 9, 26 April 1990, pp. 593–595.

[8]   Linnartz, J.P.M.G., "Slotted ALOHA Land-Mobile Radio Networks With Site Diversity," *IEE Proceedings I*, Vol. 139, No. 1, Feb. 1992, pp. 58–70.

[9]   Kuperus, F., and J. Arnbak, "Packet Radio in a Rayleigh Channel," *Electron. Lett.*, Vol. 18, No. 10, 10 June 1982, pp. 506–507.

[10]   Namislo, C., "Analysis of Mobile Radio Slotted ALOHA Networks," *IEEE J. Sel. Areas Commun.*, Vol. SAC-2, No. 4, July 1984, pp. 583–588.

[11]   Verhulst, D., M. Mouly, and J. Szpirglas, "Slow Frequency Hopping Multiple Access for Digital Cellular Radiotelephone," *IEEE J. Sel. Areas Commun.*, Vol. SAC-2, No. 4, July 1984, pp. 563–574.

[12]   Linnartz, J.P.M.G., "Spatial Distribution of Traffic in a Cellular Mobile Data Network," EUT Report 87-E-168, Eindhoven University of Technology, The Netherlands, 1987, ISBN 90–6144–168–4.

[13]   Linnartz, J.P.M.G., R. Prasad, and J.C. Arnbak, "Spatial Distribution of Traffic in a Cellular ALOHA Network," *Archiv für Elektronik und Übertragungstechnik* (AEÜ), Vol. 42, No. 1, Jan./Feb. 1988, pp. 61–63.

[14]   Lau, C., and C. Leung, "Capture Models for Mobile Packet Radio Network," *IEEE Trans. on Comm.*, Vol. 40, No. 5, May 1992, pp. 917–925.

[15]   Prasad, R., and J.C. Arnbak, "Enhanched Throughput in Packet Radio Channels With Shadowing," *Electron. Lett.*, Vol. 24, No. 16, 4 Aug. 1988, pp. 986–988.

[16]   Prasad, R., and J.C. Arnbak, "Capacity Analysis of Non-persistent Inhibit Sense Multiple Access in Channels With Multipath Fading and Shadowing," *Proc. 1989 Workshop on Mobile and Cordless Telephone Communications*, IEE, London, Sept. 1989, pp. 129–134.

[17]   Prasad, R., and J.C. Arnbak, "Effects of Rayleigh Fading on Packet Radio Channels With Shadowing," *Proc. IEEE Tencon 1989*, Bombay, India, Nov. 1989, pp. 27.4.1–27.4.3.

[18]   Linnartz, J.P.M.G., and R. Prasad, "Near-Far Effect on Slotted ALOHA Channels With Shadowing and Capture," *Proceedings IEEE Veh. Tech. Conf. 1989*, San Francisco, 3–5 May 1989, pp. 809–813.

[19]   Prasad, R., "Performance Analysis of Mobile Packet Radio Networks in Real Channels With Inhibit Sense Multiple Access," *IEE Proc. I*, Vol. 138, No. 5, Oct. 1991, pp. 458–464.

[20] Van der Plas, C., and J.P.M.G. Linnartz, "Stability of Mobile Slotted ALOHA Network With Rayleigh Fading, Shadowing and Near-Far Effect," *IEEE Trans. on Veh. Tech.*, Vol. 39, No. 4, Nov. 1990, pp. 359–366.

[21] Zdunek, K.J., D.R. Ucci, and J.L. Locicero, "Throughput of Nonpersistent Inhibit Sense Multiple Access With Capture," *Electron. Lett.*, Vol. 25, No. 1, 5 Jan. 1989, pp. 30–32.

[22] Linnartz, J.P.M.G., "Near-Far Effects in Some Random-Access Radio Systems With Fading," *1990 Bilkent Int. Conf. on New Trends in Comm., Control, and Signal Processing*, Ankara, Turkey, 2–5 July 1990, pp. 504–510.

[23] Prasad, R., "Throughput Analysis of Non-persistent Inhibit Sense Multiple Success in Multipath Fading and Shadowing Channels," *Eur. Trans. on Telecommunications* (ETT), Vol. 2, No. 3, May/June 1991, pp. 313–318.

[24] Fronczak, J., "Data Communications in the Mobile Radio Channel," EUT Report 83-E-142, Dept. of El. Eng., Eindhoven University of Technology, The Netherlands, 1983.

[25] Sinha, R., and S.C., Gupta, "Mobile Packet Radio Networks: State-of-the-Art," *IEEE Comm. Mag.*, Vol. 23, No. 3, March 1985, pp. 53–61.

[26] Sinha, R., and S.C. Gupta, "Performance Evaluation of a Protocol for Packet Radio Network in Mobile Computer Communications," *IEEE Trans. Veh. Tech.*, Vol. VT-33, No. 3, Aug. 1984, pp. 250–258.

[27] Gardiner, J.G., and A.I.A. Jabbar, "Performance of CSMA Protocols in Fading Mobile Radio Environments," *Proc. 5th IEE Int. Conf. on Mobile Radio and Personal Communications*, Warwick, U.K., 11–14 Dec. 1989, pp. 10–14.

[28] Ramamurthi, B., A.A.M. Saleh, and D.J. Goodman, "Perfect-Capture ALOHA for Local Radio Communications," *IEEE J. Sel. Areas Commun.*, Vol. SAC-5, No. 5, June 1987, pp. 806–814.

[29] Davis, D.H., and S.A. Gronemeyer, "Performance of Slotted ALOHA Random Access With Delay Capture and Randomized Time of Arrival," *IEEE Trans. on Comm.*, Vol. COM-28, No. 5, May 1980, pp. 703–710.

[30] Pluymers, R., "Computer Simulation of a Mobile Packet Radio System," *Electron. Lett.*, Vol. 24, No. 6, 17 March 1988, pp. 316–317.

[31] Zhang, K., K. Pahlavan, and R. Ganesh, "Slotted Aloha Radio Networks With PSK Modulation in Rayleigh Fading Channels," *Electron. Lett.*, Vol. 25, No. 6, 16 March 1989, pp. 413–414.

[32] Habbab, I.M.I., M. Kavehrad, and C-E.W. Sundberg, "ALOHA With Capture Over Slow and Fast Fading Radio Channels With Coding and Diversity," *IEEE J. on Sel. Areas in Comm.*, Vol. SAC-7, No. 1, Jan. 1989, pp. 79–88.

[33] Sheikh, A.U.H., Y.D. Yao, and X. Wu, "The ALOHA Systems in Shadowed Mobile Radio Channels With Slow and Fast Fading," *IEEE Trans. on Veh. Tech.*, Vol. VT-39, No. 4, Nov. 1990, pp. 289–298.

[34] Zhang, K., and K. Pahlavan, "A New Approach for the Analysis of the Slotted ALOHA Local Radio Networks," *Proc. International Conference on Communications ICC*, Atlanta, April 1990, pp. 1231–1235.

[35] Gradsteyn, I.S., and I.M. Ryzhik, *Table of Integrals, Series and Products*, 4th ed., New York: Academic Press, 1965.

[36] Jakes, W.C., Jr., ed., *Microwave Mobile Communications*," New York: John Wiley and Sons, 1974.

[37] Habbab, I.M.I., M. Kavehrad, and C-E.W. Sundberg, "ALOHA With Capture Over Rayleigh Fading Local Radio Channels," *Proc. IEEE Globecom*, Tokyo, 15–18 Nov. 1987, paper 21.6.1, pp. 818–822

[38] Zhang, K., and K. Pahlavan, "Relation Between Transmission and Throughput of Slotted ALOHA Local Packet Radio Networks," *IEEE Trans. on Comm.*, Vol. 40, No. 3, March 1992, pp. 577–583.

[39]   Eaves, R.E., and A.H. Levesque, " Probability of Block Error in Very Slow Rayleigh Fading in Gaussian Noise," *IEEE Trans. on Comm.*, Vol. COM-25, No. 3, March 1977, pp. 368–374.

[40]   Kleinrock, L., and F.A. Tobagi, "Packet Switching in Radio Channels: Part 1—Carrier Sense Multiple Access Modes and Their Throughput-Delay Characteristics," *IEEE Trans. on Comm.*, Vol. COM-23, No. 12, Dec. 1975, pp. 1400–1416.

[41]   Goodman, D.J., and A.A.M. Saleh, "The Near/Far Effect in Local ALOHA Radio Communications," *IEEE Trans. on Veh. Tech.*, Vol VT-36, No. 1, Feb. 1987, pp. 19–27.

Chapter 8
Spatial Distribution of Traffic in Mobile Slotted ALOHA Networks

Chapter 7 showed that differences in received signal power largely influence the probability of capture and thus the performance of the network. A special case has been addressed: packets were assumed to be offered to the channel according to a quasi-uniform distribution of transmitting terminals over the service area, and shadowing was not considered. These assumptions facilitated the investigation of capture or packet-success probabilities according to various models for fast and slow Rayleigh-fading channels. In this chapter, the detection process will be simplified by considering a receiver capture *threshold*, but the spatial distribution of traffic over the service area and the power fluctuations caused by shadowing, the near-far effect, and Rayleigh fading will now be treated in more detail. Also, network stability and delay will be addressed.

It has been demonstrated already, particularly in Chapters 3 and 7, that Laplace image functions, introduced in Section 2.1.5, facilitate the analysis of the threshold model. The motivation to study these images in more detail has been the two observations that (1) Laplace images are closely analogous to *characteristic functions* or *moment generating functions* and fully describe the statistical behavior of the fluctuations of the received power, and (2) Laplace transforms appear explicitly in the expression for receiver capture in a Rayleigh-fading channel. In our case, inverse transformation, which is a difficult mathematical problem [1, 2], is not required.

Section 8.1 discusses the properties of the Laplace transform of the pdf of received signal power. Expressions for the probability of successful transmission and for total channel throughput are given in Sections 8.2 and 8.3 for coherent and incoherent cumulation of interference power, respectively. Section 8.4 studies a number of special cases of the spatial distribution of offered traffic. The combined effects of Rayleigh fading, shadowing, and UHF path loss are addressed in Section 8.5. Stability and delay are addressed in Section 8.6. Section 8.7 concludes this chapter.

8.1 LAPLACE IMAGE FUNCTIONS AND INTEGRAL TRANSFORMS OF SPATIAL DISTRIBUTION

Since Laplace images of the pdf of received interference power play a key role in the subsequent analysis, a number of properties are summarized in this section. First, we introduce the image function $\psi_j(s)$, defined as the one-dimensional, one-sided Laplace image of the pdf of the *local-mean* power of the jth signal, with

$$\psi_j(s) \triangleq \int_{0-}^{\infty} \exp\{-s\bar{p}_j\} f_{\bar{p}_j}(\bar{p}_j) \, d\bar{p}_j \tag{8.1}$$

We also introduce the image function $\phi_j(s)$, defined as the Laplace transform of the *instantaneous* packet power f_{p_j}. So,

$$\phi_j(s) \triangleq \int_{0-}^{\infty} \exp\{-sp_j\} f_{p_j}(p_j) \, dp_j \tag{8.2}$$

As distinct from $\psi_j(s)$, $\phi_j(s)$ thus incorporates the effect of Rayleigh fading. The image function $\phi_\sigma(s)$, defined in Section 2.1.5 as the Laplace transform of the pdf of the power of a signal subjected to multipath fading and shadowing, is an example of (8.2).

A number of properties of these images follow.

The kth derivative of the images in the point $s = 0$ equals the k-th linear moment μ_k of the original pdf, except for a sign-factor $(-1)^k$, since

$$\phi_j^{(k)}(0) = \int_{0-}^{\infty} (-x)^k e^0 f_{\bar{p}_j}(x) \, dx = (-1)^k \mathrm{E}[\bar{p}_j^k] \triangleq (-1)^k \mu_k \tag{8.3}$$

This illustrates the relation between the Laplace transform and the *moment-generating function*. The defining integrals of the Laplace images (8.1) and (8.2) will converge (at least) for $\mathrm{Re}(s) \geq 0$, because any pdf integrated over its domain yields unity, and $0 \leq \exp\{-x\} \leq 1$ for x real and positive. The relations

$$0 \leq \phi_j(s) \leq \phi_j(0) = 1 \tag{8.4}$$

apply for $\mathrm{Re}(s) \geq 0$. Similar relations hold for $\psi_j(s)$. Further, if the pdf $f_{p_j}(p_j)$ is bounded for $p_j \geq 0$,

$$\lim_{s \to \infty} \phi_j(s) = 0 \tag{8.5}$$

Equation (8.5) shows that, mostly, the image vanishes for real s tending to infinity. The case of no interference (i.e., $f_{p_j}(p_j) = \delta(p_j)$) is an exception to this (see Section 7.4.2).

The relation between $\psi_j(s)$ and $\phi_j(s)$ follows from lemma I.

Lemma I

Assume two functions with Laplace images, $\phi_1(q)$ and $\phi_2(p)$, respectively. Let the relation between $f_1(x)$ and $f_2(w)$ be

$$f_2(w) = \int_0^\infty \frac{1}{x} \exp\left\{-\frac{w}{x}\right\} f_1(x)dx \tag{8.6}$$

The Laplace image $\phi_2(p)$ then equals

$$\phi_2(p) = \int_0^\infty \frac{1}{p} \exp\left\{-\frac{q}{p}\right\} \phi_1(q)dq \tag{8.7}$$

A typical pair of $f_1(x)$ and $f_2(w)$ are the pdfs of received local-mean and instantaneous power in a Rayleigh-fading channel, respectively.

Proof

Using the Laplace transform definition gives

$$\phi_2(w) = \int_0^\infty \int_0^\infty \frac{1}{x} \exp\left\{-\frac{w}{x} - wp\right\} f_1(x)dxdw$$

$$= \int_0^\infty \frac{1}{xp + (1/x)} f_1(x) \, dx \tag{8.8}$$

A substitution of $y = x + p^{-1}$ gives

$$\phi_2(p) = \int_{p^{-1}}^\infty \frac{1}{yp} f_1\left(y - \frac{1}{p}\right) dy \tag{8.9}$$

This can be written as a integral over q, namely,

$$\phi_2(p) = \frac{1}{p} \int_0^\infty \int_{p^{-1}}^\infty f_1\left(y - \frac{1}{p}\right) \exp\{-yq\} \, dy \, dq$$

$$= \frac{1}{p} \int_0^\infty \int_0^\infty f_1(x) \exp\left\{-q\left(x + \frac{1}{p}\right)\right\} dx \, dq \qquad (8.10)$$

The integral over x corresponds to a Laplace transform from x into q domain, so

$$\phi_2(p) = \frac{1}{p} \int_0^\infty \phi_1(q) \exp\left\{-\frac{q}{p}\right\} dq \qquad \text{Q.E.D.} \qquad (8.11)$$

We now address the relation between the image functions $\phi(s)$ and $\psi(s)$ as defined in (8.2) and (8.1), respectively. Equation (8.7) can be rewritten as

$$\phi_j(s) = \int_0^\infty \psi_j(sx) \, e^{-x} \, dx \qquad (8.12)$$

Generalizing, the kth derivative of $\phi_j(s)$ is

$$\phi_j^{(k)}(s) = \int_0^\infty x^k e^{-x} \psi_j^{(k)}(xs) \, dx \qquad (8.13)$$

In the limit $s \to 0$ and applying integral formula (3.351.3) in [3], this becomes

$$\phi_j^{(k)}(0) = \psi_j^{(k)}(0) \int_0^\infty x^k e^{-x} dx = k! \psi_j^{(k)}(0) \qquad (8.14)$$

provided that the limit operation ($s \to 0$) and integration may be interchanged. A formal proof of the existence of the limit (8.14) is not given, since the same result can be obtained by comparing the derivatives $\phi_j^{(k)}(0)$ and $\psi_j^{(k)}(0)$ [4]. Equation (8.14) is a trivial result for $k = 0$ and $k = 1$, but it shows that higher order linear moments v_k of the instantaneous power ($v_k \triangleq \mathrm{E}[p_{jj}^k]$) equal $k!$ times the linear moments μ_k of the pdf of the local-mean power ($\mu_k \triangleq \mathrm{E}[\bar{p}_j^k]$). This property was used by Prasad in [5] to compute the moments of the pdf of a signal with combined shadowing and Rayleigh fading. Furthermore, we conclude that the image $\phi_j(s)$, which at first glance seems like only an unfocused picture of $\psi_j(s)$ (blurred by the spreading of the received power caused by Rayleigh fading), does uniquely specify all moments of the mean power pdf. Moreover, as seen via a Taylor expansion of $\psi_j(s)$, $\phi_j(s)$ also uniquely specifies the pdf of the local-mean interference power

$$\psi(s) = \sum_{k=0}^{\infty} \mu_k \frac{(-s)^k}{k!} = \sum_{k=0}^{\infty} v_k \frac{(-s)^k}{(k!)^2} \tag{8.15}$$

provided that the two series converge.

8.1.1 Image Functions in Channels Without Shadowing

In channels without shadowing ($\sigma_s = 0$), the local-mean power is assumed to be normalized in the form

$$\bar{p}_j = \tilde{p}_j = r_j^{-\beta} \tag{8.16}$$

If the position of the terminal is unknown, the pdf of its mean power is found by transforming the stochastic variable from the r (normalized distance) domain into the $\bar{p}$ (normalized power) domain:

$$\left| f_{\bar{p}_j}(\bar{p}_j) d\bar{p}_j \right| = \left| \frac{2\pi r_j G(r_j)}{G_t} dr_j \right| \tag{8.17}$$

In this case, the image $\psi_j(s)$, together with the multiplicative factor G_t for the total offered traffic, uniquely describes the spatial distribution of the packet traffic offered to the channel. We do not consider any specific information on individual interfering signals, so we may assume that the pdf of received power is identical for each terminal. Therefore, we omit the index j in the notation of the images in this case.

The image $\psi(s)$ can be obtained directly from the spatial distribution of the offered traffic $G(r)$ by substituting (8.17) in the definition (8.1), so

$$G_t \psi(s) = \int_0^{\infty} 2\pi r\, G(r)\, \exp\{-sr^{-\beta}\}\, dr \tag{8.18}$$

Standard properties of the Laplace transform are summarized in Table 8.1 using the notation $G(r) \leftrightarrow G_t\psi(s)$ for a spatial distribution $G(r)$ and its image $G_t\psi(s)$.

Examples [4] of image functions of typical spatial distributions of the offered traffic are given below in Table 8.2.

It is seen that the images of the second and third example have an unbounded derivative at $s = 0$ ($\psi'(s) \to -\infty$ if $s \downarrow 0$). This is caused by unbounded mean

Table 8.1
Properties of Spatial Distibutions and Images [4]

$$G(r) \leftrightarrow G_t\psi(s)$$
$$aG(r) \leftrightarrow aG_t\psi(s)$$
$$-r^{-\beta}G(r) \leftrightarrow G_t\psi'(s)$$
$$G(r)\exp(-vr^{-\beta}) \leftrightarrow G_t\psi(s + v)$$

Table 8.2
Table of Spatial Distributions and Images [4]

Model	$G(r) =$	$G_t\psi(s) =$
Ring	$\dfrac{1}{2\pi}\delta(r - 1)$	e^{-s}
Quasi-uniform	$\dfrac{1}{\pi}\exp\left(-\dfrac{\pi}{4}r^2\right)$	$e^{-\sqrt{\pi s}}$
Uniform	$\dfrac{1}{\pi}$ for $0 < r < 1$	$e^{-s} - \sqrt{\pi s}\,\text{erfc}\sqrt{s}$
Circular belt	$\dfrac{2}{\pi r^6}\exp(-r^{-4})$	$\dfrac{1}{s + 1}$

received powers: Any spatial distribution with non-zero offered traffic infinitely close to the receiver, i.e. with $G(r) = G_\epsilon < 0$ for $r \to 0$, has unbounded moments, since

$$\mu_k = (-1)^k\psi^{(k)}(s) = \int_0^\infty 2\pi r^{1-k\beta}\frac{G(r)}{G_t}\,dr > 2\pi\frac{G_\epsilon}{G_t}\int_0^\epsilon r^{1-k\beta}dr \qquad (8.19)$$

if ϵ is small enough to assume $G(r)$ to be constant for $0 < r < \epsilon$. This integral diverges for $k \geq 1$ if $\beta \geq 2$. This result indicates that with distributions of the offered traffic that appear uniform, at least in the vicinity of the receiver, the unconditional mean-received power is unbounded ($\mu_1 \overset{\Delta}{=} E[\bar{p}_j] \to \infty$). As will be shown later, this significantly affects the estimate of the throughput of random-access nets at high traffic loads.

On the other hand, assuming all traffic to be generated beyond a distance of at least ϵ, with $0 < \epsilon \ll 1$, the moments increase no faster than exponentially with k, since

$$\mu_k = \int_\epsilon^\infty 2\pi r^{1-k\beta} \frac{G(r)}{G_t} dr \le \epsilon^{-k\beta} \int_\epsilon^\infty 2\pi r \frac{G(r)}{G_t} dr = \epsilon^{-k\beta} \qquad (8.20)$$

We conclude this section by showing that for channels without shadowing, any of the functions $G(\cdot)$, $f_{\bar{p}}(\cdot)$, $f_p(\cdot)$, $\phi(\cdot)$, or $\psi(\cdot)$ uniquely specifies all other functions, except possibly for the multiplicative factor G_t. In the case of channels without shadowing, the image $\phi(s)$ is found as

$$\phi(s) = \int_0^\infty \int_{0-}^\infty \frac{1}{\bar{p}_j} \exp\left\{ -sp_j - \frac{p_j}{\bar{p}_j} \right\} f_{\bar{p}_j}(\bar{p}_j)\, d\bar{p}_j dp_j \qquad (8.21)$$

Similar to (8.18), for channels without shadowing ($\sigma_s = 0$), the definition of $\phi(s)$ as an integral transform of the received power can also be stated in terms of an integral transform of the spatial distribution $G(r)$, namely,

$$\phi(s) = \int_0^\infty \int_0^\infty 2\pi\lambda^{\beta+1} \exp\{-(\lambda^\beta + s)w\} \frac{G(\lambda)}{G_t} d\lambda\, dw \qquad (8.22)$$

Interchanging the order of integration yields

$$\phi(s) = \frac{1}{G_t} \int_0^\infty \frac{r^\beta}{r^\beta + s} 2\pi r\, G(r)\, dr \qquad (8.23)$$

We have now established the relations illustrated in Figure 8.1, where arrows indicate definitions by the equations quoted. It can be seen that any distribution or image defines all other functions.

After this general introduction to image functions, we will now use these images to compute the probability of capture in Rayleigh-fading channels, and we will distinguish between coherent and incoherent cumulation of interference signals.

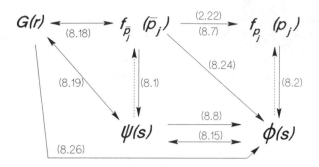

Figure 8.1 Relations between pdfs and image functions.

8.2 CAPTURE PROBABILITY FOR COHERENT CUMULATION

Coherent cumulation was discussed in Chapter 3. If interference signals add coherently, the amplitude of the joint signal of n Rayleigh-fading signals is again Rayleigh-distributed with mean $\bar{p}_t = \Sigma \bar{p}_i$. Hence, the conditional probability that a signal from terminal j captures receiver A (event A_j) is

$$\Pr(A_j|\bar{p}_j, n) = \mathscr{L}\left\{ f_{p_t}(p_t|n), \frac{z}{\bar{p}_j} \right\}$$

$$= \mathscr{L}\left\{ \int_0^\infty \frac{1}{\bar{p}_t} \exp\left(-\frac{p_t}{\bar{p}_t} \right) f_{\bar{p}_t}(\bar{p}_t|n) d\bar{p}_t, \frac{z}{\bar{p}_j} \right\} \tag{8.24}$$

Assuming a Poisson-distributed number of interfering signals and using lemma **I**, we find

$$\Pr(A_j|\bar{p}_j) = \sum_{n=0}^\infty \frac{G_t^n}{n!} e^{-G_t} \int_0^\infty \frac{\bar{p}_j}{z} \exp\left(-\frac{q\bar{p}_j}{z} \right) \psi(q|n) dq \tag{8.25}$$

where we used $\psi(s|n)$ to denote the Laplace image of the conditional pdf of the joint local-mean interference power, given the presence of n interfering signals.

This joint power is the sum of the local-mean power levels of the individual contributing signals, so the image of the pdf of the local-mean power of the joint signal $\psi(q|n)$ equals $\psi^n(q|1)$. If one substitutes $x = q\bar{p}_j/z$ and uses the series expansion of the exponential function, (8.25) becomes

$$\Pr(A_j|\bar{p}_j) = \int_0^\infty \exp\left\{ -x - G_t + G_t\psi\left(z\frac{x}{\bar{p}_j} \right) \right\} dx$$

$$= r_j^{-\beta} \int_0^\infty \exp\{\lambda r_j^{-\beta} - G_t + G_t\psi(z\lambda)\} d\lambda \tag{8.26}$$

where we used (8.16) to derive the expression on the second line of (8.26). This can be evaluated numerically with the Gauss-Laguerre integration method (see Appendix E). In this approximation, the probability of success $Q(r_j)$ is expressed as

$$Q(r_j) \triangleq \Pr(A_j|\bar{p}_j = r_j^{-\beta}) = \sum_{k=1}^{m} w_k \exp\{-L(zr_j^\beta x_k)\} + R_m \qquad (8.27)$$

Here, w_k are weight factors in the sample points x_k for an M-point quadrature, R_m denotes the remainder, and $L(s)$ is defined as

$$L(s) \triangleq -G_t + G_t\psi(s) = \int_0^\infty [\exp\{-s\lambda^{-\beta} - 1\} - 1] \, 2\pi r G(r) dr \qquad (8.28)$$

This result can be used to calculate the throughput for an arbitrary distribution of $G(r)$. Specific properties of the performance of typical channels with coherent cumulation will be addressed in later sections.

The probability of successful transmission q_n, conditional on the number of interferers n, but averaged over all expected positions of the terminal transmitting the test signal is

$$q_n = \int_0^\infty 2\pi r G(r) r^{-\beta} \int_0^\infty \exp\{\lambda r^{-\beta}\}\psi(z\lambda|n) d\lambda \, dr$$

$$= -\int_0^\infty \psi'(\lambda)\psi''(\lambda z) \, d\lambda \qquad (8.29)$$

where we used (8.18). Numerical examples will be discussed in Sections 8.2.3 and 8.2.4. In the special case of a perfect-capture receiver (i.e., if $z = 1$ (0 dB)), one finds

$$q_n = \int_0^1 \psi^n d\psi = \frac{1}{n+1} \qquad (8.30)$$

for any spatial distribution of the offered traffic.

8.2.1 Image of the Spatial Distribution of the Throughput

Analogous to the image $\psi(s)$ of received power of *attempted* packets, a Laplace image function of the received signal power of *successful* packets can be defined. For channels without shadowing, we define

$$\chi(s) \triangleq \mathcal{L}\{f_{\bar{p}_j}(\bar{p}_j | A_j), s\}$$

$$= \frac{1}{S_t} \int_0^\infty 2\pi r S(r) \exp\{-sr^{-\beta}\} dr \qquad (8.31)$$

where A_j denotes that test packet j captures receiver A. Inserting (8.26) gives

$$S_t \chi(s) = \int_0^\infty 2\pi r r^{-\beta} G(r) \Pr(c_0 | \bar{p}_0 = r^{-\beta}) \, dr$$

$$= \int_0^\infty 2\pi r r^{-\beta} G(r) \exp\{-(s + \lambda) r^{-\beta} - G_t + G_t \psi(z\lambda)\} dr$$

$$= -G_t \exp(-G_t) \int_0^\infty \psi'(\lambda + s) \exp\{G_t \psi(z\lambda)\} d\lambda \qquad (8.32)$$

The last step is based on the properties summarized in (8.18). In the special case of a perfect-capture receiver (i.e., a receiver with a capture ratio of $z = 1$ (0 dB)), the integral can be solved for $s = 0$. One finds

$$S_t = -\exp\{-G_t[1 - \psi(\lambda)]\}\big|_{\lambda=0}^\infty = 1 - \exp\{-G_t\} \qquad (8.33)$$

Since the number of packets in a time slot is assumed to be Poisson-distributed, precisely the part $1 - \exp\{-G_t\}$ of all slots is occupied by one or more packets. Thus, all nonempty slots *on the average* contribute one successful packet. However, (8.33) does not guarantee that every occupied slot contributes exactly one received packet: some occupied slots may contribute none, while other slots might, in principle, contribute more than one packet. Examples of the latter event can be easily found. For instance, Figure 8.2 depicts three signals, each of the form $v_i(t) \approx \cos(\omega_c t + \theta_i + \phi_\Delta \kappa_i)$, with $i = 1, 2,$ or 3, with small phase modulation index ($\phi_\Delta \ll 1$) to ensure coherent cumulation. In Figure 8.2, $\theta_1 \approx 0$, $\theta_2 \approx 0$, and $\theta_3 \approx \pi$ rad, so signal $v_1(t)$ and $v_2(t)$ satisfy the condition for receiver capture, because $v_2(t) - v_3(t) \approx 0$ and $v_2(t) - v_3(t) \approx 0$, respectively.

Figure 8.2 Phasor diagram of the typical event (for coherent cumulation) in which more than one signal can capture the receiver.

An asymptotic expansion for high traffic loads can be obtained by integrating by parts:

$$S_t\chi(s) = -e^{-G_t} \int_0^\infty \frac{\psi'(\lambda + v)}{z\psi'(z\lambda)} \frac{d}{d\lambda} \exp\{-G_t\psi(z\lambda)\}d\lambda \qquad (8.34)$$

This technique and conditions for its application are discussed in [6]. By repeating integration by parts of (8.34), an asymptotic expansion of $\chi(s)$ is found in the form of [4]

$$S_t\chi(s) = -\frac{\psi'(s)}{z\mu_1} - \frac{1}{G_t}\left[\frac{\psi''(s)}{z^2\mu_1^2} + \frac{\mu_2}{\mu_1^3}\frac{\psi'(s)}{z}\right] + \ldots \qquad (8.35)$$

with μ_k the kth linear moment of the local-mean power of received packets. If $s = 0$ is inserted, one finds

$$S_t = \frac{1}{z} + \frac{1}{G_t}\frac{\mu_2 z - 1}{\mu_1^2 z^2} + \ldots \qquad (8.36)$$

provided that the moments (8.3) of the pdf of received power exist. It can be seen that, for $G_t \to \infty$, $S_t \to 1/z$. This is a generalization of the limit $S_t \to 1/z$ found by Arnbak and v. Blitterswijk [7] for Rayleigh-fading channels without near-far effect. So, for any realistic spatial distribution of the offered traffic, the throughput does not decrease to zero for $G_t \to \infty$ if interfering signals indeed add coherently. The limit z^{-1} is approached increasingly rapidly if the spread μ_2 of the received signal powers is small. If moments are unbounded ($\mu_k \to \infty$) (e.g., in the case of the quasi-uniform distribution (8.18)), different limits can occur, as in [7].

8.2.2 Ring Model of Offered Traffic

If all signals arrive with identical mean power ($\bar{p}_j = 1$), say from mobiles moving on a ring around the central receiver, the image function $\psi(s)$ equals $\exp\{-s\}$. The probability q_n of successful reception (see (8.26)) becomes

$$q_n = \int_0^\infty \exp\{-(nz + 1)\lambda\}d\lambda = \frac{1}{nz + 1} \qquad (8.37)$$

which agrees with results reported by Arnbak and Van Blitterswijk [7]. Using the techniques embodied in (8.32), the total throughput is found as

$$S_t = S_t \chi(0) = -G_t e^{-G_t} \int_0^{\infty} \exp\{-\lambda + G_t e^{-z\lambda}\} d\lambda$$

$$= G_t \int_0^1 e^{-G_t x}(1 - x)^{1/z-1} dx \qquad (8.38)$$

This can be expressed in terms of the incomplete gamma function [1, 4] by applicating the integral expression (3.331.1) in [3] or in terms of Kummers function [1, 4], with $S_t = G_t M(1, 1/z + 1, -G_t)$. Moreover, the series expansion

$$S_t = G_t - \frac{G_t^2}{\frac{1}{z} + 1} + \frac{G_t^3}{\left(\frac{1}{z} + 1\right)\left(\frac{1}{z} + 2\right)} - \cdots \qquad (8.39)$$

was developed in [4]. Numerical results for S_t are in Figure 8.3.

It can be seen that at high offered traffic loads, say $G_t > 2$ ppt, coherent cumulation gives substantially higher throughput than incoherent cumulation (if the

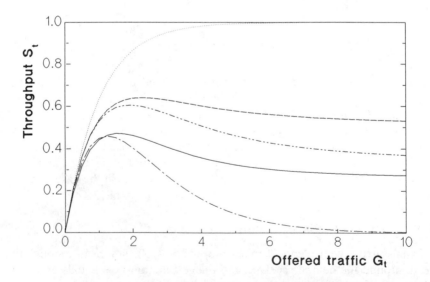

Figure 8.3 Throughput S_t versus offered traffic G_t with coherent cumulation for (——) ring model and (---) quasi-uniform offered traffic and with incoherent cumulation for (—·) ring model and (—··) quasi-uniform offered traffic. Receiver threshold $z = 4$ (6 dB). (···) One successful packet per slot ($S_t = 1 - \exp\{-G_t\}$).

receiver threshold is the same in both cases). In particular, if $G_t \rightarrow \infty$, the ring model gives $S_t \rightarrow 0$ for incoherent cumulation (see Section 8.4.1), but $S_t \rightarrow 1/z$ for coherent cumulation.

8.2.3 Quasi-Uniform Offered Traffic

The quasi-uniform distribution (see Chapter 7 and Table 8.2) of offered traffic has the image function $\psi(s) = \exp\{-\sqrt{\pi s}\}$. The probability of successful transmission is

$$q_n = \int_0^\infty \sqrt{\pi} \exp\{-\sqrt{\pi\lambda}\,(n\sqrt{z} + 1)\}\, d\lambda$$

$$= \left. \frac{e^{-t}}{n\sqrt{z} + 1} \right|_{t=0}^\infty = \frac{1}{n\sqrt{z} + 1} \tag{8.40}$$

It can be seen that (8.40) is in the same form as (8.37), though with z replaced by $\sqrt{z}$. This implies that the total throughput for a quasi-uniform spatial distribution must be in the form of (8.38), with $\sqrt{z}$ inserted for z [7].

Numerical results for Figure 8.3 show that coherent cumulation gives substantially higher throughput than incoherent cumulation. Coherent cumulation of colliding interference packets is likely to occur during the initial lockon period. However, with the modulation techniques used in practical mobile radio systems, signals are likely to add incoherently. Further, signals as depicted in Figure 8.2, with a large carrier component (because $\phi_\Delta << 1$), are relatively vulnerable to interference (see section on "minimum energy signals" in Section 4.5 of [27]). Thus, comparison between coherent and incoherent cumulation may not be fair if identical receiver thresholds are considered.

8.3 CAPTURE PROBABILITY FOR INCOHERENT CUMULATION

With coherent cumulation, individual fluctuations of carrier phases are assumed to be negligibly small during the reception of each packet. In contrast to this, this section addresses *in*coherent cumulation, which occurs as soon as modulation causes substantial phase deviations. For further discussion of incoherent cumulation, refer to Chapter 3.

Initially, a finite population of N terminals with known positions is considered. The probability $\Pr(k_{ON})$ that terminal k ($k = 1, 2, \ldots, N$) transmits during a particular time slot is assumed stationary. This probability includes the transmission of a newly arrived packet, as well as the retransmission of previously collided packets. Similar to the expression for outage probability in Chapter 3, the conditional probability that a test packet (with index j, from a terminal at distance r_j) captures the receiver is

$$\Pr(A_j|\bar{p}_j) = \prod_{\substack{k=1 \\ k \neq j}}^{N} \mathscr{L}\left\{ f_{p_k}, \frac{z}{\bar{p}_j} \right\} \tag{8.41}$$

where f_{p_k} represents the unconditional pdf of received interference power, taking into account the probability $\Pr(k_{OFF})$ that the terminal remains idle ($p_k = 0$ if k_{OFF}).

8.3.1 Vulnerability Weight Function

We continue to consider channels without shadowing in this section. If the local-mean power of the signal from each terminal is known, the pdf of the local mean-power becomes an exponential distribution. In this case,

$$\Pr(A_j|r_j, \{r_k\}_{k=1}^{N}) = \prod_{\substack{k=1 \\ k \neq j}}^{N} 1 - (1 - \mathscr{L}\{r_k^\beta \exp(-r_k^\beta p_k), zr_j^\beta\})\Pr(k_{ON})$$

$$= \prod_{\substack{k=1 \\ k \neq j}}^{N} \{1 - W(r_j, r_k)\,\Pr(k_{ON})\} \tag{8.42}$$

For case of notation we defined the weight function $W(r_j, r_k)$ $(0 < W(\cdot,\cdot) < 1)$ as [4, 8]

$$W(r_j, r_k) \triangleq \frac{zr_j^\beta}{zr_j^\beta + r_k^\beta} = 1 - \frac{1}{1 + z\left(\dfrac{r_j}{r_k}\right)^\beta} \tag{8.43}$$

This function is illustrated in Figure 8.4 for a test packet j transmitted at unity distance ($r_j = 1$) and for a receiver threshold of 0 dB ($z = 1$). Other threshold values can be considered by scaling the curve appropriately.

A signal from a distant terminal ($r_j \to \infty$) is most vulnerable to interference from terminal $k(r_k \ll r_j)$. This is represented by a weight factor close to unity ($W(r_j, r_k) \to 1$). Moreover, for channels without capture ($z \to \infty$), the weight factor goes to unity ($W(\cdot,\cdot) \to 1$) for all ranges. Then

$$\Pr(A_j|\{r_i\}_{i=1}^{N}) = \prod_{\substack{k=1 \\ k \neq j}}^{N} 1 - \Pr(k_{ON}) \tag{8.44}$$

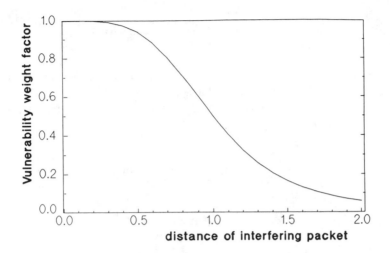

Figure 8.4 Factor $W(1,r_k)$ to weigh the vulnerability of a test packet from unity distance ($r_j = 1$) to an interfering signal from r_k. Receiver threshold $z = 1$ (0 dB).

which agrees with the well-known results for wired networks without capture [9]. On the other hand, for a test packet arriving at the receiver from a nearby terminal ($r_j \to 0$), interference from remote terminals may be neglected, since the weight factor tends toward zero ($W(r_j,r_k) \to 0$) for $r_j \ll r_k$.

We now address an infinite population of independently operating mobile terminals by letting $N \to \infty$. Moreover, the exact location of each terminal is not considered to be known. Taking account of Rayleigh fading, the pdf of the instantaneous power received from an individual interfering terminal is

$$f_{p_k}(p_k) = \int_0^\infty f_{p_k}(p_k|\bar{p}_k) f_{\bar{p}_k}(\bar{p}_k)\, d\bar{p}_k$$

$$= \int_0^\infty r_k^\beta \exp(-r_k^\beta p_k) \frac{2\pi r_k G(r_k)}{G_t}\, dr_k \qquad (8.45)$$

From this, the capture probability $q_n(r_j)$ is determined as

$$q_n(r_j) = \mathscr{L}^n \left\{ \int_0^\infty r^\beta \exp(-r^\beta p_k) \frac{2\pi r G(r)}{G_t}\, dr,\ z r_j^\beta \right\}$$

$$= \left[\int_0^\infty \int_0^\infty r^\beta \exp(-(r^\beta + z r_j^\beta)x) \frac{2\pi r G(r)}{G_t}\, dx\, dr \right]^n$$

$$= \left[\frac{1}{G_t} \int_0^\infty \frac{r^\beta}{r^\beta + zr_j^\beta} 2\pi r G(r) dr \right]^n \tag{8.46}$$

where the Lapalce transformation and the integral over r have been interchanged. Moreover, (8.46) with (8.23) verifies that $q_n(r_j) = \phi^n(zr_j^{-\beta})$.

Since $q_n(r) = q_1^n(r)$, the unconditional probability of capture $Q(r)$ is, for a Poisson distributed number of interfering packets,

$$Q(r) = \sum_{n=0}^\infty \frac{G_t^n}{n!} \exp(-G_t)q_n(r) = \exp\{-G_t(1 - q_1(r))\}$$

$$= \exp\left\{ -\int_0^\infty W(r,x)G(x)2\pi x dx \right\} \tag{8.47}$$

Thus, the expression for the probability of a successful transmission contains the interfering traffic intensity $G(r)$ multiplied by a weight factor, which is entirely determined by the mean propagation attenuation and the receiver capture ratio z. It is seen that interfering traffic from remote areas ($r_i \gg r_j$) does not play a major role since $W(r_j, r_i) \to 0$, whereas nearby interfering traffic causes more destructive collisions, and must be weighted by a factor approaching unity ($W(r_j, 0) = 1$).

If the weight function is replaced by a step function, the vulnerability circle proposed by Abramson [16] is recovered. This model [10] ignores cumulation of interference power and Rayleigh fading. However, active terminals outside this circle do introduce some interference. Accordingly, the integral in (8.47) has its upper limit at infinity, rather than at $c_z r_j$. Apparently, Rayleigh fading of the interfering packet traffic change the circle adopted by Abramson into the softer transition region described by the vulnerability weight function.

8.4 CAPTURE PROBABILITIES FOR SPECIAL CASES OF SPATIAL DISTRIBUTION

This section summarises results from an evaluation of a number of examples of the spatial distribution of the offered traffic. Incoherent cumulation of interference power is assumed.

8.4.1 Ring Distribution of Offered Traffic

For channels without near-far effect and shadowing, the local-mean power of all received signals is identical ($\bar{p}_i \equiv 1$ for all i). A similar situation arises if adaptive power is used to ensure equal local-mean power for all terminals. The probability of capture is evaluated by inserting a spatial distribution of the form

$$G(r) = \frac{G_t}{2\pi r} \delta(r - 1) \tag{8.48}$$

in (8.46). This gives

$$q_n = q_n(1) = \frac{1}{(z + 1)^n} \tag{8.49}$$

and the total throughput (see Figure 8.3) is

$$S_t = G_t \exp\left\{-\frac{z}{z + 1} G_t\right\} \tag{8.50}$$

These results were first presented in [7] and [11] with another method of derivation, thus without explicit use of the weight function.

8.4.2 Uniform Distribution With Infinite Extension

For globally, uniformly distributed offered traffic with intensity $G(r) \equiv G_0$ over a service area with infinite extension $(0 < r < \infty)$, the total offered traffic G_t is unbounded. As will now be shown, a nonzero throughput S_t is nonetheless found for $\beta > 2$. Using (8.47), the probability of a successful transmission from distance r is

$$Q(r) = \exp\left\{-2\pi G_0 \int_0^\infty \frac{zr^\beta \lambda}{\lambda^\beta + zr^\beta} d\lambda\right\} \tag{8.51}$$

After substituting $t = \lambda \rho^{-1} z^{-1/\beta}$, this gives

$$Q(r) = \exp\left\{-2\pi G_0 z^{2/\beta} \int_0^\infty \frac{t\, dt}{t^\beta + 1}\right\} \tag{8.52}$$

The integral is given as Equation (3.241.2) in [3] and in [12]:

$$Q(r) = \exp\left\{-\frac{2\pi^2 G_0 z^{2/\beta}}{\beta \sin \dfrac{2\pi}{\beta}} r^2\right\} \tag{8.53}$$

In the event of the propagation law $\beta = 4$, one finds

$$Q(r) = \exp\left\{-\frac{\pi^2}{2} G_0 \sqrt{z} r^2\right\} \tag{8.54}$$

Figure 8.7 (– –) gives the probability of capture versus distance r for globally uniform offered traffic.

For a minimum success rate Q_{MIN} to be realized at any position within the unit circle, thus for $Q(r) \geq Q(1) = Q_{MIN}$ for any $0 \leq r \leq 1$, the packet traffic per unit of area offered to the network must be bounded by

$$G_0 < -\frac{2}{\pi^2 \sqrt{z}} \ln Q_{MIN} \qquad (8.55)$$

Figure 8.5 gives the maximum allowed traffic intensity per unit of area as a function of the required Q_{MIN}.

If we also account for additive noise with power N_A (see Sections 3.3.3 and 7.4), the total throughput S_t is found as

$$S_t = \int_0^\infty 2\pi r_j G_0 Q(r_j) \exp\{-z N_A r_j^4\} dr_j$$

$$= \begin{cases} \dfrac{\pi}{2} G_0 \sqrt{\dfrac{\pi}{z N_A}} \exp\left\{\dfrac{\pi^4 G_0^2}{16 N_A}\right\} \mathrm{erfc}\left(\dfrac{\pi^2 G_0}{4\sqrt{N_A}}\right) & \text{if} \quad N_A > 0 \\[3mm] \dfrac{2}{\pi\sqrt{z}} \approx \dfrac{0.64}{\sqrt{z}} & \text{if} \quad N_A = 0 \end{cases}$$

$$(8.56)$$

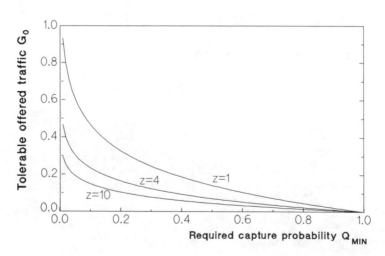

Figure 8.5 Maximum allowed offered traffic per unit of area for a given minimum probability of capture Q_{MIN}. Uniform offered traffic with infinite extension.

The total throughput is shown in Figure 8.6(a) versus the signal-to-noise ratio for terminals at normalized distance $r = 1$ $(C/N = N_A^{-1})$, with total offered traffic within the unit circle $\pi G_0 = 1$ ppt. The effect of the propagation exponent β is depicted in Figure 8.6(b) for $N_A = 0$,

$$S_t = \int_0^\infty 2\pi r G_0 \exp\left\{-\frac{\dfrac{2}{\beta}\pi^2 G_0 \dfrac{z^2}{\beta}}{\sin\dfrac{2\pi}{\beta}} r^2\right\} dr = \frac{\beta}{2\pi z^{2/\beta}} \sin\frac{2\pi}{\beta} \qquad (8.57)$$

In the special case of offered traffic with infinite extension, the throughput appears to be sensitive to the path loss law β. Moreover, for channels with free-space propagation ($\beta = 2$), the throughput even degrades to zero. This occurs because the expected number of interferers in a circular band at distance r_i from the base station is proportional to $2\pi r_i$. Since the signals from these interfering terminals are attenuated proportionally to r_i^2, the expected interference power per band decreases as r_i^{-1}. The nearest active terminal, at a distance r_0, experiences an unbounded total interference power because the integral of r_i^{-1} from r_0 to infinity diverges. Hence, for $\beta = 2$, $S_t \rightarrow 0$.

In fact, the above discussion can be extended to conclude that any form of cellular radio with regular frequency reuse is possible only because signal powers

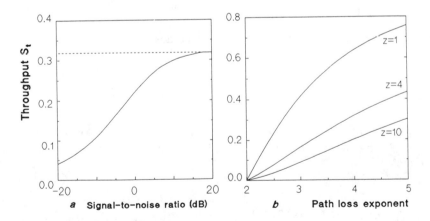

Figure 8.6 (a) Total throughput S_t versus signal-to-noise ratio at $r = 1$. (– –) Limit for noise-free channel. Globally uniform offered traffic with $\pi G_0 = 1$ ppt. Receiver threshold $z = 4$ (6 dB). Propagation exponent $\beta = 4$. (b) Total throughput S_t versus propagation exponent β. Globally uniform offered traffic. Receiver threshold $z = 1$, 4, and 10. Noise-free channel.

degrade on earth far more rapidly than in free space. For finite offered traffic loads, the effect of β is presumably much less pronounced.

A related situation has been addressed by astronomers for many centuries [13]. Although light from remote stars is heavily attenuated (i.e., proportional to r^{-2}), the solid angle covered by a remote star is also proportional to r^{-2}. Hence, the amount of light (per steradian) at the surface of any star seen by an observer on earth would be equal to the intensity of light (per steradian) seen at the surface of the sun. If the universe were infinitely large with a uniform distribution of stars, and if the universe had already existed for an indefinite period of time, in any direction one would see a star. In this case, the amount of light on earth would be tremendous. Moreover, no difference between day and night could be observed. In contrast to this, "it gets dark at night . . . ."

8.4.3 Uniform Distribution in Circular Band

If the offered traffic is uniformly distributed with intensity G_0 between two circles with radius r_1 and r_2, respectively, that is,

$$G(r) = \begin{cases} G_0 = \dfrac{G_t}{\pi(r_2^2 - r_1^2)} & r_1 < r < r_2 \\ 0 & \text{elsewhere} \end{cases} \tag{8.58}$$

the probability of a successful transmission becomes

$$Q(r) = \exp\left\{ -\pi G_0 \int_{\lambda=r_1}^{r_2} \frac{zr^\beta}{\lambda^\beta + zr^\beta} d\lambda^2 \right\} \tag{8.59}$$

For $\beta = 4$, the integral can be expressed in terms of the arctangent function. Using Equation (4.4.34) in [1] or (1.625.9) in [3] to subtract the resulting arctan terms, one obtains

$$Q(r) = \exp\left\{ -\sqrt{z}\pi r^2 G_0 \arctan\left(\frac{\sqrt{z}r^2(r_2^2 - r_1^2)}{zr^4 + r_1^2 r_2^2} \right) \right\} \tag{8.60}$$

Numerical results are shown in Figure 8.9 by broken lines (---). In the limit for increasing offered traffic ($G_t \to \infty$), the probability $Q(r)$ tends toward zero for all r

$(0 < r_1 < r < r_2)$. The total throughput also reduces to zero $(S_t \rightarrow 0)$ for any positive (and fixed) inner boundary r_1 $(r_1 > 0)$. Moreover,

$$\lim_{r_1 \rightarrow 0} \lim_{G_r \rightarrow \infty} S_t = 0 \qquad (8.61)$$

However, as will be shown in the next section,

$$\lim_{G_r \rightarrow \infty} \lim_{r_1 \rightarrow 0} S_t = \frac{2}{\pi \sqrt{z}} \qquad (8.62)$$

As was already seen in (8.56), the limit (8.62) also holds for an infinitely extended distribution of offered traffic $(r_2 \rightarrow \infty)$.

8.4.4 Uniform Spatial Distribution Within Unit Cell

The uniform spatial distribution within the unit cell, namely,

$$G(r) = \begin{cases} G_0 = \dfrac{G_t}{\pi} & 0 < r < 1 \\ 0 & \text{elsewhere} \end{cases} \qquad (8.63)$$

is a special case of the distribution addressed in the previous section. Inserting $r_2 = 1$ and taking the limit $r_1 \rightarrow 0$, the throughput for homogeneous offered traffic within the unit cell is found to be

$$Q(r) = \exp\left\{ -\sqrt{z} r^2 G_t \arctan\left(\frac{1}{\sqrt{z} r^2}\right) \right\} \qquad (8.64)$$

For the minimum success rate Q_{MIN} required at the cell boundary, the packet traffic offered to the net must be bounded by

$$G_0 < \frac{-\ln Q_{MIN}}{\pi \sqrt{z} \arctan\left(\dfrac{1}{\sqrt{z}}\right)} \qquad (8.65)$$

Since $\arctan \pi/4$ for near-perfect capture $(1 < z < 4)$ this is slightly less than twice the maximum allowed traffic in the event of uniform spatial distribution with infinite extension ($r_2 \to \infty$, as in Figure 8.5).

If the offered traffic G_t is increased without limit, the probability of successful reception for any individual terminal with finite (nonzero) distance to the base station necessarily tends toward zero. Nonetheless, packets from the immediate vicinity of the receiver continue to contribute to the throughput. The probability C_{i+1} of correct reception of one out of $i + 1$ packets (see Figure 8.13) can be found from

$$C_{i+1} = (i + 1) \int_0^1 q_i(r) 2r \, dr$$

$$= (i + 1) \int_0^1 \left[\int_0^1 \frac{2x^5 dx}{x^4 + zr^4} \right]^i 2r \, dr$$

$$= \frac{(i + 1)}{\sqrt{z}} \int_0^{\sqrt{z}} \left[1 - y \arctan \frac{1}{y} \right]^i dy \tag{8.66}$$

For large i, small values of y offer the principal contribution to the integral. We now apply integration by parts to evaluate the asymptotic behavior for $i \to \infty$:

$$\lim_{a \downarrow 0} \int_a^1 \frac{d\{f(y)\}^{i+1}}{f'(y)} = -\lim_{a \downarrow 0} \frac{(f(a))^{i+1}}{f'(a)} + \frac{\{f(1)\}^{i+1}}{f'(1)} + \lim_{a \downarrow 0} \int_a^1 \frac{\{f(y)\}^{i+1}}{\{f'(y)\}^2} f''(y) dy \tag{8.67}$$

where, for convenience of notation, $f(y) \triangleq 1 - \arctan(y^{-1})$. For large i, the integral and the term for $y = 1$ on the right-hand side vanish and we find $C_{i+1} \to 2/(\pi\sqrt{z})$. For large G_t, the probability of less than i_e packets in any finite interval of time can be made arbitrarily small. Thus, for slotted ALOHA, the throughput at high traffic loads must approach $S_t \to 2/(\pi\sqrt{z})$. This agrees with (8.56). Lau and Leung showed in [14] that this limit also occurs in channels without Rayleigh fading.

8.4.5 One-Dimensional Uniform Distribution

A slotted ALOHA net supporting packet traffic transmitted by vehicles located on a highway is addressed with intensity $(1/2)G_0$ packets per slot per unit of distance, passing by the base station with angle θ_0. The offered traffic per unit of area is

$$G(r, \theta) = \frac{1}{2} G_0 [\delta(\theta - \theta_0) + \delta(\theta - \theta_0 + \pi)] \tag{8.68}$$

So, by averaging over a circular ring with radius r ($0 \leq \theta < 2\pi$), $G(r)$ is found to be

$$G(r) = \int_0^{2\pi} \frac{G(r,\theta)}{2\pi r}\, dr = \frac{1}{2\pi r} G_0 \tag{8.69}$$

Then, if $\beta = 4$,

$$Q(r) = \exp\left\{ -G_0 \int_0^{\infty} \frac{zr^4}{x^4 + zr^4}\, dx \right\} = \exp\left\{ -\frac{1}{4} \sqrt{2}\pi G_0 \sqrt[4]{zr} \right\} \tag{8.70}$$

Numerical results are shown in Figure 8.7 by $(\cdot\!-\!\cdot)$ It can be seen that, compared to the two-dimensional globally uniform distribution (see (8.54) and $(-\,-)$ in Figure 8.7), packets from relatively remote terminals still have a significant probability of capture. The total throughput becomes

$$S_t = \int_0^{\infty} G_0 \exp\left\{ -\frac{\sqrt{2}}{4} \pi G_0 \sqrt[4]{zr} \right\} dr = \frac{2\sqrt{2}}{\pi \sqrt[4]{z}} \approx \frac{0.90}{\sqrt[4]{z}} \tag{8.71}$$

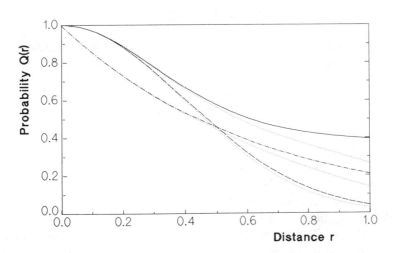

Figure 8.7 Probability of successful transmission $Q(r)$ versus distance **r** for $(-\,-)$ two-dimensional uniform distribution with infinite extension, (———) uniform distribution in unit circle, and $(\cdot\!-\!\cdot)$, one-dimensional uniform distribution with infinite extension. Receiver threshold $z = 4$ (6 dB). Total traffic offered within unit circle is 1 ppt. $(\cdots)$ Corresponding results for noisy channel ($N_A = 0.1$)

for any positive, nonzero value of G_0. Since we confine ourselves to realistic narrowband receivers with $z > 1$, this throughput is higher than the throughput with uniform offered traffic in two dimensions (compare (8.56)).

8.4.6 Nonuniform Distributions

Exactly or quasi-uniform offered traffic was studied in the previous sections. Such distributions may be reasonable in a number of cases. For instance, if vehicle speeds are sufficient to ensure that the position of a terminal becomes uncorrelated between each retransmission, a uniform offered packet traffic may be reasonable. Moreover, in certain vehicle location systems or telemetry applications, a routine status report lost in a collision may not need to be retransmitted. This leads to a uniform distribution if the participating terminals are uniformly distributed around the base station. However, major deficiencies of the uniform distribution are that (1) uniform distributions allow an incidental transmitter to be unrealistically close to the receiver, which seriously hinders realistic assessment of the throughput, and (2) uniform distributions ignore the fact that retransmissions are likely to arrive predominantly from areas with poor propagation. The first aspect is addressed in the next section concerning log-normal spatial distributions, and the second aspect will be considered in Section 8.6.3 and in Chapter 9.

8.4.7 Log-Normal Spatial Distribution

Linnartz and Prasad suggested in [15] that the uniform spatial distribution be approximated by a two-parameter *log-normal spatial distribution* in the form of

$$G(r) = \frac{G_t \beta}{(2\pi)^{3/2} r^2 \sigma_d} \exp\left\{ -\frac{\beta^2}{2\sigma_d^2} \ln^2 \frac{r}{r_m} \right\} \tag{8.72}$$

with r_m the logarithmic mean distance. We normalize distances by taking the parameter $r_m = 1$. The parameter σ_d defines the spatial spread of the offered traffic.

Basically, this approximation has two effects:

1. It smooths the transition from high offered traffic intensity inside the cell to zero offered traffic outside the cell.
2. The density of users unrealistically close to the receiver is sharply diminished.

Moreover, the distribution (8.72) produces a log-normal pdf of the area-mean power, so study of the effects of combined shadowing and near-far effects becomes mathematically tractable [15]. Figure 8.8 illustrates this spatial distribution for $\beta = 4$ and various σ_d. Initially, we assess an appropriate choice of σ_d to approximate the exact uniform distribution. To this end, the logarithmic moments of the uniform and

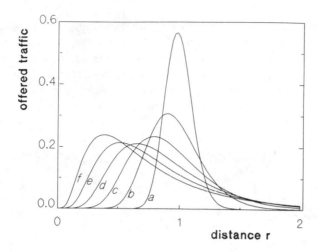

Figure 8.8 Normalized log-normal spatial distribution of offered traffic $G(r)/G_t$ for $\beta = 4$ for various spatial spreads σ_d: $s_d = 2, 4, \ldots, 12$ dB, curve (a) to (f), respectively.

the log-normal distribution are compared. Analogous to definitions in Chapter 2, the kth logarithmic moment μ_k of the mean received signal power is defined as

$$\mu_k \triangleq \int_0^{\infty} [-\ln \bar{\bar{p}}_j]^k f_{\bar{p}_j}(\bar{\bar{p}}_j) \bar{\bar{p}}_j$$

$$= \frac{1}{G_t} \int_0^{\infty} [-\beta \ln r]^k 2\pi r G(r) dr \qquad (8.73)$$

provided that the integral exists. The logarithmic variance is found from $\sigma_d^2 = \mu_2 - \mu_1^2$. For the uniform spatial distribution, the logarithmic moments (8.73) exist, namely,

$$\mu_k = (-\beta)^k \int_0^1 \ln^k r \, r \, dr = (-\beta)^k \frac{k!}{2^k} \qquad (8.74)$$

This is in contrast to the *linear* moments of the exactly uniform distribution, which diverges, as seen from (8.19). Equation (8.74) shows that the logarithmic standard deviations of the uniform distribution and the log-normal distribution become equal if one takes $\sigma_d = \beta/2$. For UHF groundwave propagation over a smooth, idealized earth ($\beta = 4$), the logarithmic standard deviation of the uniform distribution equals $\sigma_d = 2$ ($s_d = 8.68$ dB).

As discussed in Section 2.1.2, the log-normal shadowing for the model by Egli [16] was reported to be on the order of $s_s = 12$ dB. Remarkably, since we found

$\sigma_d < \sigma_s$ for all realistic values of β, other parameters (e.g., obstacles to propagation) would appear to have a more pronounced influence on the standard deviation of the signal power fluctuations than the distance between transmitter and receiver.

Prasad and Arnbak [17] presented the throughput for ALOHA networks over channels with log-normal fading and developed an approximate technique for channels with combined shadowing and Rayleigh fading [5]. In contrast to this, we will adopt here a formally exact technique similar to the derivation of outage probabilities as presented in Chapter 3. Using the weight function with (8.72), the throughput of slotted ALOHA with Rayleigh fading and shadowing is found to be

$$S_t = \int_0^\infty \frac{G_t\beta}{r_j\sigma_d\sqrt{2\pi}} \exp\left\{ -\frac{\beta^2 \ln^2 r_j}{2\sigma_d^2} - \int_0^\infty \frac{zr_j^\beta}{r^\beta + zr_j^\beta} \frac{G_t\beta}{\sqrt{2\pi}r\sigma_d} \exp\left(\frac{\beta^2 \ln^2 r}{2\sigma_d^2}\right) dr \right\} dr_j$$

(8.75)

Numerical results shown in Figure 8.9 have been obtained from the Hermite polynomial method (Appendix E). The computation algorithm is similar to the calculation of signal outages in macrocellular nets (Section 3.2).

It can be seen that at high traffic loads, say $G_t > 3$ ppt, the assumption of log-normally distributed offered traffic leads to a substantially lower estimate of the throughput than uniform offered traffic within $0 < r < 1$. Although for the main part of coverage area the near-far effect does not lead to large power fluctuations

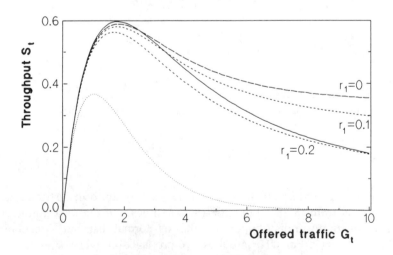

Figure 8.9 Throughput S_t of slotted ALOHA in a Rayleigh-fading channel with (——) a log-normal spatial distribution ($\sigma_d = 2$), (– –) exactly uniform distribution for $r_1 < r < 1$, and (---) exactly uniform within the unit circle $0 < r < 1$ ($r_1 = 0$). Receiver threshold $z = 4$ (6 dB) and ($\cdots$) no capture ($S_t = G_t\exp\{-G_t\}$).

compared to shadowing, the possibility that occasionally a transmitter may be unrealistically close to the receiver substantially influences the average throughput.

8.5 CAPTURE PROBABILITY IN CHANNELS WITH RAYLEIGH FADING, SHADOWING, AND NEAR-FAR EFFECT[1]

In order to investigate combined Rayleigh fading, shadowing, and near-far effects,
the weight-function method is now extended. The probability of capture $q_n(r)$, given
the area-mean power $\bar{p}_j = r^{-\beta}$ of the test packet and given the presence of n interferers, is

$$q_n(r) = \int_0^\infty f_{\bar{p}_j}(\bar{p}_j | \bar{\bar{p}}_j = r^{-\beta}) \Pr(A_j | n, \bar{p}_j) d\bar{p}_j$$

$$= \int_0^\infty \frac{1}{\sqrt{2\pi}\sigma_s \bar{p}_j} \exp\left\{\frac{1}{2\sigma_s^2} \ln^2(\bar{p}_j r^\beta)\right\} \phi^n\left(\frac{z}{\bar{p}_j}\right) d\bar{p}_j \qquad (8.76)$$

The Laplace image $\phi(s)$ of the pdf of the power received from an interfering mobile
terminal experiencing combined groundwave path loss, shadowing, and Rayleigh
fading is

$$\phi(s) = \frac{1}{G_t} \int_0^\infty \int_0^\infty \frac{1}{1 + s\bar{p}_i} \frac{2\pi r G(r)}{\sqrt{2\pi}\sigma_s \bar{p}_i} \exp\left\{\frac{\ln^2(\bar{p}_i r^\beta)}{2\sigma_s^2}\right\} dr \, d\bar{p}_i \qquad (8.77)$$

This can be written as an integral transform of the spatial distribution by interchanging the order of integration. Rewriting the integral over $\bar{p}_i$ using the substitution
proposed in (2.23) with $\bar{\bar{p}}_i = r_i^{-\beta}$, the image (8.77) can be interpreted as weighing
the vulnerability of the test packet to interference at distance r_i, that is,

$$G_t\phi(s) = \int_0^\infty 2\pi r_i W_\sigma(s, r_i) f_r(r_i) dr_i \qquad (8.78)$$

where the vulnerability weight function $W_\sigma(\cdot)$ $(0 < W_\sigma(\cdot) < 1)$ is defined by

$$W_\sigma(r_j, r) \triangleq \frac{1}{\sqrt{\pi}} \int_{-\infty}^\infty \frac{\exp(-x^2)dx}{1 + zr_j^\beta r^{-\beta}\exp(\sqrt{2}\sigma_s x)} = \phi_\sigma(zr_j^\beta r^{-\beta}) \qquad (8.79)$$

[1]Sections 8.5 and 8.6: Portions reprinted, with permission, from IEEE Transactions on Vehicular
Technology, Vol. VT-39, No. 4, pp. 359–366, November 1990. © 1990 IEEE.

where $\phi_\sigma(\cdot)$ has been defined in Section 2.1.5 as the Laplace image of the pdf of the received power from a Suzuki-fading signal. For channels without shadowing $(\sigma_s \to 0)$, the simpler weight function defined in (8.43) is recovered. Figure 8.10 compares the weight function for channels including shadowing with the case of no shadowing $(\sigma_s \to 0)$.

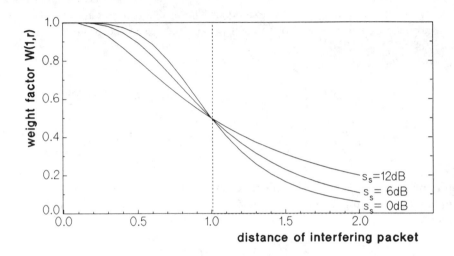

Figure 8.10 Weight factor $W_o(1,r)$ and (———) approximation by a step function, describing the vulnerability of a test packet (at $r_j = 1$) to interference at distance r_k. Shadowing is $s_s = 0$, 6, and 12 dB. UHF groundwave propagation with $\beta = 4$.

For uniform spatial density of offered traffic, the Laplace image of the unconditional pdf of received interference power p_i becomes

$$\phi(s) = \int_0^1 2r_i\phi_\sigma(sr_i^{-\beta})dr_i = \int_0^1 \frac{2r_i}{\sqrt{\pi}} \int_{-\infty}^\infty \frac{r_i^4 \exp(-y^2)\,dy}{r_i^4 + s\exp(\sqrt{2}\sigma_s y)}\,dr_i \quad (8.80)$$

After substitution of

$$t \triangleq \frac{r_i^\beta}{s\exp\{\sqrt{2}\sigma_s y\}} \quad (8.81)$$

the image becomes, for $\beta = 4$,

$$\phi(s) = \frac{1}{2\sqrt{\pi}} \int_{-\infty}^{\infty} \int_{0}^{t_{max}} \frac{\sqrt{t}}{1+t} dt \sqrt{s} \exp\left\{-y^2 + \frac{1}{2}\sqrt{2}\,\sigma_s y\right\} dy \quad (8.82)$$

with t_{max} defined as

$$t_{max} \triangleq \frac{1}{s} \exp\{-\sqrt{2}\sigma_s y\} \quad (8.83)$$

Using Equation (14.115) in [18], the image function becomes

$$\phi(s) = \frac{1}{\sqrt{\pi}} \int_{-\infty}^{\infty} \left[1 - \sqrt{s} \exp\left\{\frac{1}{2}\sqrt{2}y\sigma_s\right\} \arctan\left(\sqrt{\frac{1}{s}} \exp\left\{-\frac{1}{2}\sqrt{2}y\sigma_s\right\}\right) \right] e^{-y^2} dy$$

$$(8.84)$$

which is to be inserted in (8.76). In the latter, the integral over the local-mean power $\bar{p}_j$ of the test packet is rewritten, using logarithmic integration variables x and y in the form of (2.23). Hence,

$$q_n(r) = \frac{1}{\sqrt{\pi}} \int_{-\infty}^{\infty} \exp(-x^2) \left[\frac{1}{\sqrt{\pi}} \int_{-\infty}^{\infty} f(x,y) \exp(-y^2) dy \right]^n dx \quad (8.85)$$

where for shortness of notation we introduced $f(x,y)$, defined as

$$f(x,y) \triangleq 1 - \sqrt{z}r^2 \exp\left\{\frac{\sqrt{2}}{2}\sigma_s(y-x)\right\} \arctan\left\{\frac{1}{zr^4}\exp\left\{\frac{1}{2}\sqrt{2}\sigma_s(x-y)\right\}\right\} \quad (8.86)$$

It can be seen that in the event of combined near-far effect, shadowing, and Rayleigh fading, capture probabilities $q_n(r)$ can be expressed as a two-dimensional integral.

Capture probabilities $\{C_i\}$ with $C_i = iq_{i-1}$ ($i = 1, 2, \ldots$) (thus unconditional on the location of the particular mobile transmitting the test packet) will be used in Section 8.6 to study the stability of the network in its entirety. To this end, probability (8.85) is averaged over the pdf of the unknown area-mean power of the test packet. The capture probability q_n, given that n packets interfere, is

$$q_n = \frac{2}{\sqrt{\pi}} \int_{0}^{1} \int_{-\infty}^{\infty} r \exp(-x^2) \left[\frac{1}{\sqrt{\pi}} \int_{-\infty}^{\infty} f(x,y) \exp(-y^2) dy \right]^n dx\, dr \quad (8.87)$$

264

Numerical results for this capture probability can be obtained using the Hermite polynomial method (Appendix E) for both infinite integrals in (8.87). Results are included later in Figure 8.13 for various propagation models. The nonzero limit of C_i for $i \to \infty$ is due to the unrealistic traffic assumed in the immediate vicinity of the base station ($r \to 0$).

8.6 DYNAMIC PERFORMANCE OF MOBILE SLOTTED ALOHA NETS

It has been discussed in Chapter 6 that data packets experience delays because of a number of queuing mechanisms. Here we address the delay caused by the fact that packets unsuccessfully offered to the random-access channel have to be retransmitted after a random waiting time. During the analysis of throughput S_t and the probability of success $Q(r)$, it was assumed that unsuccessful packets are rescheduled, so that each new attempt is seen as a new contribution to the offered traffic G_t. Unsuccessful packets remain in the buffer of the terminal for retransmission. In this section we address a queuing model based on a Markov Chain. This model offers the possibility of evaluating the effect that after an unsuccessful transmission attempt, the terminal retransmits the packet with a minimum random waiting time to reduce the delay. If multiple terminals have packets for retransmission, destructive collisions may continue to occur. The backlog, packet delay, and also the stability of the network are studied from a Markov Chain.

A well-known procedure for evaluation of the stability of the ALOHA channel [19–21] considers a finite population of N terminals. Each terminal has a buffer of length 1 to store the last packet, which may need retransmission. The local behavior of such a terminal is modeled by a tristate Markov chain (see Figure 8.11).

In the origination state (O), a packet is generated (for transmission in the next available time slot) with probability P_o. In the transmission state (T), the terminal is busy with either transmitting a new packet or retransmitting a collided packet. The mobile terminal returns to the O-state if it receives a positive acknowledgment at the end of the time slot during which a packet is transmitted; otherwise, it enters the retransmission or *backlogged* state (R). From the latter state, retransmission (i.e., transition into the T-state) occurs with probability P_r. A backlogged terminal is blocked

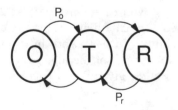

Figure 8.11 Markov chain model for a terminal.

in the sense that no new messages can arrive when the terminal is in state R. This tristate model implies the absence of a buffer for more than one packet in a mobile terminal.

The global behavior of the entire slotted ALOHA network is also modeled by means of a finite Markov chain (Figure 8.12). The state of the network represents the number of terminals m in the state R at the start of the time slot. This is also called the *system backlog*. With a population of N terminals, we have $N + 1$ states. During a time slot, i ($i = j + k$) packets are transmitted whenever j terminals in the O-state and k terminals in the R-state change into the T-state at the same time. The probability of a transition into another state may be obtained from the probability distribution of j and k and the capture probability C_{j+k} that one of the packets survives the collision. Here, j and k are binomial random variables. For a given backlog, j and k are mutually independent.

These investigations take into account that the system backlog m and the probability of successful transmission are statistically correlated: with a high system backlog, frequent retransmissions occur (if $P_r > P_o$), and collisions are very likely. Hence, the probability of success generally reduces with increasing system backlog (if $C_k < C_m$ for $k > m$), which may lead to saturation or instability of the network if terminals in backlog employ too-short retransmission waiting times (too-high P_r). If the number of terminals N is finite, the network depicted in Figure 8.12 may become bistable; that is, the network oscillates between low and high backlog states [19–21].

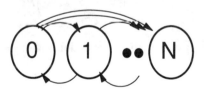

Figure 8.12 Markov chain model for the ALOHA network.

8.6.1 Measures of Dynamic Performance

The throughput S_m in state m is found from the probability of successfully transmitting a packet in a time slot, and is determined by averaging the probability of i ($i = j + k$) colliding packets, given a backlog of m terminals; that is,

$$S_m = \sum_{k=0}^{N-m} \binom{N-m}{N-m-k} P_o^k (1 - P_o)^{N-m-k} \sum_{j=0}^{m} \binom{m}{j} P_r^j (1 - P_r)^{m-j} C_{j+k} \quad (8.88)$$

Stability is studied from the expected drift d_m in each state, defined as the difference between the expected input traffic $(N - m)P_o$ and the expected output traffic S_m in that state, that is,

$$d_m = (N - m)P_o - S_m \qquad (8.89)$$

expressed in states per time slot. In a simulation to be described in Section 8.6.4, this drift is obtained directly as the average motion towards another (higher or lower) state, when the network is in state m. The network is expected to operate in a state near an equilibrium point; that is, where the expected drift (8.89) crosses zero with negative derivative [19–20].

The average total network throughput S_t is found by averaging (8.88) over the state probabilities π_m, namely,

$$S_t = \sum_{m=0}^{N} \pi_m S_m \qquad (8.90)$$

where the probabilities $\{\pi_m\}$ that the network is in state m $(m = 0, 1, \ldots, N)$ can be obtained recursively from the probabilities of transition from states m_1 to m_2, which depend on P_o, P_r, $\{C_i\}$, N, m_1, and m_2 [18, 22].

In a well-functioning network in equilibrium, the channel throughput S_t must equal the newly generated traffic, so

$$S_t = \left(N - \sum_{m=0}^{N} m\pi_m \right) P_o \qquad (8.91)$$

For stable (and bistable) systems, the access delay can be evaluated from Little's formula [23]

$$D_a = \frac{E[m]}{S_t} = \frac{\displaystyle\sum_{m=0}^{N} m\pi_m}{\displaystyle\sum_{m=0}^{N} S_m \pi_m} \qquad (8.92)$$

The stability of mobile ALOHA channels including Rayleigh fading and shadowing, in addition to near-far effects, will be reported in the next section.

8.6.2 Network Stability

Namislo [21] reported drifts, stability, and delays for a network with capture in terms of N, P_o, $\{C_i,\}$ and P_r. The probabilities of capture were obtained from Monte-Carlo simulation, taking into account path losses; fading and shadowing were not considered. Van der Plas and Linnartz included the effects of Rayleigh fading and shadowing in [12]. Figure 8.13 summarizes the probability that one out of i packets survives a collision for various propagation models. Inserting the capture probabilities $\{C_i\}$ in (8.88) and (8.89) gives the expected dynamic behavior of the network. In Figure 8.14, the dynamic behavior of the mobile network is shown for various propagation models in terms of the expected drift (8.89). A network with N ($N = 100$) mobile terminals, uniformly spread over coverage area, is assumed. The assumed probability of packet generation is $P_o = 0.0055$, and the probability of retransmission is $P_r = 0.08$.

If capture never occurs in the presence of other packets ($C_i \equiv 0$ for $i = 2, 3, \dots$), entire saturation ($E[m] \approx N = 100$) leads to extraordinarily high delay ($D_a \approx 4.6 \times 10^5$ slots). For the traffic parameters studied here, network performance is not essentially improved by Rayleigh fading only; for example, if adaptive power control is applied to compensate slow power variations: steady-state operation is found at a backlog of nearly $E[m] \approx 100$ terminals, leading to an average delay of $D_a \approx 7000$ slots. For Rayleigh fading combined with shadowing or near-far effects, a single equilibrium occurs, with low average delay on the order of 10 to 15 time slots. Even if the near-far effect is ignored (i.e., all terminals are located at an identical distance from the central receivers), as in curve (c), the network is stable and terminals experience little backlog. Therefore, it may be concluded that for the reported traffic

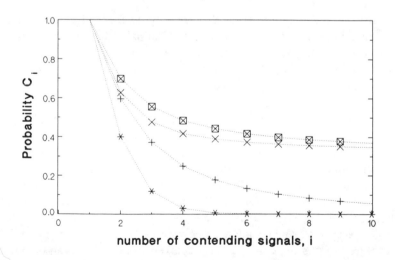

Figure 8.13 Probability that one out of i packets successfully captures the base station, for ($-$) Rayleigh fading, ($+$) Rayleigh fading and shadowing ($\sigma_s = 1.36$ or 6 dB), ($\times$) near-far effect and Rayleigh fading, and ($\boxtimes$) near-far effect, Rayleigh fading, and shadowing. Receiver threshold $z = 4$ (6 dB). Path loss law $\beta = 4$, uniform spatial distribution of offered traffic.

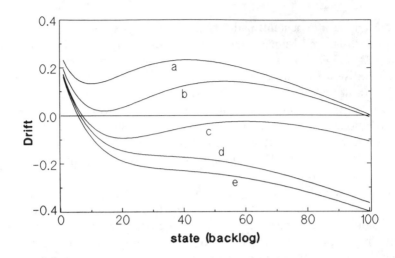

Figure 8.14 (a) Drift for a network of $N = 100$ users for a receiver without capture, and (b) for a receiver with threshold of 6 dB ($z = 4$) for a channel Rayleigh fading only, (c) shadowing ($\sigma_s = 1.36$ or 6 dB) and Rayleigh fading, (d) near-far effect and Rayleigh fading, (e) near-far effect, shadowing ($\sigma_s = 1.36$), and Rayleigh fading.

parameters, the stability of the network does not rely on terminals unrealistically close to the base station.

The drift at high backlog ($m \uparrow N$) is mainly determined by the probability of capture realized in the event of many colliding terminals (C_∞). With Rayleigh fading, upfades are rare, so the probability that one out of the i Rayleigh-fading signals is sufficiently stronger than the joint interference is low, as compared to other propagation mechanisms. Consequently, the drift from full saturation d_{100} is almost zero in curve (b). The log-normal distribution, due to shadowing, exhibits a higher probability of upfades of the signal power, resulting in a negative drift of $d_{100} \approx -0.1$ away from the situation with all terminals in backlog. The uniform spatial distribution, with nonzero traffic in the immediate vicinity of the receiver, has a fast recovery from states with high backlog ($d_{100} \approx -0.4$). In this event, $-d_N$ and S_N roughly equal the capture probability C_∞, since C_i is almost constant for large i.

Figure 8.15 gives the region for P_o and P_r where the network with $N = 100$ terminals is bistable. A conservative receiver threshold of 10 dB ($z = 10$) is considered, so these results may be somewhat pessimistic. The bistability area for ALOHA without capture ($C_i \equiv 0$ for $i = 1, 2, \ldots$) is taken from Onozato and Noguchi [24]. Areas mapping the stability of a slotted ALOHA network with capture were derived in [26] for the case that the population of terminals is divided into a group transmitting at high power and a group transmitting at low power. Van der Plas and Linnartz [12] estimated the area of bistability by trial and error with the technique

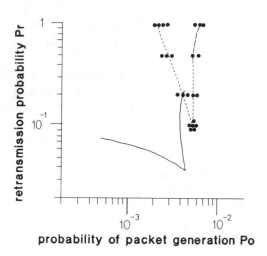

Figure 8.15 Bistability area for a network with $N = 100$ users without capture [19] and $(--)$ with imperfect capture ($z = 10$) in mobile channels. The latter area has been estimated by trial in 23 points (●).

used to obtain drift curves as in Figure 8.14. A uniform spatial distribution, shadowing ($s_s = 6$ dB), and Rayleigh fading are considered. When receiver capture occurs, the mobile network exhibits bistability at substantially higher packet traffic loads, even for a pessimistic receiver threshold of 10 dB ($z = 10$).

It appears that the network is always stable, irrespective of the retransmission probability P_r, if $P_o < P_{o,max}$, with roughly $P_{o,max} \approx 2 \cdot 10^{-3}$. This agrees with the observation by Ghez, Verdú, and Schwartz [22], that for

$$N_{p_0} < \lim_{n \to \infty} C_{n+1} \tag{8.93}$$

the channel is stable if the probability of capture C_{n+1} is independent from slot to slot. However, this may be a strong assumption if the retransmission waiting time is small ($P_r \uparrow 1$). If packets are retransmitted from the same location, they are received with the same power as the initial transmission attempt. If the same set of data packets collides again with the same powers for all signals involved, interference is likely to cause packet loss during all successive collisions.

It has been shown in Section 8.4 that, for noise-free channels with uniform offered traffic without shadowing and $\beta = 4$,

$$\lim_{G_r \to \infty} S_t = \lim_{n \to \infty} C_{n+1} = \frac{2}{\pi \sqrt{z}} \tag{8.94}$$

whereas for channels with shadowing, C_∞ is believed to be at least as large as $2/\pi\sqrt{z}$ (see Figure 8.13). For $z = 10$ and $N = 100$, this indicates that the net is stable as long as $P_o < 2/(100\pi\sqrt{10}) \approx 0.002$. The region of bistability in Figure 8.15 is relatively small: the transition through this area from a single (low-backlog) equilibrium into saturation occurs for relatively small increments of the packet generation probability P_o. In contrast to this, since the bifurcation set of the mobile net is almost parallel to the P_r-axis, a change in the retransmission probability will have less effect on the stability of the net. For a packet generation probability up to $P_o = 0.002$, even a persistent retransmission schedule ($P_r = 1$) leads to a stable network. Decreasing the generation probability P_o always has a positive effect on the network performance, since the drift towards lower backlog increases for all n.

Simulations [25] indicated that reducing the probability of retransmission P_r has a *positive* effect on the network performance in saturated networks (increasing throughput, decreasing backlog, and delay), but a *negative* effect in stable, unsaturated networks (lower throughput, increasing backlog, and delay). In bistable nets, appropriate reduction of P_r can remove bistability, but this measure may not sufficiently relieve the backlog and packet delay: Figure 8.15 suggests that a relatively drastic reduction of P_r may be required. Reducing P_r in a bistable network can result in a stable network, but with relatively high backlog [25].

By reducing the probability of retransmission, the effective time each terminal spends in the origination mode is also reduced, which indirectly resulted in a low input traffic load. This would suggest that mobile channels might as well be managed by directly controlling the input traffic, for instance, by limiting the number of terminals N that are allowed to be signed on simultaneously. Simulations gave the impression that the area where the network is bistable is decreased further if z becomes smaller.

8.6.3 Influence of Near-Far Effect on Stability

Packet transmissions considered so far were assumed to be uniformly distributed over the coverage area. In this event, the probability of correct reception of one out of $n + 1$ colliding packets, C_{n+1}, has some nonzero limit for $n \to \infty$ (see Section 8.4.3). As an alternative, the log-normal spatial distribution (8.72) is now considered. This allows more realistic modeling of the packet traffic very close to the central receiver. The true uniform distribution is compared with a log-normal distribution with $\sigma_d = 2$ ($s_d = 8.68$ dB). The state probability π_m and average drift d_m are depicted in Figure 8.16 for a receiver threshold of $z = 10$ (10 dB). The traffic parameters ($P_o = 0.0055$, $P_r = 0.08$) and the pessimistic receiver threshold ($z = 10$) were chosen such that

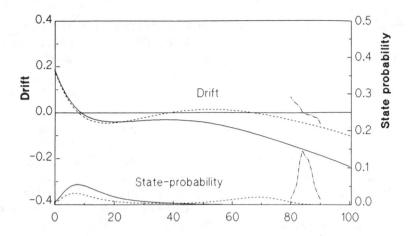

Figure 8.16 Expected drift and state probability for a slotted ALOHA network with $N = 100$ users and a receiver threshold of $z = 10$, (——) according to a uniform distribution, ($\cdots$) a log-normal spatial distribution with $\sigma_d = 2$ ($s_d \approx 8.68$ dB), and (–·–) a simulation. Probability of a new packet generation $P_o = 0.0055$. Probability of retransmission $P_r = 0.08$.

the two spatial models studied give different assessments of the stability of the network with $N = 100$ terminals. While the assumption of a uniform (continuous) distribution of the offered traffic suggests a stable network with little backlog, the log-normal model predicts bistable performance.

Van der Plas [25] took into account the effect that most retransmissions originate from the boundary of the service area. Initially, the software simulation program randomly distributes N ($N = 100$) terminals over the area $0 < r < 1$ and estimates the corresponding local-mean powers for every terminal according to $\bar{p}_j = r_j^{-4}$. Shadowing is ignored in this experiment ($\sigma_s = 0$). All terminal states are modeled as a tristate variable. In the O-state, the program performs a random experiment to simulate generation of a packet with probability P_o. In the R-state, permission for retransmission of a previously collided packet is granted with probability P_r. In both cases, the terminal enters the T-state, where the transmission and multipath propagation of a packet is simulated by the random generation of an instantaneous received power p_j according to an exponential distribution with local-mean $r_j^{-\beta}$. By accounting for all the received interference packets, the program determines whether any of the colliding packets is strong enough to capture the receiver. Figure 8.16 indicates that both the uniform and the log-normal distributions lead to optimistic estimates of the network performance.

Because of the near-far effect, the average time a terminal is in the retransmission mode increases with distance. Since $P_o < P_r$, the traffic offered per unit of area increases with distance: This has a disadvantageous influence on the network performance: colliding signals are more often received from far away and hence more with nearly equal (low) power levels. Consequently, the probability C_i that the power of one of these signals sufficiently exceeds the joint interference is relatively low.

8.7 CONCLUSION

The capture-ratio model and Laplace transforms of received signals have proved a most fruitful method to assess capture probabilities and throughput of mobile slotted ALOHA nets. Results confirmed observations previously reported in the literature that the throughput of the ALOHA channel is substantially enhanced by receiver capture. The throughput of a number of spatial distributions has been compared, including a ring model, quasi- and exactly uniform distributions in one and two dimensions, and a log-normal spatial distribution. The way terminals are distributed over the coverage area has a substantial influence on the throughput of the net. Also, the effect of combined Rayleigh fading, shadowing, and near-far effect has been studied. A general impression from these results is that the larger the differences in received power, the larger the probability that one packet is received correctly in a collision of many competing packets. This enhances the throughput and the stability of the network.

Uniform distribution of offered traffic, although frequently adopted in the literature, gives an optimistic assessment of the network behavior under high offered traffic loads. This popular model fails in two respects:

1. It allows terminals to be infinitely close to the base station. For high traffic loads, stability may therefore rely on terminals unrealistically close to the base station.
2. Another effect is the interaction between the attempted traffic and the system performance: almost all retransmissions are clustered in distant parts of the coverage area with poor propagation. Since these transmissions may arrive with almost equal low power, mutual collisions usually result in the loss of all packets involved.

Stability and delay performance have been investigated from a Markov chain model. The calculations reveal that bistability of the considered mobile networks occurs only for a very limited range of transmission parameters. The transition from low-backlog to saturation (i.e., the transition through the area of bistability) is very abrupt. This suggests that, in contrast to the case for wired networks, radio protocol design should focus on continuously ensuring sufficient throughput and acceptable delay, for instance by exercising slow centralized control over the number of terminals allowed to sign on. In mobile ALOHA networks, the issues of dynamically repairing instability and collision resolution are presumably less critical compared to networks without capture. However, the near-far effect may not always be modeled by a continuous uniform spatial density of the traffic.

Analysis of various alternative distributions revealed that throughput and delay are less enhanced by capture than suggested by the theoretical investigations of quasi-uniform spatial distributions. Similarly, the estimation of stability areas made in Section 8.6 may be further refined by investigating the drift at high backlog from a nonuniform and dynamically changing spatial distribution of the backlogged terminals. To this end, the spatial distribution of offered traffic would have to be assessed as a function of the number of terminals in backlog.

An effect similar to the one in item 2 is believed to occur in networks with stationary terminals, such as in narrowband indoor wireless office communication. The main cause of fading, transmitter mobility, is not present for terminals at a fixed location. Transmitters accidentally placed near a multipath null generate the main part of all retransmissions. Simulation revealed that area-mean powers remaining fixed for each terminal due to virtually constant propagation distances degrades the performance for mobile networks. Presumably, constant shadow and multipath attenuation during retransmissions in fixed networks further degrades overall performance.

REFERENCES

[1] Abramowitz, M., and I.A. Stegun, eds., *Handbook of Mathematical Functions*, New York: Dover, 1965.

[2] Doetsch, G., *Introduction to the Theory and the Application of the Laplace-Transform: A Textbook for Students in Mathematics, Physics and Engineering* (in German), 2nd ed., Vol. 24 of Textbooks and Monographs from the Area of Exact Sciences: Mathematical Series, Birkhäuser, Basel, 1970.

[3] Gradshteyn, I.S., and I.M. Ryzhik, *Table of Integrals, Series and Products*, 4th ed., New York: Academic Press, 1965.

[4] Linnartz, J.P.M.G., "Spatial Distribution of Traffic in a Cellular Mobile Data Network," EUT Report 87-E-168, Eindhoven University of Technology, Feb. 1987, ISBN 90-6144-168-4.

[5] Prasad, R., and J.C. Arnbak, "Effects of Rayleigh Fading on Packet Radio Channels With Shadowing," *Proceedings IEEE Tencon 1889*, Bombay, India, Nov. 1989, pp. 27.4.1–27.4.3.

[6] Bender, C.M., and S.A. Orszag, Chapter 6.3 in *Advanced Mathematical Methods for Scientists and Engineers*, New-York: MacGraw-Hill, 1978.

[7] Arnbak, J.C., and W. van Blitterswijk, "Capacity of Slotted-ALOHA in a Rayleigh Fading Channel," *IEEE J. Sel. Areas Comm.*, Vol. SAC-5, No. 2, Feb. 1987, pp. 261–269.

[8] Linnartz, J.P.M.G., R. Prasad, and J.C. Arnbak, "Spatial Distribution of Traffic in a Cellular ALOHA Network," *Electronics and Communication* (Archiv für Elektronik und Übertragungstechnik), Vol. 42, No. 1, Jan./Feb. 1988, pp. 61–63.

[9] Tanenbaum, A.S., *Computer Networks*, 2nd ed., London: Prentice-Hall International Editions, 1989.

[10] Abramson, N., "The Throughput of Packet Broadcasting Channels," *IEEE Trans. on Comm.*, Vol. COM-25, No. 1, Jan. 1977, pp. 117–128.

[11] Verhulst, D., M. Mouly, and J. Szpirglas, "Slow Frequency Hopping Multiple Access for Digital Cellular Radiotelephone," *IEEE J. Sel. Areas Comm.*, Vol. SAC-2, 1984, pp.563–574.

[12] Van der Plas, C., and J.P.M.G. Linnartz, "Stability of Mobile Slotted ALOHA Network With Rayleigh Fading, Shadowing and Near-Far Effect," *IEEE Transactions on Vehicular Technology*, Vol. 39, No. 4, November 1990, pp. 359–366.

[13] Harrison, E., *Darkness at Night*, Harvard University Press, 1987.

[14] Lau, C.T., and C. Leung, "Capture Models for Mobile Packet Radio Networks," *in Proc. International Conference on Communications ICC, Atlanta, April 1990, pp. 1226–1230. Extended version: IEEE Trans. on Comm.*, Vol. 40, No. 5, May 1992, pp. 917–925.

[15] Linnartz, J.P.M.G., and R. Prasad, "Near-Far Effect on Slotted ALOHA Channels With Shadowing and Capture," *Proc. IEEE Veh. Tech. Conf. 1989*, San Francisco, 3——5 May 1989, pp. 809–813.

[16] Egli, J.J., "Radio Propagation Above 40 MC/s Over Irregular Terrain," *Proc. IRE*, Vol. 45, No. 10, Oct. 1957, pp. 1383–1391.

[17] Prasad, R., and J.C. Arnbak, "Enhanced Throughput in Packet Radio Channels With Shadowing," *Electron. Lett.*, Vol. 24, No. 16, 4 Aug. 1988, pp. 986–988.

[18] Spiegel, M.R., *Mathematical Handbook of Formulas and Tables*, Schaum's Outline Series, McGraw-Hill, New York, 1968.

[19] Carleial, A.B., and M.E. Hellman, "Bistable Behavior of ALOHA-Type Systems," *IEEE Trans. on Comm.*, Vol. COM-23, No. 4, Apr. 1975, pp. 401–410.

[20] Kleinrock, L., and S.S. Lam, "Packet Switching in a Multiaccess Broadcast Channel-Performance Evaluation," *IEEE Trans. on Comm.*, Vol. COM-23, No. 4, Apr. 1975, pp. 410–423.

[21] Namislo, C., "Analysis of Mobile Radio Slotted ALOHA Networks," *IEEE J. Sel. Areas Comm.*, Vol. SAC-2, No. 4, July 1984, pp. 583–588.

[22] Ghez, S., S. Verdú, and S.C. Schwartz, "Stability Properties of Slotted ALOHA With Multipacket Reception Capability," *IEEE Trans. on Automatic Control*, Vol. 33, No. 7, July 1988, pp. 640–649.

[23] Little, J.D.C., "A Proof of the Queuing Formala $L = \lambda W$," *Operations Research*, Vol. 9, 1961, pp. 383–387.

[24] Onozato, Y., and S. Noguchi, "On the Thrashing Cusp in Slotted Aloha Systems," *IEEE Trans. on Comm.*, Vol. COM-33, No. 11, Nov. 1985, pp. 1171–1182.

[25] Van der Plas, C., "Stability of Mobile Slotted ALOHA Networks," Delft University of Technology, The Netherlands, M.Sc. E.E. Thesis, No. 1–68250–289, September 1989.

[26] Onozato, Y., J. Liu, and S. Noguchi, "Stability of a Slotted ALOHA System With Capture Effect," *IEEE Trans. on Veh. Tech.*, Vol. VT-38, No. 1, Feb. 1989, pp. 31–36.

[27] Wozencraft, J.M., and I.M. Jacobs, *Principles of Communication Engineering*, New York: John Wiley and Sons, 1965.

Chapter 9
Frequency Reuse in Wide-Area
Packet-Switched Networks

In present mobile voice networks, high spectrum efficiency has been achieved by extensive frequency reuse. In the late 1970s, cellular engineering was developed to accommodate the rapid growth of mobile telephony. Since in numerous mobile communication systems, mostly stereotyped and short messages are sent, data packet transmission can provide more efficient use of the available bandwidth. The possibility of frequency reuse in wide-area random-access networks is studied in this chapter by extending the results in Chapter 8 to include the effect of intercell interference. Study of wide-area data networks with cellular frequency reuse requires distinction between the inbound channel, which is randomly accessed by a large population of transmitters, and the outbound channel, where the base station multiplexes data packets for various receiving mobile terminals [1]. Here we confine ourselves to the inbound contention-type channel.

In this case, a large number of terminals can operate within one cell and inbound messages are sent to the base station over a common radio channel according to an appropriate random-access scheme. The access to the radio channels has a bursty character, and packets lost due to interference are automatically retransmitted. In these networks, therefore, the need to continuously safeguard the traffic in one cell from excessive interference by cochannel transmissions in other cells is less demanding than in mobile CW telephony. Thus, frequency reuse distances and cell cluster sizes may be smaller. Moreover, a packet-switched cellular radio network might exploit the diversity resulting from packet reception by base stations outside the particular cell in which the terminal is located. Two methods to increase the system capacity are considered in this chapter: dense cellular frequency reuse and site diversity.

Most of the recent studies of the performance of mobile ALOHA networks address the case of a single base station, whereas spatial frequency reuse in adjacent areas and site diversity has received relatively little attention. Most existing wide-area data networks use a frequency reuse pattern similar to the cellular structure for

CW telephony. In other words, the mechanism by which time-frequency resourses are dynamically assigned for transmission of data packets is separated from the method for assigning corresponding coverage areas. One example is the Nordic Mobitex system [2] for public mobile data communication. Mobitex was mainly developed for short data messages. The channel access protocol resembles reservation ALOHA, but a cellular frequency reuse pattern is used with reuse distances determined by the required protection ratio for analog emergency voice transmissions. Later generations of the network supported only packet data communication. Another example is the *Packet Reservation Multiple Access* (PRMA) system proposed by Goodman et al. [3, 4] for mixed traffic. PRMA combines aspects of framed ALOHA and TDMA. It allows a user to reserve a time slot in successive frames when periodically transmitting packets or voice segments, and combines aspects of framed ALOHA and TDMA. Performance analyses mainly focused on the traffic within one particular cell (e.g. [5]), simplifying the spatial aspects of frequency reuse.

Few research papers on frequency reuse in data networks are known to the author. Hafez and Nehme [6] addressed spatial spectrum reuse in a mobile data network using a single inbound frequency. Wilson and Rappaport [7] addressed multichannel CSMA schemes in a cellular communication environment. Further, wide-area frequency reuse has been studied for military multihop packet radio nets. Such networks mostly use distributed frequency management without any distinction between mobile terminals and base stations. In contrast to this, civil (public) networks mostly use fixed base stations and centralized frequency management.

As we will show in this chapter, data and voice communication require different methods for efficient spatial reuse of the available radio spectrum. The possible application of slotted ALOHA in wide-area packet data networks is addressed. Section 9.1 describes wide-area data networks and a number of assumptions made for analysis. The model for the mobile radio channel and for receiver capture is a combination of models used earlier (e.g., in Chapters 3 and 8). Section 9.2 summarizes the relevant expressions and establishes a notation appropriate to multireceiver ALOHA networks. Cellular ALOHA networks are addressed in Section 9.4 on the assumption that each base station only accepts packets from terminals within its cell, although interference may be caused by packets in other cells. Section 9.5 accounts for the effect that most retransmissions of previously unsuccessful packet transmissions occur in areas with poor propagation to the common receiver. The analysis is extended to site diversity in Sections 9.6 to 9.9. Section 9.10 concludes this chapter.

9.1 MODEL FOR CELLULAR DATA NETWORKS

In contrast to cellular CW communication, in a cellular packet data network, each inbound channel supports intermittent traffic from multiple users. For an infinite population of terminals, data packets are transmitted with a spatial distribution $G(x)$, where x), is the position in the cell area. The location x_j of mobile terminal j and the location x_A), of base station A determine the propagation distance a_j. For the packet traffic offered to receiver A, we use the simpler notation $G(a_j)$. The cellular frequency reuse pattern is depicted in Figure 9.1.

In the analysis of hexagonal cellular networks, it is common practice to normalize the length of the sides of each hexagon (and the radius) to unity. Normalization of sides of the hexagon to unit length in Figure 9.1 is in contrast to the normalization adopted in Chapter 8, where the radius of a circular cell was normalized to unity. A hexagonal cell with sides of unit length has surface area $(3\sqrt{3})/2$. This corresponds to the surface area of a circular cell with radius $R = 3^{3/4}/\sqrt{2\pi} \approx 0.91$.

If the offered traffic per unit area is constant throughout the entire network, the traffic offered from a cell to its base station A has the spatial distribution

$$G(a_j) = \begin{cases} G_0 = \dfrac{1}{\pi R^2} G_c & 0 < a_j < R \\[2mm] 0 & \text{elsewhere} \end{cases} \tag{9.1}$$

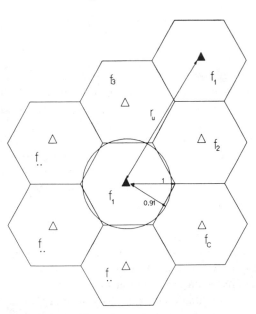

Figure 9.1 Cellular frequency reuse in a mobile ALOHA network. Hexagonal cells are approximated by circular cells of identical surface area.

with $G(a_j)$ expressed in packets per packet time per unit of area, where a_j is the distance to base station A, and G_c is the total traffic offered per cell, expressed in packets per packet time. Besides $G(a_j)$, interfering packet signals from cochannel cells may also be received at the base station.

If C different carrier frequencies are used in each cluster of cells, the frequency reuse distance is $r_u = \sqrt{3C}$, normalized to the radius of a hexagonal cell (see (1.1)). Interfering signals from cochannel cells arrive from distances in a subinterval of $r_u - 1 < a_i < r_u + 1$. The exact distribution depends on the specific layout of the hexagons, particularly on i and j in Section 1.5.1.1. The main body of the distribution is $\langle r_u - R, r_u + R \rangle$, If adjacent cells use different inbound frequencies, intercell (cochannel) interference arrives only from distant terminals. If the frequency reuse distance r_u is relatively large ($r_u \gg 1$, $a_i \gg 1$), all interfering signals from cochannel cells arrive with almost the identical mean power $\bar{p}_{A_i} \approx r_u^{-\beta}$. This approximation will be used here also for relatively small reuse distances.

In the following, we ignore interference from higher-order tiers of cochannel cells, we ignore propagation delays and we assume global slot synchronization. The packet traffic interfering with a test packet has the spatial distribution

$$G(a_i) = G_0[U(a_i) - U(a_i - R)] + 6 \frac{G_c}{2\pi r_u} \delta(a_i - r_u) \qquad (9.2)$$

where $U(x)$ is the unit step function ($U(x) = 1$ if $x > 0$ and zero otherwise).

Random access for mobile terminals by unslotted or slotted ALOHA is considered. If, in practical applications, relatively long messages are to be transmitted or if messages arrive according to some predictable pattern, other access techniques may be more appropriate, but the evaluation of mixed traffic is outside the scope here. For slotted wide-area networks, we assume that time slot synchronization is perfect throughout the entire network and known to all terminals; propagation delays and technical difficulties of implementation are thus ignored. Despite guard times required in practical networks, all data packets are assumed to be of uniform duration, equal to the slot length, which is taken as the normalized unit of time. (The duration of a packet, expressed in seconds, is denoted by T_s.) The feedback channel for acknowledgments of correctly received packets is assumed to be perfect. This implies that transmissions by the base stations are coordinated to guarantee that messages to a mobile terminal experience no harmful interference. The network protocol requires the presence of a suitable control network interconnecting all base stations. Here, we do not address the design of this wired network and of the supporting protocols.

Global slot synchronization might prove feasible in networks of limited size employing a relatively low bit rate; for instance, $r_b = 1200$ b/s [2, 8]. If the bit rate is higher, as is increasingly the case in mobile data nets, propagation delays will become the same order of magnitude as the duration of a data packet. For instance, a call request by a mobile station in the GSM system ($r_b \approx 270$ kb/s) is made by means of slotted ALOHA. During the telephone call, in TDMA operation, timing advance and retard is possible for each mobile to ensure slotted arrival of blocks of bits. In contrast to this, isolated data packets are offered to the channel in random-access networks, so providing feedback information on variable propagation delays is not feasible.

If the bit rate in a random-access channel is increased, packets' duration becomes shorter. The effective throughput can ultimately be severely impaired by the necessary guard times to ensure that, despite random propagation delays, received packets fit within prescribed time slots. In Section 9.3, it will be verified that the throughput advantages of slotted ALOHA over pure, unslotted ALOHA vanish if required guard times exceed the duration of a packet. That section also discusses that the results derived in this chapter for slotted ALOHA can be adapted to pessimistically approximate the throughput for pure ALOHA by considering interfering traffic with doubled intensity.

Time slotting may, nonetheless, be required to allow implementation of collision resolution schemes to ensure the stability of the network. If propagation delays become very large, a practical solution may be to split the inbound channel into a number of parallel channels, each with smaller bit rates and correspondingly longer packets.

9.2 CHANNEL AND CAPTURE MODEL

UHF path loss and Rayleigh fading are considered, whereas shadowing is ignored for ease of analysis. The received packet power from two successive transmissions of the terminal is assumed to be entirely uncorrelated because of Rayleigh fading. Note that this assumes that the terminal is moving. We will distinguish between the case in which, during a retransmission waiting time, the mobile terminal moves sufficiently far to assume uncorrelated distances a_j during each retransmission ($T_9 \ll T_8$), and, alternatively, the case in which, for each packet, the distance a_j remains constant until a retransmission attempt is eventually successful ($T_8 \ll T_9$).

The threshold model for receiver capture (see Chapters 7 and 8) will be used. Thus, a packet is assumed to capture the receiver in a base station if and only if its instantaneous power exceeds the instantaneous joint interference power by a certain margin (factor) z, called the receiver threshold. During the reception of a packet, the received signal power levels (thus, also, the C/I ratio) are assumed to be constant. We ignore propagation delays and guard times. Thus, we assume $T_4 \ll T_5 \ll T_6$. Although the threshold model simplifies the detection process in the receiver, we found in Chapter 7 that the capture-ratio model may be considered appropriate if the probability that the C/I ratio is in the transition range ($z_1 < C/I < z_2$) is relatively low; z_1 is the C/I ratio at which the probability of successful reception becomes significant, and z_2 the C/I ratio at which the probability of excessive bit errors in the packet becomes negligible. The accuracy of the capture-ratio model depends highly on the probability that during a certain time slot the C/I ratio is in the transition range, which is found to be the throughput for z_2 minus the throughput for z_1. Typically, this transition was found for C/I ratios ranging from 0 to 6 dB ($z_1 \approx 1$; $z_2 \approx 4$). Numerical results are presented for $z = 4$. We verified in [9] that, for the situation addressed in this chapter, the variations in received signal powers may be sufficiently high to ensure that the critical probability $\Pr(z_1 < C/I < z_2)$ is relatively small. However, the effect of the choice of z is more significant than was seen in Section 7.9 for $Q(r)$. The accuracy of the model decreases somewhat if the offered traffic G_c becomes large ($G_c >> 1$).

The notation adopted here is a combination of the notation introduced in Chapters 3 and 8. Results relevant to the analysis here are briefly summarized in this section. In a random-access net, the distinction between wanted and interfering signals vanishes. We consider one particular signal, called the test signal, which is denoted by index j. The event in which the C/I ratio of a packet signal from terminal j at receiver A is above the receiver threshold z is denoted by A_j. The probability of capture $\Pr(A_j|\bar{p}_{A_j})$ can be expressed in the form of (8.42). Three cases of contaminating signals will be considered: (1) Interference caused by terminals occasionally transmitting a packet in the same cell, terminals occasionally transmitting a packet in a cochannel cell, and the receiver noise floor. This probability is assumed stationary for each terminal. (3) The receiver noise floor, caused by additive noise with power N_A, is described by the pdf $\delta(p_n - N_A)$, with Laplace image $\exp\{-sN_A\}$. We denote the probability that the signal from terminal j sufficiently exceeds the noise floor at receiver A with P_{NA}.

9.2.1 Summary of Weight-Function Technique

For a finite population of N terminals, transmitting with probability $\Pr(k_{ON})$ with $k = 1, 2, \ldots, N$, the probability $\Pr(A_j|\{a_j\})$ that a test packet transmitted by terminal j captures the receiver in base station A, given all distances a_i ($i = 1, 2, \ldots, N$) and a_j, was shown to be (Equation (8.43))

$$\Pr(A_j|\{a_i\}_{i=1}^{N}) = P_{NA} \prod_{i=1, i \neq j}^{N} [1 - W(a_j, a_i) \Pr(i_{ON})] \qquad (9.3)$$

with the vulnerability weight function $W(a_j, a_i)$ ($0 \leq W(\cdot, \cdot) \leq 1$) defined as $W(a_j, a_i) \triangleq za_j^\beta (za_j^\beta + a_i^\beta)^{-1}$.

For an infinite population, the probability of successfully receiving signal j from a known distance a_j in the presence of a Poisson-distributed number of n interfering signals ($j \notin \{1, 2, \ldots, n\}$) with unknown positions was expressed in (8.48) as

$$\Pr(A_j|a_j) = P_{NA} \exp\left\{ -\int_0^\infty 2\pi a_i \, W(a_j, a_i) \, G(a_i) da_i \right\} \qquad (9.4)$$

The throughput per unit of area $S(a_j)$, defined as the average number of successful packets per unit area per time slot, is obtained from $S(a_j) = G(a_j)\Pr(A_j|a_j)$. The total throughput at receiver A, denoted by S_A and expressed in successful packets per packet time, is obtained by polar integration of $S(a_j)$.

9.3 GUARD TIMES AND UNSLOTTED ALOHA

This section refines the discussion on propagation delays and guard times that was initiated in Section 9.1. We assume a guard time T_g required to ensure synchronized arrival of packets. For a network with cell radius of D meters, $T_g = 2D/c$, while for slot synchronization of interfering cells, a larger T_g is required. The duration of each slot has to be at least $T_s + T_g$, with T_s and T_g expressed in seconds. We consider the situation that the offered packet traffic λ_c per *second* is independent of the guard time. Consequently, in a system where guard times are required, the offered traffic G_{tg} expressed in packets per *time slot*, is larger than the offered traffic per slot G_t in a theoretical network without guard times; that is,

$$\lambda_c = \frac{G_{tg}}{T_s + T_g} = \frac{G_t}{T_s} \qquad (9.5)$$

If required guard times become excessively large, say if $2D > T_s c$, unslotted ALOHA might be considered. Exact analysis of unslotted ALOHA is complicated by the fact that the number of interfering signals changes during the reception of a packet. We now propose a model that somewhat overestimates the effect of interference in pure ALOHA. A test packet is assumed to capture the receiver if the received power exceeds z times the total power accumulated from *all* signals present during at least a part of the duration of reception of the test packet. For an infinite population of terminals, the number of packets overlapping with the test packet is Poisson-distributed with mean $2G_t$, so the expression presented in the next sections may be used if the contending or interfering traffic per cell G_c is replaced by $2G_c$, and, similarly, if the traffic per area $G(x)$ is replaced by $2G(x)$. This also corresponds to the case $G_{tg} = 2G_t$ in (9.5). Hence, the probability of capture can be estimated from

$$\Pr(A_j|n,\bar{p}_{A_j}) > P_{UA} \triangleq P_{NA} \prod_{i=1}^{n} \mathscr{L}\left\{f_{p_{A_i}}, \frac{z}{\bar{p}_{A_j}}\right\} = P_{NA} \prod_{i=1}^{n_1} \mathscr{L}\left\{f_{p_{A_i}}, \frac{z}{\bar{p}_{A_j}}\right\} \prod_{i=n_1+1}^{n} \mathscr{L}\left\{f_{p_{A_i}}, \frac{z}{\bar{p}_{A_j}}\right\}$$

$$(9.6)$$

where $1, 2, \ldots, n_1$ denote the interfering packet signals present at the instant of arrival of the test packet j, and $n_1 + 1, n_1 + 2, \ldots, n$ denote packets present at the end of the test packet. One, may conclude that unslotted ALOHA becomes favorable if T_g approaches or exceeds T_s.

9.4 SPECTRUM EFFICIENCY FOR GLOBALLY UNIFORM OFFERED TRAFFIC

If cellular frequency reuse is employed in mobile data networks, the outage probability (as expressed in Chapter 3) may not be the most appropriate performance measure. Spectrum-efficient design requires more appropriate criteria. Data mostly require highly reliable communication without end-to-end outages, but are significantly less vulnerable to delay than voice communication. Data packets lost during an RF signal outage are, if necessary for the application, retransmitted. In a cellular data network, the spectrum efficiency is affected by the choice of the cluster size in two mutually conflicting ways. The bandwidth available per cell or per base station decreases proportionally to the cluster size if the total bandwidth allocated to the system is fixed. On the other hand, small cluster sizes lead to large mutual interference between cells. This in turn leads to poor reception, requiring many retransmissions, so the effective throughput may be reduced if the cluster size is chosen too small. A first step to understanding the spectrum efficiency of a mobile ALOHA network uses a criterion based on the attempted traffic that can be supported with a certain prescribed probability of capture. Analogous to assumptions in Chapters 6, 7, and 8, it is assumed that the offered traffic is known. A uniformly distributed offered traffic is reasonable, for instance, if the packet traffic load is kept relatively low to ensure that few packets are lost in collisions. If vehicle speeds are sufficiently large to ensure that the positions of a terminal become uncorrelated between each retransmission, uniform offered packet traffic also may be a reasonable assumption. Moreover, in certain vehicle location systems or telemetry applications, a routine status report lost in a collision may not need to be retransmitted. For instance, the proposed SOCRATES system, which is intended for road traffic management, gathers information on urban traffic flows that is randomly transmitted by vehicles participating in the road traffic [10]. This leads to uniform offered traffic if the road traffic is uniformly distributed around the location of the base station.

Cellular frequency reuse is considered. A base station is assumed to accept packets only from terminals inside its cell. Packets from co-channel cells are considered interference. The following measure of the spatial bandwidth (or spectrum) efficiency of a cellular data network, expressed in bits per second per hertz per base station, is proposed:

$$SE_1 \triangleq \frac{\lambda_c L}{B_c} = \frac{\lambda_c L}{CB_T} \tag{9.7}$$

where L is the packet size in bits, λ_c is the maximum offered traffic expressed in packets per second ($\lambda_c = G_c/T_s$) that can be supported with a satisfactory grade of service (i.e., with a probability of success larger than Q_{MIN} for a worst-case location of the mobile terminal ($a_j = R$)). B_c denotes the bandwidth occupied by the cellular network with C frequencies, each of bandwidth B_T. For a given type of modulation, η_r can be defined as the bit rate per occupied bandwidth ($\eta_r = r_b/B_T$). So, the spectrum efficiency SE_1 can be expressed as

$$SE_1 \triangleq \frac{G_c}{LT_b} \frac{L}{CB_T} = \pi R^2 G_0 \frac{\eta_r}{C} \tag{9.8}$$

We assume that the parameters B_c, η_r, L, and z are known and that $R \approx 0.91$. The spectrum efficiency, or, equivalently, the arrival rate λ_c that can be supported, is to be optimized. Both the spectrum efficiency and the probability of success are functions of G_0 and C. Thus, efficient spectrum use requires that the total traffic offered within the cell ($G_c = \pi R^2 G_0$) and the tolerated co-channel interference from outside the cell are optimized jointly. The resulting optimum cluster size C_{OPT} then determines the system parameters B_T, r_b, and T_b. This two-dimensional optimization problem is in contrast to the case for optimum spectrum efficiency in cellular telephony [11], where the traffic within each cell consists of only one signal. In the latter case, efficient spectrum efficiency only requires a one-dimensional optimization.

We now use our weight-function technique to express the influence of G_0 and C on the probability of success. The interfering packet traffic has the spatial distribution (9.2). Using (9.4), the probability of capture is found as

$$\Pr(A_j|a_j) = P_{NA} \exp\left\{-\int_0^R 2\pi\lambda W(a_j,\lambda)G_0 d\lambda - 6G_c \frac{za_j^\beta}{za_j^\beta + r_u^\beta}\right\} \tag{9.9}$$

The worst-case probability of successful access occurs at $a_j = R \approx 0.91$, with, for $\beta = 4$,

$$\Pr(A_j \mid a_j = R) = \exp\left\{ -zN_A R^4 - \pi\sqrt{z}R^2 G_0 \arctan\left(\frac{1}{\sqrt{z}R^2}\right) - 6\frac{zR^4 R^2 \pi G_0}{zR^4 + r_u^4} \right\} \quad (9.10)$$

where we used (8.67). The maximum value of G_0 that satisfies $\Pr(A_j \mid a_j = R) > Q_{\text{MIN}}$ is found from

$$G_{0,\text{MAX}} = \frac{-\ln Q_{\text{MIN}} - zN_A R^4}{\pi\sqrt{z}R^2 \arctan\left(\dfrac{1}{\sqrt{z}R^2}\right) + \dfrac{6z\pi R^6}{zR^4 + 9C^2}} \quad (9.11)$$

with $Q_{\text{MIN}} < \exp\{-zN_A R^4\}$. Equation (9.11) is inserted into the definition of spectrum efficiency (9.7):

$$SE_1 = \frac{\eta_r}{C} \frac{-\ln Q_{\text{MIN}} - zN_A R^4}{\sqrt{z}\arctan\left(\dfrac{1}{\sqrt{z}R^2}\right) + 6\dfrac{zR^4}{zR^4 + 9C^2}} \quad (9.12)$$

Thus, SE_1 depends on Q_{MIN}, N_A, C, z, and η_r. Figure 9.2 portrays Equation (9.12)

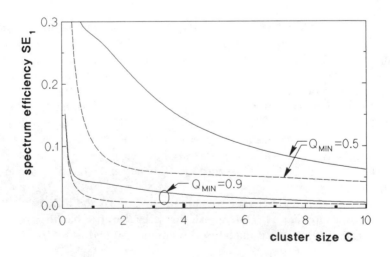

Figure 9.2 Spatial bandwidth efficiency SE_1, in bits/s/Hz per base station, versus cluster size C for a minimum probability of capture $Q_{\text{MIN}} = $ and 0.9 and 0.5. $\eta_r = 1$ bits/s/Hz. Receiver threshold: (——)$z = 4$ (6 dB) and (— —) $z = 100$ (20 dB). Noise-free channel ($N_A = 70$). Path loss law $\beta = 4$.

versus cluster size C for $\eta_r = 1$ b/s/Hz and receiver thresholds of $z = 4$ (6 dB) and $z = 100$ (20 dB). The criterion Q_{MIN} is taken equal to 0.5 and 0.9 for a noise-free exchannel ($N_A = 0$). For a noisy channel, the curves represent the case $Q_{MIN} = 0.5 \exp\{-(0.91)^2 z N_A\}$ and $Q_{MIN} = 0.9 \exp\{-(0.91)^2 z N_A\}$, respectively.

As the numerical examples of (9.12) suggest, optimum spectrum efficiency is thus achieved if all cells use the same inbound frequency. This can be understood intuitively from earlier results for uniform offered traffic within a single cell and with infinite extension. If $C = 1$, the case for infinitely extended uniform offered traffic (8.59) applies: $G_{0,max} = 2\pi^{-2} z^{-1/2} Q_{MIN}$.

Alternatively, if one takes $C > 1$ ($C = 3, 4, 7, 9, \ldots$), the bandwidth per cell is reduced by a factor C. In the most optimistic scenario, the effect of co-channel interference from other cells vanishes. According to (8.68), in this case the offered traffic $G_{0,MAX}$ could be approximately doubled. However, for doubled G_0 and increased C, the arrival rate λ is multiplied by a factor $2/C$, (because $\lambda = G_0/T_s = G_0 r_b/L = G_0 \eta_r B_c/(CL)$). Thus, any value of $C \geq 2$ is likely to lead to lower tolerable offered traffic than in the case with $C = 1$.

Hence, in the ALOHA network studied here, the reduction of interference from cochannel cells does not justify the choice of different inbound frequencies for adjacent cells. In practical networks with mixed traffic, the cluster size may still have to be chosen larger to ensure reliable operation during transfer of longer messages for priority traffic. In this case, the results in Figure 9.2 can be of assistance to estimate the improved spectrum efficiency from separating the channels for inbound (randomly arriving) short packets and reserved transfer of longer messages.

9.5 TRAFFIC TO BE OFFERED TO ACHIEVE UNIFORM THROUGHPUT[1]

In contrast to the previous case of a known distribution of the *offered* packet traffic, in practical random access networks, most retransmissions can be expected to occur in areas with poor propagation to the base stations. We now assume that retransmissions are always transmitted from the same distance as that from which the first transmission was attempted. This corresponds to a prescribed throughput per unit of area. A reasonable assumption is uniform throughput, with spatial distribution

$$S(a_j) = \begin{cases} S_0 = \dfrac{1}{\pi R^2} S_t & 0 < a_j < R \\ \\ 0 & \text{elsewhere} \end{cases} \tag{9.13}$$

[1]Sections 9.5 through 9.9: Portions reprinted, with permission, from IEE Proceedings I, Vol. 139, No. 1, Feb. 1992, pp. 58–70. © 1992, IEE.

We address the effect of interference from outside the coverage area of the base station in more detail. Analogous to (9.8), the spectrum efficiency can be defined as

$$SE_2 \triangleq \frac{S_A \eta_r}{C} \qquad (9.14)$$

where S_A is the maximum uniform throughput in the form of (9.13) that can be achieved. Prescribed throughput and hexagonal frequency reuse are considered by rewriting (9.4) and (9.2) to be

$$G(a_j) = \frac{S_0}{\Pr(A_j | a_j)}$$

$$= S_0 \exp\left\{ + zN_A a_j^\beta + \int_0^\infty 2\pi a_i W(a_j, a_i)(G(a_i) + 6G_I(a_i))da_i \right\} \qquad (9.15)$$

where $G(a_i) = G(a_j)$ denotes the offered traffic within the cell. Interference from a cochannel cell is denoted by $G_I(a_i)$. It is also assumed that the total traffic offered within each cochannel cell is identical to the traffic G_c offered in the cell served by receiver A. A solution for the offered traffic is obtained by means of an iterative computational technique with (9.15): each new estimate of $G(a_j)$ on the left-hand side of (9.15) is computed by inserting the previous estimate of $G(a_i)$ and $G_I(a_i)$ in the integral, with $G_I(a_i) = G_c \delta(a_j - r_u)/(2\pi r_u)$. The traffic per cell G_c is obtained by polar integration of $G(a_j)$ from 0 to R $\approx$ 0.91. Analogous to the second member of the right-hand side of (9.2), $G_I(a_i)$ is then found from $G_c \delta(a_j - r_u)/(2\pi r_u)$. If a single solution of (9.15) exists and numerical (rounding) errors are negligible, the iteration scheme converges to the solution $G(a_i) = G(a_j)$ from a first estimate $G(a_i) = S_0$.

Figure 9.3 gives the traffic $G(a_j)$ to be offered to achieve uniform throughput of $S_A = 0.4$ ppt per cell for various cluster sizes ($C = 4, 9$, and 100), although for small cluster sizes, approximation (9.2) may loose its accuracy because it simplifies the pdf of interference power from cochannel cells. Tables 9.1 and 9.2 give the expected number of required retransmission attempts for terminals at the boundary of the cell and averaged over all positions in the cell ($0 < a_j < R$) to achieve $S_A = 0.4$ and 0.2 ppt, respectively. Also, the effect of a noise floor $N_A = 0.1$, which corresponds to a worst-case local-mean C/N ratio of 11.6 dB for a terminal at $a_j = R \approx 0.91$, is considered.

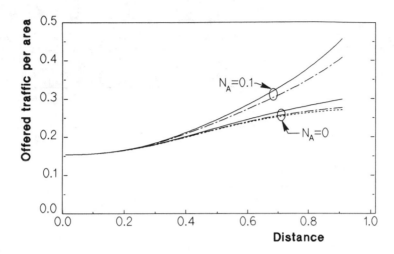

Figure 9.3 Traffic per unit of area $G(a_j)$ to be offered to achieve throughput $S_t = 0.4$ ppt with uniform distribution. Receiver threshold $z = 4$ (6 dB). Cluster size $C = 4$ (——), 9 (—·—) and 100 (– –). Effect of noise: $N_A = 0$ and 0.1. Path loss law $\beta = 4$.

Table 9.1

Expected Number of Required Transmission Attempts in a Cellular ALOHA Data Network for Various Cluster Sizes C and Noise Levels N_A. Throughput per cell is $S_t = 0.4$ ppt. Receiver threshold $z = 4$ (6 dB).

C	$r_u = \sqrt{3C}$	N_A	G_t	Number of Attempts Average	Number of Attempts Fringe of Cell
1	1.73	0	No solution found		
1	1.73	0.1	No solution found		
4	3.46	0	0.64	1.61	1.94
4	3.46	0.1	0.80	2.00	2.97
9	5.20	0	0.62	1.56	1.79
9	5.20	0.1	0.75	1.88	2.68
100	17.3	0	0.62	1.54	1.76

Table 9.2

Expected Number of Required Transmission Attempts in a Cellular ALOHA Data Network for Various Cluster Sizes C and Noise Levels N_A. Throughput per Cell is $S_t = 0.2$ ppt. Receiver Threshold $z = 4$ (6 dB).

C	$r_u = \sqrt{3C}$	N_A	G_t	Number of Attempts Average	Number of Attempts Fringe of Cell
1	1.73	0	0.29	1.45	1.96
1	1.73	0.1	0.35	1.76	2.96
4	3.46	0	0.24	1.21	1.28
4	3.46	0.1	0.27	1.36	1.74
9	5.20	0	0.24	1.19	1.25
9	5.20	0.1	0.27	1.35	1.69
100	17.3	0	0.24	1.19	1.25
100	17.3	0.1	0.27	1.34	1.68

UHF groundwave propagation with $\beta = 4$ is considered. For a receiver threshold of 6 dB, a fade margin of 5.6 dB remains at $a_j = R \approx 0.91$. Interestingly, the effect of intercellular interference is relatively small, even for $C = 1$ or 4, compared to the effect of noise. The observation that noise may substantially diminish the performance in mobile ALOHA channels is in contrast to observations in [12], where a prescribed offered traffic was assumed. It can be seen in Figure 9.3 that packets from the remote terminals are much more likely to be lost in noise, which causes a significant increase in the number of retransmission attempts at larger ranges. It was seen in [11] that poor reception at the fringe of the cell may also threaten the stability of the network. A higher fade margin appears desirable.

For a cluster size of $C = 1$ and a receiver threshold of $z = 4$, the maximum uniform throughput of slotted ALOHA is $S_A \approx 0.25$ ppt for $N_A = 0.1$ and $S_A \approx 0.3$ ... 0.35 ppt for a noise-free channel ($N_A = 0$). It turned out from our computations that the traffic to be offered for the cases $C = 4$ with $S_A = 0.4$ ppt and $N_A = 0.1$ is almost identical with the case $C = 1$ with $S_A = 0.2$ ppt and $N_A = 0.1$. In both cases, the expected number of required transmission attempts is about three for terminals near the fringe of the cell. A slightly higher required throughput caused divergence of the iteration scheme (9.15). The spectrum efficiency is thus about $SE_2 \approx 0.2\eta_r$ for $C = 1$ and $SE_2 \approx 0.1\eta_r$ for $C = 4$. For larger cluster sizes, the spectrum efficiency SE_2 decreases further. It can also be concluded that, if spectrum utilization is optimized, distant (terminals may have to perform many retransmissions. SE_2 can be seen to be of the same order of magnitude as SE_1 with $Q_{\mathrm{MIN}} = 0.5$.

9.6 MODEL FOR TWO-BRANCH SITE DIVERSITY

It has been illustrated that relatively short frequency reuse distances can be tolerated without substantial sacrifices to the throughput of each cell. In the extreme case $C = 1$, adjacent cells use the same frequency and slot synchronization, so packets transmitted near the boundary of the two cells can be received successfully at more than one base station. This offers the perspective of site diversity. Although some results are available for antenna (or micro) diversity at a single base station (e.g., [13]), the effect of applying multiple receiver sites has received little attention in technical literature. Chang [1] studied the throughput of a random-access network with multiple receiver sites by extensive evaluation of expected events of collisions. Here, the vulnerability-weighing technique is extended to produce analytical expressions for the throughput of slotted ALOHA with site diversity. In the next two subsections (9.6.1 and 9.6.2), results for site diversity in Chapter 3 are simplified by ignoring the effect of shadowing ($\sigma_s = 0$). This allows study in Section 9.7 of the spatial distribution of an infinite population of terminals.

9.6.1 Activity of Interfering Terminals

The probability $\Pr(k_{ON}|A_j)$ that terminal k has transmitted a packet, given the a posteriori knowledge A_j that a packet from terminal j captures base station A, was derived in Chapter 3 from Bayes' rule:

$$\Pr(k_{ON}|A_j,\{a_k\}) = \frac{1 - W(a_j,a_k)}{1 - W(a_j,a_k)\Pr(k_{ON})} \Pr(k_{ON}) \qquad (9.16)$$

where, in contrast to (3.35), shadowing has been ignored ($\sigma_s = 0$).

Figure 9.4 agrees with the intuitive understanding that the probability of an active interferer in the range $0 < a_k < {}^\beta\sqrt{z}\, a_j$ must be very small. In cellular telephone networks, it is generally *a priori* unlikely that this range contains an interferer, because reuse distances are usually relatively large. It was seen in Chapter 3 that considering a posteriori probabilities of active interferers is relevant only if substantial shadowing is present and relatively small reuse distances are employed. In random-access networks, however, (9.16) singles out a significant area for interfering contenders, which may be relevant even if no shadowing occurs $\sigma_s = 0$.

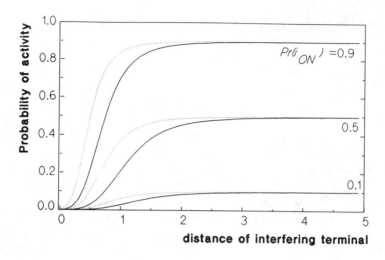

Figure 9.4 A posteriori probability $P(k_{ON}|A_j)$ that an interfering terminal k at distance a_k transmitted a packet, given that a packet from terminal j at distance $a_j = 1$ captures receiver A in the same slot. The a priori probability of transmission is $\Pr(k_{ON}) = 0.1$, 0.5, and 0.9. Receiver threshold (--) $z = 1$ and (——) $z = 4$.

9.6.2 Capture Probability for Finite Population of Terminals

The positions of the N terminals are denoted by $\underline{x}_i$. Since terminal i is at normalized distances a_i to base station A and at distance b_i to base station B, $|\underline{x}_i - \underline{x}_A| = a_i$ and $|\underline{x}_i - \underline{x}_B] = b_i$ (see Figure 3.5). Similar to (9.3), the conditional probability that the packet from terminal j captures base station B given that it also captures base station A is found to be

$$\Pr(B_j|A_j,\{\underline{x}_i\}_{i=0}^N) = P_{NB} \prod_{k=1,k\neq j}^{N} 1 - W(b_j,b_k)\,\Pr(k_{ON}|A_j,a_j,a_k)$$

$$= P_{NB} \prod_{k=1,k\neq j}^{N} 1 - \frac{1 - W(a_j,a_k)}{1 - W(a_j,a_k)\Pr(k_{ON})}\,W(b_j,b_k)\Pr(k_{ON}) \quad (9.17)$$

where P_{NB} is the probability that the received signal fails to exceed the noise floor at receiver B. The probability that a packet from terminal j captures at least one of the two base stations (A or B), given the position of each terminal i ($i = 1, 2, \ldots,$ N), equals

$$\Pr(A_j \vee B_j | \{x_i\}_{i=1}^N) = \Pr(A_j | \{x_i\}) + \Pr(B_j | \{x_i\}) - \Pr(A_j | \{x_i\})\Pr(B_j | A_j, \{x_i\}) \quad (9.18)$$

Inserting (9.3) and (9.17) into (9.18) gives

$$\Pr(A_j \vee B_j | \{x_i\}_{i=1}^N) = P_{NA} \prod_{i=1, i \neq j}^N 1 - W(a_j, a_i)\Pr(i_{ON})$$
$$+ P_{NB} \prod_{i=1, i \neq j}^N 1 - W(b_j, b_i)\Pr(i_{ON})$$
$$- P_{NA}P_{NB} \prod_{i=1, i \neq j}^N 1 - W_{AB}\Pr(i_{ON}) \quad (9.19)$$

where we introduced the joint weight function

$$W_{AB} \triangleq W(a_j, a_k) + W(b_j, b_k) - W(a_j, a_k)W(b_j, b_k)$$
$$= 1 - \frac{a_k^\beta b_k^\beta}{(za_j^\beta + a_k^\beta)(zb_j^\beta + b_k^\beta)} \quad (9.20)$$

Here, the factor W_{AB} weighs the disturbance caused by an interfering packet signal from position x_k to a reception of a data packet by terminal j at the two base stations A and B simultaneously. For a large receiver threshold ($z \to \infty$) or for an interferer relatively close to either of the base stations ($a_k \ll a_j$ or $b_k \ll b_j$), the weight factor tends toward unity. This represents the fact that the interference signal is very likely to disturb duplicated reception of the test packet from terminal j at both stations. For a remote interferer ($b_k, a_k \to \infty$), W_{AB} tends toward zero: despite interference from k, the test packet is likely to be received correctly at both base stations simultaneously.

9.7 CAPTURE PROBABILITY FOR INFINITE POPULATION OF TERMINALS

For a packet transmitted from a position x_j in the presence of n interfering packets, the probability of capturing at least one of the two base stations is found by rewriting

(9.19) for a population of n interfering terminals known to transmit and averaging over all possible positions of the n interferers. So,

$$\Pr(A_j \vee B_j|\underline{x}_j,n,\{i_{\text{on}}\}) = \iint_{\text{area}} \cdots \iint_{\text{area}}$$
$$\Pr(A_j \vee B_j|\underline{x}_j,\{\underline{x}_i,i_{\text{ON}}\}_{i=1}^n,n)f_{\underline{x}_1}(\underline{x}_1)\cdots f_{\underline{x}_n}(\underline{x}_n)d\underline{x}_1\cdots d\underline{x}_n \quad (9.21)$$

Since $\Pr(i_{\text{ON}}|i_{\text{ON}}) = 1$ for $i = 1, 2, \cdots, n$, (9.19) and (9.21) give

$$\Pr(A_j \vee B_j|n,\underline{x}_j,\{i_{\text{ON}}\}) = \iint_{\text{area}} \cdots \iint_{\text{area}} \left\{ P_{NA}\prod_{i=1}^n [1 - W(a_j,a_i)] + P_{NB}\prod_{i=1}^n \right.$$
$$\left. \cdot [1 - W(b_j,b_i)] - P_{NA}P_{NB}\prod_{i=1}^n [1 - W_{AB}] \right\} f_{\underline{x}_1}(\underline{x}_1)\cdots f_{\underline{x}_n}(\underline{x}_n)d\underline{x}_1\cdots d\underline{x}_n \quad (9.22)$$

Since each of the integration variables $\underline{x}_i$ occurs only in one factor of each of the products, product and integration may be interchanged. Hence,

$$\Pr(A_j \vee B_j|n,\underline{x}_j,\{i_{\text{ON}}\}) = P_{NA}\left[\iint_{\text{area}} \{1 - W(a_j,a_i)\}f_{\underline{x}_i}(\underline{x}_i)d\underline{x}_i\right]^n$$
$$+ P_{NB}\left[\iint_{\text{area}} \{1 - W(b_j,b_i)\}f_{\underline{x}_i}(\underline{x}_i)d\underline{x}_i\right]^n - P_{NA}P_{NB}\left[\iint_{\text{area}} \{1 - W_{AB}\}f_{\underline{x}_i}(\underline{x}_i)d\underline{x}_i\right]^n$$
$$(9.23)$$

For n Poisson-distributed, this becomes a sum of exponential functions, namely,

$$\Pr(A_j \vee B_j|x_j) = \sum_{n=0}^{\infty} \frac{G_t^n}{n!} e^{-G_t} \Pr(A_j \vee B_j|n,\underline{x}_j,\{i_{\text{ON}}\})$$

$$= P_{NA} \exp\left\{ -\iint_{\text{area}} W(a_j,a_i)G(\underline{x}_i)\,d\underline{x}_i \right\}$$

$$+ P_{NB} \exp\left\{ -\iint_{\text{area}} W(b_j,b_i)G(\underline{x}_i)\,d\underline{x}_i \right\}$$

$$- P_{NA}P_{NB} \exp\left\{ -\iint_{\text{area}} W_{AB}G(\underline{x}_i)\,d\underline{x}_i \right\} \quad (9.24)$$

with $G(\underline{x}_i)$ the offered traffic per unit of area at $\underline{x}_i$. Equation (9.24) offers a mathematical expression for the probability of capturing at least one receiver, given an arbitrary spatial distribution of the offered traffic, described by $G(\underline{x}_i)$. The expression contains three terms: the first and second terms are in the form of (9.4) for the individual receivers A and B; the third term corresponds to successful reception at both receivers simultaneously.

9.8 CAPTURE PROBABILITY FOR GLOBALLY UNIFORM OFFERED TRAFFIC

We assume uniform offered packet traffic of $G(x) \equiv G_0$ for all x. The normalized distance between the two base stations is R_d. The first two terms in (9.24) are in the analytic form of (8.59). Using the substitutions $r = a_k$ and $b_k^2 = r^2 + R_d^2 - 2R_d r \cos\phi$, the third term in (9.24) equals, for a noise-free system,

$$\Pr(A_j \wedge B_j | x_j) =$$
$$\exp\left\{ -\int_0^{2\pi} \int_0^\infty \left[1 - \frac{r^4(r^2 + R_d^2 - 2rR_d \cos\phi)^2}{(za_j^4 + r^4)\{zb_j^4 + (r^2 + R_d^2 - 2rR_d \cos\phi)^2\}} \right] G_0 r\, dr\, d\phi \right\} \quad (9.25)$$

which has been evaluated numerically.

Figures 9.5 and 9.6 give numerical results of the probability of capturing at least one of two geographically separated base stations as a function of the two-dimensional position of the transmitting terminal. Base station A is located at $(-1,0)$ and B is located at $(+1,0)$. Figure 9.7 presents the cross section of Figure 9.5 along the axis through both base stations. The exact result of the probability $\Pr(A_j \vee B_j | x_j)$ of capture at one of the two receivers is compared to the approximation $Q_M(x_j)$, calculated by

$$Q_M(x_j) \triangleq \Pr(A_j | x_j) + \Pr(B_j | x_j) - \Pr(A_j | x_j)\Pr(B_j | x_j) \quad (9.26)$$

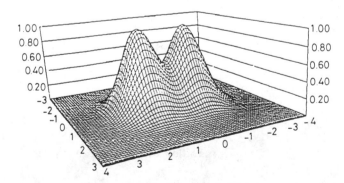

Figure 9.5 Probability of capture $\Pr(A_j \vee B_j | x_j)$ as a function of the position of the terminal. Base stations are located at $(1,0)$ and $(-1,0)$. The receiver threshold is 6 dB ($z = 4$). The offered traffic per unit of area is $G_0 = 0.1$ ppt.

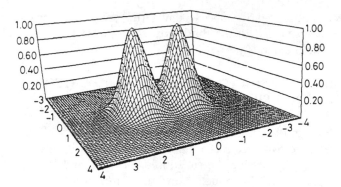

Figure 9.6 Probability of capture $\Pr(A_j \vee B_j|x_j)$ as a function of the position of the terminal. Base stations are located at (1,0) and (−1,0). The receiver threshold is 6 dB ($z = 4$). The offered traffic per unit of area is $G_0 = 0.2$ ppt.

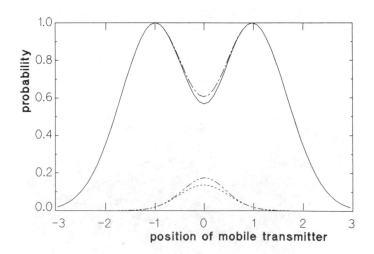

Figure 9.7 Probability of successful reception in at least one receiver $\Pr(A_j \vee B_j|x_j)$ and at both receivers simultaneously $\Pr(A_j \wedge B_j|x_j)$ for a terminal located on the line through the base stations at (1,0) and (1,0). The total offered traffic per unit of area is $G_0 = 0.1$ ppt. The receiver threshold is 6 *dB* ($z = 4$). (——) Reception at one receiver; (—·—) approximation; (−··) reception at both receivers; (····) approximation.

Also, the probability $\Pr(A_j \wedge B_j)$ of successful reception at both receivers simultaneously is compared to $\Pr(A_j) \cdot \Pr(B_j)$. These approximations correspond to the case that the interference level at the two receivers would be uncorrelated at both sites, which underestimates $\Pr(A_j \wedge B_j)$ and overestimates $\Pr(A_j \vee B_j)$. At the traffic load studied ($G_0 = 0.1$ ppt), probability (9.25) of reception at both base stations simultaneously is seen to be underestimated by about 20% in the worst case; that is, for a terminal located halfway between the two receivers ($\underline{x}_j = (0,0)$). The effect on the probability of capture $\Pr(A_j \vee B_j | \underline{x}_j = (0,0))$ is less than 5%.

9.9 TOTAL THROUGHPUT AND APPROXIMATION TECHNIQUES

The total throughput of the two base stations is given by the integral

$$S_{A \vee B} = \int\!\!\int_{\text{area}} G_0 \Pr(A_j \vee B_j | \underline{x}_j) \, dx_j \qquad (9.27)$$

which, taking into account the three terms in (9.24), may also be written as

$$S_{A \vee B} = S_A + S_B - S_{A \wedge B} \qquad (9.28)$$

The common throughput of the two base stations can be computed from

$$S_{A \wedge B} = S_A \int_0^\infty f_{b_j}(b_j | A_j) \Pr(B_j | b_j, A_j) \, db_j \qquad (9.29)$$

This might be approximated by

$$S'_{A \wedge B} \triangleq S_A \int_0^\infty f_{b_j}(b_j | A_j) \Pr(B_j | b_j) \, db_j \qquad (9.30)$$

The approximation $S'_{A \wedge B}$ ignores that, during each slot, the number of relevant interferers and the interference power level at receivers A and B are dependent, and the vulnerabilities of the test packet to these interferers are correlated at the two receivers. This causes B_j and A_j to be mutually correlated, as is considered in (9.29).

The exact total throughput depicted in Figure 9.8 is obtained by four-fold integration, arising from combining (9.25) with (9.27), in order to take due account of the correlated interference experienced at the two base stations. We now address the approximate and much simpler technique based on (9.30) to estimate the common throughput $S'_{A \wedge B}$.

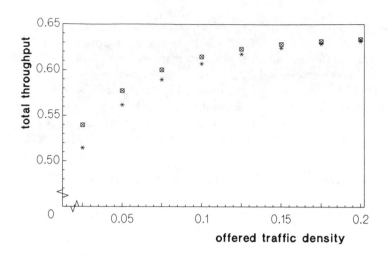

Figure 9.8 (*) Exact total throughput of two base stations versus the traffic G_0 per unit of area, and (⊗) approximation. The receiver threshold is $z = 4$ (6 dB) and the receiver separation is $R_d = 2$. UHF path loss law $\beta = 4$.

As derived in Chapter 8, the probability of capture in a noise-free channel with infinitely extended uniform offered traffic $(G(a_i) = G_0)$ and $\beta = 4$, is

$$\Pr(A_j|a_j) \triangleq Q(a_j) = \exp\left\{-\frac{\pi^2}{2}G_0\sqrt{z}a_j^2\right\} \quad (9.31)$$

and the total throughput per receiver is $S_t = 2/(\pi\sqrt{z})$. The a posteriori pdf of the propagation distance a_j of a correctly received packet is Rayleigh-distributed, with pdf

$$f_{a_j}(a_j|A_j) = \frac{2\pi a_j S(a_j)}{S_A} = \frac{2\pi a_j G_0 \Pr(A_j|a_j)}{S_A} = \frac{a_j}{\sigma^2}\exp\left\{-\frac{a_j^2}{2\sigma^2}\right\} \quad (9.32)$$

with variance

$$\sigma^2 = \frac{1}{\pi^2\sqrt{z}G_0} \quad (9.33)$$

We now consider distances to base station B, which has a distance R_d from base station A. Transforming the Rayleigh-distributed a_j into the statistical variable of the distance b_j to base station B, one obtains the Rician distribution

$$f_{b_j}(b_j|A_j) = \frac{b_j}{\sigma^2} \exp\left\{ -\frac{R_d^2 + b_j^2}{2\sigma^2} \right\} I_0\left(\frac{R_d b_j}{\sigma^2}\right) \tag{9.34}$$

where $I_0(\cdot)$ is the zeroth-order modified Bessel function of the first kind. Inserting (9.34) and using capture probabilities in the form of (9.4) to describe the probability of capture at B, (9.30) becomes

$$S'_{A \wedge B} = \frac{2}{\pi \sqrt{z}} \int_0^\infty \exp\left\{ -\frac{b_j^2}{2\sigma^2} \right\} \frac{b_j}{\sigma^2} \exp\left\{ -\frac{R_d^2 + b_j^2}{2\sigma^2} \right\} I_0\left(\frac{R_d b_j}{\sigma^2}\right) db_j \tag{9.35}$$

The approximation (9.30) is seen to allow an analytical solution, namely, by substituting

$$x^2 \triangleq 2b_j^2 \quad \text{and} \quad y^2 \triangleq \frac{1}{2} R_d^2 \tag{9.36}$$

one finds

$$S'_{A \wedge B} = \frac{1}{\pi \sqrt{z}} \exp\left(-\frac{R_d^2}{4\sigma^2} \right) \tag{9.37}$$

The approximate total throughput $S'_{A \vee B}$ of the two base stations is found from

$$S'_{A \vee B} = S_A + S_B - S_{A \wedge B} \tag{9.38}$$

The above approximate joint throughput (9.38) (indicated by $\otimes$ in Figure 9.8) has been compared with the results found from numerical integration of the exact expression (9.27) with (9.24) and (9.25) (indicated by * in Figure 9.8). Comparison with exact results confirmed that the approximation technique underestimates the number of packets that capture both receivers simultaneously; however, it is noted that the accuracy of the approximation is better than a few percent for sufficiently spaced receivers, say for $R_d \gg \sigma$.

9.10 CONCLUSION

The performance of wide-area slotted ALOHA networks has been addressed using the receiver threshold model. A measure for the spectrum efficiency of a cellular network with cluster size C has been proposed. This measure is particularly relevant to some information gathering systems that do not necessarily need retransmission of lost packets, such as in the SOCRATES system [10] for road traffic data. It appears for uniform offered traffic that $C = 1$ (i.e., the design where all cells use the same frequency) gives the highest spectrum efficiency. If $C > 1$, the bandwidth per receiving base station decreases accordingly, which appears to limit the throughput more significantly than the compensation achieved from reduced interference from cochannel cells. The spectrum efficiency was found to depend largely on the required minimum probability of successful access. If a probability of access of about 0.5 is acceptable for remote terminals, the spectrum efficiency can be on the order of 0.3 b/s/Hz per receiver for robust and efficient types of modulation ($z = 4$; $\eta_r = 1$ b/s/Hz).

Systems that require retransmission of each lost packet have also been considered. The spatial distribution of the packet traffic to be offered to the channel in order to achieve a uniform throughput has been evaluated. Corresponding spectrum efficiencies were found to be on the order of 0.2 b/s/Hz per receiver or less for robust and efficient types of modulation ($z = 4$; $\eta_r = 1$ b/s/Hz). This is significantly larger than the spectrum efficiency of cellular telephone networks with narrowband channels and usual cluster sizes of about $C = 9$, which cannot accommodate more than η_r/C b/s/Hz per receiver.

We have seen that the effect of noise can be large: many packets from remote terminals are lost in noise, so these terminals are likely to perform many retransmissions, which increases the number of collisions and degrades the total system performance. Nonetheless, it appears that if high spectrum utilization is to be achieved, remote terminals have to perform many retransmissions. Thus, considering the observations in Section 8.6.4, spectrum-efficient networks in which lost packets are retransmitted from the same location may require some measures to ensure stability.

Results confirm that in the packet-switched ALOHA networks, frequency reuse distances can be substantially smaller than in circuit-switched networks for CW radio telephony. This suggests that if frequency reuse patterns are designed for mixed traffic (e.g., to support circuit-switched voice for emergency calls [2]), spectrum usage during normal packet-switched operation is far from optimum [14]. Moreover, results motivate the use of wide-area networks with contiguous frequency reuse, thus using the entire available bandwidth in the entire area rather than avoiding the use of the same channel in adjacent cells, as is conventional in cellular telephony networks. In this case, packets can be received outside the cell where the transmitting terminal is located. This type of site diversity may enhance the system throughput.

A technique has been proposed to compute the probability of successful reception of a data packet if site diversity is employed. A two-receiver configuration has been studied analytically. Numerical results have been obtained for the capture probability and the network throughput for the case of a uniform spatial distribution of the total packet traffic offered to the network. It can be concluded that site diversity, which is inherent to the optimum cluster size $C = 1$ in a random-access network, significantly increases the probability of successful access for the terminals near the boundary between two cells.

REFERENCES

[1]   Chang, C.M., "Multisite Throughput of a Mobile Data Radio Link," *IEEE Trans. on Veh. Tech.*, Vol. VT-39, No. 3, August 1990, pp. 190–204.

[2]   Berntson, G., "Mobitex—A New Network for Mobile Data Communications," *Ericsson Review*, No. 1, 1989, pp. 33–39.

[3]   Goodman, D.J., "Cellular Packet Communications," *IEEE Trans. on Comm.*, Vol. 38, No. 8, Aug. 1990, pp. 1272–1280.

[4]   Goodman, D.J., R.A. Velanzuela, K.T Gayliard, and B. Ramamurthi, "Packet Reservation Multiple Access for Local Wireless Communications," *Proc. 39th IEEE Veh. Tech. Conf.*, Philadelpia, June 1988, pp. 701–706

[5]   Sastry, K., and R. Prasad, "Speech Packet Dropping Probability for Speech/Data Transmission Using Packet Reservation Multiple Access," *European Transactions on Telecommunications*, Vol. 2, No. 5, Sep./Oct. 1991, pp. 25–28.

[6]   Hafez, H.M., and G.H. Nehme, "Data Transmission Over Cellular Radio Systems Using a Single Radio Frequency," *Proc. Eurocon*, Stockholm, June 1988, pp. 367–370.

[7]   Wilson, N.D., and S.S. Rappaport, "Cellular Mobile Packet Radio Using Multiple-Channel CSMA," *IEE Proc. F (GB)*, Vol. 132, No. 6, Oct. 1985, pp. 517–526.

[8]   Technical Characteristics and Test Conditions for Non-speech and Combined Analogue Speech/Non-speech Radio Equipment With Internal or External Antenna Connector, Intended for the Transmission of Data, for Use in the Land-Mobile Service (DRAFT), European Telecommunications Standards Institute, I-ETS [A] version 3.3.2, Valbonne, France, 1990.

[9]   Linnartz, J.P.M.G., "Slotted ALOHA Land-Mobile Radio Networks With Site Diversity," *IEE Proceedings I*, Feb. 1992, Vol. 139, No. 1, pp. 58–70.

[10] Cattling, I., and F. Op de Beek, "Socrates: A Cellular Radio System for Traffic Efficiency and Safety" (in French), *Navigation*, Vol. 39, No. 153, Jan. 1991, pp. 22–23.

[11] Prasad, R., and A. Kegel, "Improved Assessment of Interference Limits in Cellular Radio Performance," *IEEE Trans. on Veh. Tech.*, Vol. VT-40, No. 2, May 1991, pp. 412–419.

[12] Zhang, K., and K. Pahlavan, "A New Approach for the Analysis of the Slotted ALOHA Local Radio Networks," *Proc. Int. Conf. on Comm. ICC*, Atlanta, April 1990, pp. 1231–1235.

[13] Lau, C., and C. Leung, "Throughput of a Slotted ALOHA Channel With Multiple Antennas and Receivers," *Electron. Lett.*, Vol. 24, No. 25, 8 Dec. 1988, pp. 1540–1542.

[14] Arnbak, J.C., "Radio Data Networks: Some Differences With Digital Mobile Telephony," *Proc. Nordic Radio Symposium*, Aalborg, Denmark, June 1–4 1992, pp. 115–118.

Chapter 10
Discussion and Conclusions

This book has addressed a number of technical problems in modern mobile communication. The increased possibility of exchanging information with mobile users will render new services feasible, which may have a large impact on business activities and possibly also on society at large. Mobility enhances the need for exchange of information, and communication can be used to enhance the efficiency of mobility.

Efficient use of the limited radio spectrum resource is of key importance in the design and planning of mobile radio systems. Spectrum efficiency was previously studied in many other types of communication systems. Often a purely noise-limited situation was assumed where the quality of reception was determined mainly by the local signal-to-noise ratio at the fringe of the service area. Modulation was generally chosen such that the occupied bandwidth per transmitter was as small as possible. Digital modulation techniques with a high bit rate per occupied bandwidth have been developed (large r_b/B_T). Many of these bandwidth-efficient techniques are also applicable in trunk connections between telephony switches, either using wired or radio relay technology, or cable nets for distribution of radio and television broadcast signals to the home (CATV). A standard measure of the spectrum efficiency of voice links is expressed in erlangs per megahertz.

In present cellular radio systems, bandwidth efficiency is greatly improved by the dense reuse of available frequencies in the service area. Most mobile networks are interference-limited rather than noise-limited. This, however, requires careful consideration of the effect of mutual interference between cochannel users. Therefore, not only the occupied bandwidth, but also the size of the area in which it cannot be reused, is of importance. Hence, spectrum-efficient design must consider the spatial pattern of the communication infrastructure. The *spatial* bandwidth efficiency of mobile telephone networks is expressed in erlangs per megahertz per square kilometer. In the economic study [1], these three relevant dimensions, time, area, and spectral bandwidth, have been used to define the unit of spectrum resource: the *TAS package*.

Accordingly, we addressed the effects of interference, given a certain spatial layout of the network. Propagation channel impairments, such as multipath reception and shadowing, occasionally threaten the performance of mobile communication networks. Fluctuations of the received power necessitate the inclusion of fade margins in the maximum tolerable mean signal-to-interference ratio. Thus, the spatial bandwidth efficiency depends on the propagation characteristics of the mobile channel.

In Chapter 1, the design, planning, and control of radio networks to accommodate the required number of users within the allocated bandwidth was discussed in the relatively broad scope of a layered system model. Chapters 2 through 9 were intended to contribute to the understanding of various factors at the lower OSI layers that determine the performance of spectrum-efficient narrowband mobile networks. This concluding chapter summarizes the results most relevant to the spatial bandwidth efficiency of mobile radio networks.

10.1 DESIGN AND PLANNING OF CELLULAR VOICE NETWORKS

Limitations caused by propagation effects are of major concern in mobile radio. Most of the situations covered in this book addressed generic system design. Propagation effects were treated in a stochastic way. Idealized hexagonal or circular cell structures have been considered. In practical systems planning, however, tailor-made solutions for particular propagation environments and for specific spatial distributions of the teletraffic intensity may be required to achieve sufficient bandwidth efficiency. In this case, terrain data is usually necessary. If the area-mean power can be forecasted from this data with sufficient accuracy, margins to overcome propagation uncertainties can be kept reasonably small (see Section 3.4.5). Hence, the availability of reliable propagation models is of key importance to efficient cellular design.

UHF field measurements of path losses in rural areas have been reported and analyzed in Sections 2.2 and 2.3. The regression method adopted proved useful in determining shortcomings of the various models studied. The accuracy of the propagation prediction model could be improved by refining the model step by step (Sections 2.3.1 to 2.3.6). The combined occurrence of ground reflections, local scatters, and propagation by diffraction renders exact forecasting of propagation losses impossible. Two empirical methods for estimating the path loss, namely, (1) linear addition of losses as they would occur if the propagation mechanisms occurred independently, and (2) an empirical method proposed by Blomquist [2], have been verified and some empirical refinements have been proposed.

One counterintuitive conclusion drawn from the propagation measurements was that macrocellular networks may be more spectrum-efficient in open area than in an area with dense vegetation: because of ground reflections, the signal power was found to decay relatively fast in open area (see Figure 2.2). In wooded areas, a larger separation of cochannel users may thus be required to avoid harmful interference caused by signals propagating by diffraction over treetops. This particularly applies if base stations are more or less randomly located in the area. In practice, the network designer may, however, exploit the effects of terrain irregularities, such as shielding by hills and buildings, which significantly increases the possibilities for frequency reuse.

The performance of cellular networks for voice communication is often expressed in terms of the probability of a signal outage. An analytical technique based on integral transforms was presented in Chapter 3 for the computation of outage probabilities in multiuser mobile radio networks. Numerical results for idealized hexagonal cell structures confirmed the accuracy of previously proposed approximate techniques. It is a common frustration in scientific research that exact computations require longer computer calculations than approximations. In contrast, the exact method presented in Chapter 3 for Rayleigh-fading (macrocellular) channels turned out to be faster than other approximate techniques. This method allows faster evaluation of solutions for realistic cell layouts, so it increases the probability that the selected solution is close to optimum.

The performance analysis was extended to two-branch site (or macro) diversity in a cellular telephone network (see Sections 3.3 and 3.4). In channels with shadow fading, site diversity substantially enhances the quality of the end-to-end circuit. For a prescribed network performance, this implies that higher interference power levels may be tolerated with site diversity, which improves the spectrum efficiency of the network.

Next to improving design and planning of cell layouts, fast computational techniques for calculating outage probabilities may be advantageous in real-time *dynamic channel assignment* (DCA) algorithms or in simulation of the performance algorithms for decentralized management and handovers initiated by mobile terminals. Moreover, if the signals only attenuate slowly with increasing distance (e.g., if the propagation exponent is close to $\beta = 2$), multiple tiers of interfering signals may need to be accounted for. This requires efficient computation techniques.

10.2 PERFORMANCE OF DIGITAL CELLULAR NETWORKS

The recently developed more efficient voice-coding algorithms render spectrum-efficient digital cellular telephony feasible. In such networks, voice data is often split in frames or blocks. A block of data is likely to be received correctly unless a fade is experienced during the reception of the block. In Chapter 4, the average nonfade duration has been studied analytically for the case of a Rayleigh-fading signal in the presence of multiple interfering signals with Rayleigh fading. The results showed that in vehicular networks, the probability of a block being received within a nonfade interval is often significantly smaller than the outage probability. This suggests that blocks should be short compared to the time constant of the fading. The numerical results in Chapter 4 quantify this observation that was previously concerned with noise-limited rather than real interference-limited networks (see also Section 7.5).

For digital communication, the average bit error rate is also a relevant measure of the performance. Previous analyses of the BER in fading channels with cochannel interference appeared too complicated for use in studies of system performance. A simplified model was proposed in Chapter 5. Numerical results and closed-form asymptotic expressions were generated for the local-mean and area-mean BER in microcellular and macrocellular networks. Results of the probability of successful reception of a block of bits have been derived. The model also allowed analysis of the performance of random-access networks (Chapter 7). The analysis showed that increased shadowing has a degrading effect on the average BER. Rician K-factor (ratio of specular to scattered signal power) also appeared to have a substantial influence.

In stationary (i.e., nonfading) AWGN channels, the BER rapidly decreases if signal power is increased. This possibility is lost in mobile channels with fading. Even for high signal-to-interference-plus-noise ratios, the average BER remains relatively poor. This effect, plus the observation in Sections 4.9 and 4.10 that the average duration of fade and nonfade intervals is on the order of many hundreds or thousands of bits in typical mobile radio networks, leads to the conclusion that mobile channels require special error control. Bit interleaving and powerful error-correction coding may not suffice.

10.3 PERFORMANCE OF CELLULAR DATA NETWORKS

At the end of the 1980s, first-generation cellular telephone networks were in operation in many countries. The digital pan-European GSM system is due to be implemented, but practical introduction of this second-generation system has been delayed.

The behavior and the performance of cellular telephone networks have been determined by researchers during the past two decades. Results, or at least approximations, are now known for a number of performance aspects. This book particularly addressed interference-limited, rather than noise-limited, cases.

On the other hand, for mobile *data* communication, only a few systems are in operation and the discussions about efficient designs are still being conducted. Two aspects have been addressed, namely, (1) the performance of the access schemes, as determined by mutual competition by terminals inside the cell (Chapters 6 to 8), and (2) cochannel interference by signals from other cells (Chapter 9).

10.3.1 Effect of Competing Packets Inside the Cell

Standard results for the performance of wired data communication are not appropriate for mobile data systems. One of the problems is that, in fading channels, the performance experienced by individual mobile terminals may differ widely from the mean performance averaged over all terminals. Two types of near-far effects have been distinguished: effects caused by the fact that propagation attenuation increases with distance (Chapters 7 to 8), and effects caused by propagation delays (Chapter 6). The latter effects have been studied for ISMA networks with propagation delays in the inbound random access and feedback channel. It was found in Section 6.7.6 that the approximation in terms of fixed (i.e., distant-independent) propagation delays leads to somewhat pessimistic results for the data throughput. For a worst-case investigation of an ISMA network, the delay in dissemination of the channel status is to be based on the two-way, rather than the one-way, propagation time.

For reasonable packet traffic loads, the near-far effect of varying propagation delays appeared smaller than that of varying attenuation. A large number of channel aspects play a role in the conditional probability of successful access for a signal received from a certain distance, such as the type of modulation and coding, packet duration, speed of the terminal, and depth of shadow fading. A number of simplified models, each dealing with only a limited number of effects, have been investigated and compared in Chapter 7. In mobile slotted ALOHA networks, packets from nearby terminals have a substantially higher probability of successful reception than those from remote terminals. This effect is much smaller in networks employing ISMA, though nonpersistent ISMA does not necessarily improve the overall throughput compared to slotted ALOHA (see Section 7.8).

Chapters 7 and 8 discussed that the throughput of the ALOHA channel is enhanced by receiver capture. However, the way in which terminals are distributed over the coverage area has a great deal of influence on the throughput of the net. Various spatial distributions have been compared in Chapter 8, and the effect in which almost all retransmissions occur in distant parts of the coverage area with poor propagation was addressed in Section 9.5.

Stability and delay performance of mobile ALOHA networks have been investigated with a Markov chain model. The calculations reveal that bistability occurs only for a very limited range of traffic parameters. In contrast to the case for wired networks, this suggests that protocol design may focus on continuously assuring sufficient throughput and acceptable delay, for instance, by exercising slow centralized control over the number of mobile terminals allowed to sign on. However, effective techniques exist to ensure stability, such as dynamic frame-length ALOHA or stack algorithms, and appear relatively easy to implement. These may ensure the required performance in a wider range of traffic situations. Moreover, these techniques may be relevant if frequent retransmissions by a few terminals at the fringe of the cell threaten the stability of the entire network.

Section 8.6 showed that mobile slotted ALOHA may remain stable even without more sophisticated retransmission schemes. This was a motivation to study system behavior using steady-state analysis of the access protocol. This simplification allowed further extension of the study. For instance, the impact of noise, fade/non-fade durations, and cellular frequency reuse were computed in Chapters 7 and 9.

The steady-state spatial distribution of the packet traffic to be offered to the channel in order to achieve a uniform throughput has been determined in Section 9.5. It was found that the effect of noise can be substantial. Since packets from remote terminals are more prone to noise, they are likely to perform more retransmissions, which increase the number of collisions and degrade the total system performance.

10.3.2 Effect of Interference From Cochannel Cells

Besides the performance of isolated random-access networks, wide-area networks with frequency reuse have also been addressed, with a model recognizing that most retransmissions occur at the fringe of each cell. Results in Chapter 9 showed that in ALOHA networks, frequency reuse distances can be made substantially smaller than in networks for CW radio telephony. The computational results even motivate the use of a wide-area packet-switched network with contiguous frequency reuse; that is, using the entire available bandwidth in the entire area with cochannel receiving base stations at different locations, rather than splitting the bandwidth in a number of channels, as in cellular telephone networks. In this case, packets can be received outside the cell where the transmitting terminal is located. This type of site diversity enhances the system throughput (see Section 9.9). This conclusion is in sharp contrast to the spatial design of present mobile data networks, which appear to be based on cell layouts with cluster sizes optimized for circuit-switched voice communication. In mobile data networks, optimum spatial design requires very dense frequency reuse, while all transmission impairments are to be covered by appropriate (retransmission) protocols and coding. Design for reliable data communication by means of reducing interference between cells did not appear to be a spectrum-efficient solution.

10.4 INTEGRATED NETWORKS FOR MOBILE SERVICES

Throughout this book, our models and methods of analysis were distinctively different for circuit-switched (cellular) and packet-switched (random-access) networks. This distinction is in agreement with the fact that many existing mobile networks offer only a single transport service. Even though many telematic applications require a combination of multiple (voice and) data transport services, new systems with incompatible radio transmission standards continue to be introduced: cellular telephone networks did not appear very suitable for bursty data communications. On the other hand, the existing (packet-switched) mobile data networks do not support voice communication. Also, the current ideas for a communication architecture for Intelligent Vehicle Highway Systems (see Chapter 1) depend greatly on separate radio communication standards for different services. Moreover, new mobile radio systems are often incompatible with existing networks offering similar services. For instance, as pointed out in Chapter 1, the specifications for Telepoint or DECT microcellular telephone networks differ from the standards for vehicular cellular networks.

The needs for spectrum-efficient design and minimizing the power consumption in the mobile terminal appear to be relevant reasons for this. The analyses in the previous chapters showed a fundamental difference between optimum frequency reuse strategies for circuit-switched (CW) and packet-switched data communications. This suggests that an integrated network for mobile services would presumably not be the most spectrum-efficient solution, unless fundamentally new access techniques are developed. To ensure high spectrum efficiency, such access schemes cannot be limited to dynamically assigning time-frequency resources in some more or less efficient manner, as in most currently known integrated voice and data protocols, but also have to take into account the distinct spatial limitations to tolerable interfering traffic. As shown in Chapter 9, data transmissions may not need to be continuously safeguarded from possibly interfering transmissions occurring in adjacent areas, whereas for voice communication over narrowband channels, relatively large reuse distance are required.

Regarding battery power consumption, integration of various transport services requires that each participating terminal continuously receives the entire multiplexed data signal, whereas only a small portion of the data may be relevant to a particular terminal. Especially for terminals in standbye mode, such schemes can be prohibitively power-inefficient.

It may be concluded that the relevant design issues of wired (trunk) networks and wireless radio networks are essentially different with respect to their potential for integration of multiple transport services. Integration on a single cabled network significantly reduces the total installation costs, which is the dominant cost factor in wired networks. The available capacity of wired links often far exceeds the required transport capacity, particularly in optical communication. Hence, in wired networks the use of additional transmission bandwidth for signaling and the application of power-consuming signal processing techniques are economically feasible. In contrast to this, in mobile and personal radio networks, spectrum (TAS) resources and battery power are the most critical design issues. However, it may be argued from an economic point of view [1, 3] that in integrated networks, spectrum resources can be exchanged more flexibly between different services. This may lead in the long run to a more efficient and possibly fairer distribution of the available spectrum resources.

10.5 NARROWBAND VERSUS WIDEBAND COMMUNICATION

The discussion in this book was limited to narrowband mobile communication over fading groundwave radio channels. Wideband communication would require fundamental reconsideration of a number of topics [4–6]. For signals with a bandwidth exceeding the coherence bandwidth of the propagation environment, significant intersymbol interference will occur and techniques to combat these effects, such as adaptive channel equalization, must be considered.

Also, narrowband and wideband radio links react differently to motion of the antenna. The former experience fades during movement when bursts of bit errors occur. On the other hand, for a stationary terminal, the channel properties remain frozen. In wideband communication, however, deep fades of the received power over the entire bandwidth of the signal are very unlikely. After appropriate equalization, the bit error rate will be nearly stationary and ergodic, irrespective of whether the terminal is in motion or stationary. This also suggests that in networks with only limited terminal mobility, such as indoor wireless systems, wideband communication can ensure fairer channel performance for all users, though at the cost of increased system complexity.

Spread-spectrum and *code division multiple access* (CDMA) techniques are currently proposed for indoor and microcellular networks. However, broadband signals received from various users are often optimistically assumed to arrive with identical mean power. If this is not the case, the performance will be severely limited [7]. Because of path loss and shadowing, adaptive power control may be required to fulfill this restriction. This, however, may not be practical in a wide-area network with multiple cooperating receivers at different locations.

In the random-access networks that take advantage of feedback information on the channel state, such as ISMA, the propagation delay relative to the packet duration should be relatively small to ensure high performance. In wideband networks, the round-trip delay may become too large compared to the duration of packets to effectively employ CSMA or ISMA if the range of the cell is large (see Chapter 6).

Wideband communication generally requires substantially more signal processing; for example, for channel equalization, spread-spectrum decoding, or simply to receive a few blocks of data embedded in a continuous signal with a high symbol rate. This may be a disadvantage for handheld terminals.

10.6 MATHEMATICAL TECHNIQUES EMPLOYED FOR ANALYSES

The capture-ratio or threshold model (Chapters 3, 4, 8, and 9) produced convenient mathematical expressions for the probability of successful reception. As our discussion showed, this simplification is acceptable in the case of *narrowband* mobile channels. Since large fluctuations of the received signal power occur, the accuracy of the results may not depend critically on a detailed consideration of the behavior in a small range of the C/I ratio near the threshold. This is in contrast to recently reported observations for CDMA and spread-spectrum communication [4, 5].

Since narrowband mobile propagation is of a stochastic nature, the study of the performance of mobile communication systems relies heavily on probability theory. It appeared that the threshold model can be analyzed with the Laplace transform (also related to the characteristic function) of the probability density functions of the received interference power. The Laplace transform is widely used in electrical engineering [8] because it changes differential equations describing signals in electronic circuits into algebraic equations which are generally easier to solve. In the situation covered here, the lure of the Laplace transform is its ability to map the complicated operation of convolution into multiplication. If the interfering power from multiple terminals cumulates incoherently, the probability density of the joint interference power is the convolution of the pdfs of the interference received from the individual terminals. Laplace transformation, however, changes this into a multiplication of image functions.

Moreover, the exponentially distributed power received from a Rayleigh-fading signal was seen to match with the kernel of the Laplace image (see Section 3.2). This observation paved the way for relatively convenient mathematical expressions for outage probabilities in macrocellular telephony. The Laplace image of the Suzuki pdf was introduced and its properties have been described. The technique of Laplace transforms was also used to address the performance of random-access networks. In particular, capture probabilities in slotted ALOHA were addressed in Chapter 8. In this case, a Poisson-distributed number of colliding packets leads to expressions that also include the exponential function. Interestingly, the vulnerability circle, proposed by Abramson [9] to describe the effect of colliding packets, could be modified to take exact account of Rayleigh fading and cumulation of interference signals. It thus transpired that, because of the matching mathematical properties of Rayleigh fading, Laplace transforms, and Poisson distributions, the intensity of contending packet traffic may simply be multiplied by a propagation-dependent factor to describe the effect of causing the interference to a selected test packet. In this way, the technique was refined in Chapter 8 by interpreting the Laplace image functions as integral transforms of the spatial distribution of the offered packet traffic. In Section 8.3.1, the kernel of this integral transform was called the vulnerability weight function.

10.7 SCOPE OF FUTURE RESEARCH

In this section, further research topics relevant to the design of future personal communication networks, but not yet extensively addressed in technical literature, are reported.

Although the first experimental spread-spectrum cellular networks have recently been demonstrated, the relative advantages of narrowband, wideband and spread-spectrum transmission are not yet fully assessed. Since many new systems are being developed and standards are drafted for future global or regional networks, the need for better insight into the relative performance is acute.

Frequency management in cellular systems becomes increasingly dynamic [10, 11]. This allows assignment of channels according to actual demand and spatial distribution of active terminals, counteracting traffic variations caused by unpredictable events like traffic jams and road accidents and dynamically assigning spectrum resources according to daily patterns of mass movement. Decentralized management and handovers initiated by the mobile terminals are alternative techniques that also receive attention in the development of new systems. To the knowledge of this author, only a few results exist for the performance, the spectrum efficiency, or the stability of cellular systems with decentralized frequency management or with DCA.

Particularly from the results in Chapter 9, it appeared that further development of efficient strategies for frequency reuse in wide-area packet-switched mobile data networks may largely enhance spectrum efficiency. The inclusion of a refined model for mixed traffic in the analyses and development of a technique for dynamic control of frequency reuse according to the traffic pattern near each base station are recommended.

Refinement of the technique to study stability of networks with *nonergodic* channels is also left as a recommendation. If, because of the near-far effect, some terminals always transmit over relatively poor channels and other terminals always transmit over relatively good channels, the network is significantly less stable than if time-average and ensemble-average capture probabilities were identical, as in networks with ergodic channels. An appropriate and mathematically tractable technique for studying the stability of networks with different channels for different users has not yet been developed.

The expressions derived in Chapter 3 for the a posteriori probability that an interferer was active, despite the fact that the test signal captured the receiver, may be an interesting starting point for a number of the aforementioned topics. These expressions might be relevant particularly to the study of stability of random-access networks with near-far effect, Bayesian control algorithms for random-access channels [12], throughput of collision resolution algorithms, terminal behavior in decentralized frequency management, and spectrum-efficient wide-area packet data networks.

The analyses have been limited to cochannel interference to study the performance-versus-spectrum efficiency of radio networks. This type of interference occurs if multiple users compete for a specific resource, such as TAS-package management. Therefore it is called *scarcity interference* in the economic study of frequency assignments [3]. This in contrast to *pollution interference*, such as manmade noise or spurious emissions. Since controlling pollution requires resources, such as signal processing circuits, and consumes battery power, it may lead to less efficient use of the primarily assigned TAS package. The goal of minimizing pollution interference, as currently enforced by most frequency regulatory agencies, is not necessarily the most economic approach. As a general rule, in problems of pollution control, an efficient level of pollution is positive (i.e., nonzero) [3]. Extension of the models proposed in this book to include the effects of adjacent channel interference and spurious emissions is recommended.

Efficient design of practical systems for personal communication is not limited to addressing only the performance and spectrum efficiency: the power consumption of the mobile terminal becomes an increasingly important design aspect. The performance of a telecommunication system has traditionally been studied as a function of the transmitted energy per bit relative to the natural background noise (E_b/N_0). In handheld terminals, however, only a fraction of the battery power is consumed by the transmitting amplifier. A complete study would also require generic modeling of the power consumption of typical *digital signal processing* (DSP) circuits, studied in combination with the models for propagation, modulation, coding, and traffic characteristics described in the previous chapters. Such models become particularly relevant to the design of personal communication networks, because the size and weight of the battery is a critical issue in the design of a handheld terminal.

REFERENCES

[1] De Vany, A.S., R.D. Eckert, C.J. Myers, D.J. O'Hara, and R.C. Scott, "A Property System for Market Allocation of the Electromagnetic Spectrum: A Legal Economic-Engineering Study," *Stanford Law Review*, Vol. 21, 1969, pp. 1499–1561.

[2] Blomquist, A., and L. Ladell, "Prediction and Calculation of Transmission Loss in Different Types of Terrain," *AGARD Conf. Proc.*, No. 144, paper 32, Electromagnetic wave propagation involving irregular surfaces and inhomogeneous media.

[3] Veljanovski, C., ed., *Freedom in Broadcasting*, Chapter 8: "Radio Spectrum Management: An Economic Critique of the Trustee Model," J. Fountain, Institute of Economic Affairs, London, March 1989, Hobart Paperback 29, ISBN 0–255–36218–8

[4] Tarr, J.A., J.E. Wieselthier, and SJ A. Ephremides, "Packet-Error Probability Analysis for Unslotted FH-CDMA Systems With Error Control Coding," *IEEE Trans. on Comm.*, Vol 38, No. 11, Nov. 1990, pp. 1987–1993.

[5] Kavehrad, M., and P.J. McLane, "Performance of Low-Complexity Channel Coding and Diversity for Spread Spectrum Indoor, Wireless Communication," *AT&T Technical Journal*, Vol. 64, No. 8, Oct. 1985, pp. 1927–1965.

[6] Gilhousen, K.S., J.M. Jacobs, R. Padovani, A.J. Viterbi, L.A. Weaver, Jr., C.E. Wheatley III, "On the Capacity of a Cellular CDMA System," *IEEE Trans. on Veh. Tech.*, Vol. 40., No. 5, May 1991, pp. 303–312.

[7] Prasad, R., A. Kegel, and M.G. Jansen, "Effect of Imperfect Power Control on a Cellular Code Division Multiple Access System," *Electron. Lett.*, Vol. 28., No. 9, 23 April 1992, pp. 848–849.

[8] Nahin, P.J., "Behind the Laplace Transform," *IEEE Spectrum*, Vol. 28, No. 3, March 1991, pp. 60.

[9] Abramson, N., "The Throughput of Packet Broadcasting Channels," *IEEE Trans. Comm.*, Vol. COM-25, No. 1, Jan. 1977, pp. 117–128.

[10] Falciasecca, G., M. Frullone, G. Riva, and A.M. Serra, "Comparison of Different Hand-over Strategies for High Capacity Cellular Mobile Radio Systems," *Proc. 39th IEEE Veh. Tech. Conf.*, San Francisco, 1–3 May 1989, pp. 122–127.

[11] Whiting, P.A., "Alternate Routing Strategies at Cell Boundaries," *Proc. 5th Int. Conf. on Mobile Radio and Personal Communication*, IEE, Warwick, U.K., 11–14 Dec. 1989, pp. 84–88.

[12] Bertsekas, D., and R. Gallager, *Data Networks*, Section 4.2.3, London: Prentice Hall Int., Inc., 1987.

Appendix A
Variance of Noise Samples in
a Correlation Receiver

This appendix calculates the variance of the noise sample in the decision variable (5.6). As seen in Figure 5.1, the received signal is correlated with the locally generated signal $h(t) = 2/T_b \cdot \cos \omega_c t$. For ease of analysis, it is assumed that $\omega_c T_b = 2\pi M$. In practice, the carrier frequency ω_c will be many orders of magnitude larger than the bit rate ($\omega_c T_b \gg 1$), and the inaccuracy with M noninteger is negligible. The noise component $y(t)$ at the output of the integrator is offered to the sampling device. This signal $y(t)$ is in the form of

$$y(t) = \frac{2}{T_b} \int_{t-T_b}^{t} \cos \omega_c t \, n(t) \, dt \tag{A.1}$$

The autocorrelation is defined as $R_y(t, t + \tau) \triangleq E[y(t) \cdot y(t + \tau)]$. If the AWGN signal $n(t)$ is offered to the receiver, one obtains

$$R_y(t, t + \tau) = E\left[\int_{t-T_b}^{t} n(t_1)h(t_1)dt_1 \int_{t-T_b}^{t} n(t_2 + \tau)h(t_2 + \tau)dt_2 \right]$$

$$= \int_{t-T_b}^{t} \int_{t-T_b}^{t} E[n(t_1)n(t_2 + \tau)]h(t_1)h(t_2 + \tau)dt_1 dt_2$$

$$= \int_{t-T_b}^{t} \int_{t-T_b}^{t} R_n(t_2 + \tau - t_1)h(t_1)h(t_2 + \tau)dt_1 dt_2 \tag{A.2}$$

For AWGN with one-sided spectral power density N_0, the autocorrelation is $R_n(t, t + \tau) = 1/2\, N_0\delta(\tau)$. Thus, using the sifting property for the integral over t_2,

$$R_y(t,t + \tau) = \frac{N_0}{2} \int_{t-T_b}^{t} h(t_1)h(t_1 - \tau)dt_1 \qquad (A.3)$$

So, $R_y(0) = N_0/T_b$. This result is used in (5.7) as the variance of the noise sample.

Appendix B
Nonstationary (Poisson) Arrivals

This appendix addresses stochastic arrival processes with a quasi-stationary arrival rate $\lambda(t)$, that is, $\lambda(t)$ is assumed constant during an infinitesimal interval of duration Δt, but may fluctuate during a longer observation interval T. Nonetheless, the number of arrivals in a certain interval T is Poisson-distributed. The expected number of arrivals equals the *mean* arrival rate multiplied by the duration of the interval (see (6.7) and (6.8)). A proof is given in Appendix B.2. Initially, the familiar mathematical relation

$$\lim_{J \to \infty} \left(1 + \frac{\xi}{J} \right)^J = e^\xi \tag{B.1}$$

is generalized.

B.1 LEMMA

Let $\xi(\tau)$ be a bounded function, with $\xi(\tau) \leq M$ for any τ in $[0, a]$. Then

$$\lim_{J \to \infty} \prod_{j=0}^{J-1} \left(1 + \xi\left(j\frac{a}{J}\right)\frac{a}{J} \right) = \exp\left\{ \int_0^a \xi(\tau)d\tau \right\} \tag{B.2}$$

Proof

For ease of notation, the proof is given for the special case $a = 1$, but no essential differences occur for other values of a. We define M as the maximum of $\xi(\tau)$, so

$$M \stackrel{\Delta}{=} \max_{x \in [0,1]} \xi(x).$$

The interval $\langle 0,1 \rangle$ of the domain of $\xi(\tau)$ is partitioned into J subintervals of equal length. To derive an expression for (B.1), we consider the graph of $y = 1/x$ in Figure B.1. The shaded area is estimated by the two rectangular areas. For any j ($j \in \{0, 1, \ldots, J - 1\}$), one may write the inequalities

$$\frac{J}{J + M} \frac{\xi\left(\frac{j}{J}\right)}{J} \leq \frac{\xi\left(\frac{j}{J}\right)}{J + \xi\left(\frac{j}{J}\right)} \leq \ln\left(1 + \frac{\xi\left(\frac{j}{J}\right)}{J}\right) \leq \frac{\xi\left(\frac{j}{J}\right)}{J} \tag{B.3}$$

After summing from $j = 0$ to $J - 1$, and applying the exponential function, one finds

$$\exp\left\{\frac{J}{J + M} \sum_{j=0}^{J-1} \frac{\xi\left(\frac{j}{J}\right)}{J}\right\} \leq \prod_{j=0}^{J-1}\left(1 + \frac{\xi\left(\frac{i}{J}\right)}{J}\right) \leq \exp\left\{\sum_{j=0}^{J-1} \frac{\xi\left(\frac{j}{J}\right)}{J}\right\} \tag{B.4}$$

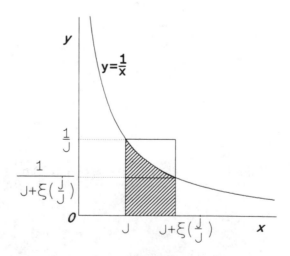

Figure B.1 Enclosure of the graph $y = 1/x$ considered in the proof of the Lemma.

In the limit for $J \to \infty$, (B.4) implies that

$$\lim_{J\to\infty} \prod_{j=0}^{J-1} \left(1 + \frac{\xi\left(\frac{j}{J}\right)}{J}\right) = \exp\left\{\int_0^1 \xi(\tau)d\tau\right\} \quad \text{Q.E.D.} \tag{B.5}$$

B.2 NONSTATIONARY ARRIVAL PROBABILITIES

Let $N_{\langle t,t+T\rangle}$ denote the random variable that records the number of arrivals of a non-stationary Poisson process during a time interval $\langle t, t + T \rangle$. The Poissonian assumptions [1] can be rewritten for the time-variant arrival rate $\lambda(t)$ ($\lambda(t) > 0$):

$$P_{N_{\langle t,t+\Delta t\rangle}}(n = 0) = 1 - \lambda(t)\Delta t + o(\Delta t)$$

$$P_{N_{\langle t,t+\Delta t\rangle}}(n = 1) = \lambda(t)\Delta t + o(\Delta t)$$

$$P_{N_{\langle t,t+\Delta t\rangle}}(n \geq 2) = o(\Delta t) \tag{B.6}$$

where $\lambda(t)$ is considered constant in $\langle t, t + T \rangle$. The interval $\langle t, t + T \rangle$ is partitioned into J subintervals, with $j_k (0 \leq j_k \leq J - 1)$ the sequence number of the subinterval where the kth arrival is located. The probability of n arrivals during $\langle t, t + T \rangle$ is

$$P_{N_{\langle t,t+T\rangle}}(n) = \lim_{J\to\infty} \sum_{j_1=0}^{J-1} \sum_{j_2=0}^{J-1} \cdots \sum_{j_n=0}^{J-1} \text{Pr}\left\{\begin{array}{l} \text{1 arrival in } j_1\Delta t, \\ \text{and 1 arrival in } j_2\Delta t, \\ \vdots \\ \text{and 1 arrival in } j_n\Delta t \end{array}\right\} \tag{B.7}$$

By inserting the first-order approximations of probabilities of arrival during infinitesimal periods, one finds

$$P_{N_{\langle t,t+T\rangle}}(n) = \lim_{J\to\infty} \frac{1}{n!} \sum_{j_1=0}^{J-1} \cdots \sum_{j_n=0}^{J-1} \prod_{k=1}^{n} \mu(j_k\Delta t)\Delta t \prod_{\substack{l=1 \\ l\neq j_1\cdots j_k}}^{N-1} 1 - \lambda(l\Delta t)\Delta t$$

$$= \lim_{J\to\infty} \frac{1}{n!} \sum_{j_1=0}^{J-1} \cdots \sum_{j_n=0}^{N-1} \prod_{k=1}^{n} \frac{\lambda_k(j_k\Delta t)\Delta t}{1 - \lambda(j_k\Delta t)\Delta t} \prod_{l=0}^{J-1} 1 - \mu(j_l\Delta t)\Delta t \tag{B.8}$$

Using the lemma (B.2), the latter product is written as the exponent of an integral, namely,

$$P_{N_{(t,t+T)}}(n) = \frac{e^{-\bar{\lambda}T}}{n!} \lim_{J \to \infty} \sum_{j_1=0}^{J-1} \cdots \sum_{j_n=0}^{J-1} \prod_{k=1}^{n} \frac{\lambda(j_k \Delta t)\Delta t}{1 - \lambda(j_k \Delta t)\Delta t} \qquad (B.9)$$

where the average arrival rate is defined as in (6.8). For $J \to \infty$, thus $\Delta t \to 0$, each summing in (B.8) goes into an integration, namely,

$$P_{N_{(t,t+T)}}(n) = \exp\frac{\{-\bar{\lambda}T\}}{n!} \int_0^T \int_0^T \lambda(t_1) \cdots \lambda(t_n)dt_1 \cdots dt_n$$

$$= \frac{(\bar{\lambda}T)^n}{n!} \exp\{-\bar{\lambda}T\} \qquad (B.10)$$

which proves (6.7) and (6.8).

REFERENCE

[1]   Kleinrock, L., *Queueing Systems, Vol. 1: Theory*, New York: John Wiley and Sons, 1975; and *Vol. 2: Computer Applications*, New York: John Wiley and Sons, 1976.

Appendix C
Further Discussion of Suzuki Image Function

In Section 2.1.5, the Laplace transform of a Suzuki pdf was found to be

$$\phi_\sigma(s) \triangleq \frac{1}{\sqrt{\pi}} \int_{-\infty}^{\infty} \frac{\exp(-x^2)dx}{1 + s \exp(\sqrt{2}x\sigma_s)} \tag{C.1}$$

A general closed-form expression for $\phi_\sigma(s)$ with arbitrary σ_s and s has not been found. However, the image function has a number of interesting mathematical properties.

C.1 COMPLEMENTARY PROPERTY

$$\phi_\sigma\left(\frac{1}{s}\right) = \frac{1}{\sqrt{\pi}} \int_{-\infty}^{\infty} \frac{s \exp(\sqrt{2}\sigma_s y)}{1 + s \exp(\sqrt{2}y\sigma_s)} e^{-y^2} dy = 1 - \phi_\sigma(s) \tag{C.2}$$

where x has been substituted by $-y$ in (C.1). It follows that $\phi_\sigma(1) = 1/2$ for any σ_s [1].

C.2 SHIFT OF THE ARGUMENT s

By substituting $y \triangleq x + (1/2)\cdot\sqrt{2}\cdot\sigma_s$, Meertens and Olde Daalhuis (at CWI, Amsterdam) [2] rewrote (C.1) as

$$\phi_\sigma(s) = \frac{1}{\sqrt{\pi}} \int_{-\infty}^{\infty} \frac{\exp\left\{-y^2 + \sqrt{2}\,\sigma_s y - \frac{1}{2}\sigma_s^2\right\} dy}{1 + s \exp\{\sqrt{2}\,\sigma_s y - \sigma_s^2\}} \tag{C.3}$$

Multiplying enumerator and denominator by $\exp\{\sigma_s^2 - \sqrt{2} \cdot y\sigma_s\}$ gives

$$\phi_\sigma(s) = \frac{e^{\sigma_s^2/2}}{s\sqrt{\pi}} \int_{-\infty}^{\infty} \frac{\exp\{-y^2\}\, dy}{\frac{1}{s} e^{-\sqrt{2}\sigma_s y + \sigma_s^2} + 1} = \frac{e^{\sigma_s^2/2}}{s} \phi_\sigma\left(\frac{e^{\sigma_s^2}}{s}\right) \tag{C.4}$$

Together with $\phi_\sigma(s^{-1}) = 1 - \phi_\sigma(s)$, this gives the property

$$\phi_\sigma(s) = 1 - s e^{\sigma_s^2/2} \phi_\sigma(s e^{\sigma_s^2}) \tag{C.5}$$

C.3 FINITE SERIES

Repeated application of (C.5) gives the series

$$\phi_\sigma(s) = \sum_{m=0}^{M-1} (-1)^m s^m \exp\left\{\frac{1}{2} m^2 \sigma_s^2\right\} + R_M \tag{C.6}$$

with the remainder R_M in the form of

$$R_M = (-1)^M s^M \exp\left\{\frac{1}{2} M^2 \sigma_s^2\right\} \phi_\sigma(s e^{M\sigma_s^2}) \tag{C.7}$$

The remainder R_M does not tend toward zero for $M \to \infty$, but may be small if M is small and $s \to 0$. The first- or second-order terms proved useful in determining the approximate behavior of $\phi_\sigma(\cdot)$ in Chapter 3.

C.4 SPECIAL CASE

In the special case that $s = \exp\{-M_0 \sigma_s^2\}$ with $M_0 = 0, 1, \cdots$, the series contains a finite number of elements, namely,

$$\phi_\sigma(\exp\{-M_0\sigma_s^2\}) = \sum_{m=0}^{M_0} (-1)^m \exp\left\{-n M_0 \sigma_s^2 + \frac{1}{2} m^2 \sigma_s^2\right\}$$

$$+ \frac{1}{2} (-1)^{M_0} \exp\left\{-\frac{1}{2} M_0^2 \sigma_s^2\right\} \tag{C.8}$$

Results for $s = \exp\{+M_0\sigma_s^2\}$ are found from $\phi_\sigma(s^{-1}) = 1 - \phi_\sigma(s)$. For instance,

$$\phi_\sigma(e^{\sigma_s^2}) = \frac{1}{2}\exp\left\{-\frac{1}{2}\sigma_s^2\right\}$$

$$\phi_\sigma(e^{2\sigma_s^2}) = \exp\left\{-\frac{3}{2}\sigma_s^2\right\} - \frac{1}{2}\exp\{-2\sigma_s^2\} \quad \text{etc.} \tag{C.9}$$

REFERENCES

[1]   Linnartz, J.P.M.G., "Site Diversity in Land-Mobile Cellular Telephony Network With Discontinuous Voice Transmission," *European Transactions on Telecommunications*, Vol. 2., No. 5, Sep./Oct. 1991, pp. 471–480.

[2]   Meertens, L., and A. Olde Daalhuis, Center for Mathematics and Computer Sciences (CWI), Amsterdam, private communication in response to a question posted on the electronic news service "usenet sci.math."

Appendix D
Other Integrals Relevant to the Analysis of Mobile Radio Nets

$$\int_0^\infty \exp\{-ax^2 + bx + c\}dx = \frac{1}{2}\sqrt{\frac{\pi}{a}}\exp\left\{\frac{b^2 - ac}{a}\right\}\text{erfc}\left(\frac{b}{\sqrt{a}}\right) \qquad (D.1)$$

The Laplace image of the pdf of the power a Nakagami-fading signal is

$$\mathcal{L}\left\{\frac{x^{m-1}e^{-ax}}{(m-1)!}, s\right\} \triangleq \int_0^\infty \frac{x^{m-1}e^{-ax}}{(m-1)!}\exp\{-sx\}dx = \frac{1}{(s+a)^m} \qquad (D.2)$$

For coherent detection in a Rayleigh-fading channel, the local-mean BER is expressed closed form, using

$$\int_0^\infty \text{erfc}(\sqrt{bx})\exp\{-ax\}dx = \frac{1}{a}\sqrt{\frac{b}{a+b}} \qquad (D.3)$$

The Marcum Q-function $Q(a,b)$ (see Section 3.2.1) can be expressed as a series of Bessel functions, with [1]

$$Q(a,b) = \begin{cases} \exp\left\{-\dfrac{a^2 + b^2}{2}\right\}\displaystyle\sum_{n=0}^\infty \left(\dfrac{a}{b}\right)^n I_n(ab), & a < b \\[4mm] 1 - \exp\left\{-\dfrac{a^2 + b^2}{2}\right\}\displaystyle\sum_{n=1}^\infty \left(\dfrac{b}{a}\right)^n I_n(ab), & a > b \end{cases} \qquad (D.4)$$

An integral relevant to the outage probability in microcellular networks with Rician fading is

$$\int_0^\infty x^{2n-1} \exp\left\{-\frac{c^2}{2}x^2\right\} Q(b,ax)dx = \frac{2^{n-1}(n-1)!}{c^{2n}}$$

$$\cdot \left[1 - \frac{a^2}{c^2+a^2} \exp\left\{-\frac{b^2}{2}\frac{c^2}{c^2+a^2}\right\} \sum_{k=0}^{n-1} \left(\frac{c^2}{c^2+a^2}\right)^k L_k\left(-\frac{b^2}{2}\frac{a^2}{c^2+a^2}\right)\right] \quad \text{(D.5)}$$

with the Laguerre polynomial L_k defined as

$$
\begin{aligned}
(n+1)L_{n+1}(x) &= (2n+1-x)L_n(x) - nL_{n-1}(x) \\
L_0(x) \quad &= \quad 1
\end{aligned}
\quad \text{(D.6)}
$$

REFERENCE

[1]   Van Trees, H.L., *Detection, Estimation and Modulation Theory*, John Wiley and Sons, 1968.

Appendix E
Numerical Evaluation of Integrals

Many numerical results have been obtained with the Gauss-Laguerre and Hermite quadrate method.

E.1 THE HERMITE POLYNOMIAL METHOD

$$\int_{-\infty}^{\infty} f(x) \exp(-x^2)dx = \sum_{l=1}^{k} w_1 f(x_l) + R_k \qquad (E.1)$$

The weight factors w_l at the sample points x_l for a 12-point integration ($k = 12$) are given in Table E.1. The remainder R_k equals

$$R_k = \frac{k!\sqrt{\pi}}{2^k(2k)!} f^{(2k)}(\xi) \qquad (E.2)$$

Table E.1
Abscissas and Weight Factors for a 12-point Hermite Quadrature [E.1]

$\pm x_i$	w_i
0.314240	$5.701352 \cdot 10^{-1}$
0.947788	$2.604923 \cdot 10^{-1}$
1.597683	$5.160799 \cdot 10^{-2}$
2.279507	$3.905391 \cdot 10^{-3}$
3.020637	$8.573687 \cdot 10^{-5}$
3.889725	$2.658552 \cdot 10^{-7}$

for a certain $\xi \in (-\infty, \infty)$. For other values of k, we refer to Mathematical handbooks such as [E.1].

E.2 THE LAGUERRE POLYNOMIAL METHOD

The Laguerre method calculates integrals on the form of the Laplace transform:

$$\int_0^\infty f(x) \exp(-x)dx = \sum_{l=1}^k w_l f(x_l) + R_k \qquad (E.3)$$

The weight factors w_l at the sample points x_l for a 6-point integration ($k = 6$) are given in Table E.2. The remainder R_k equals

$$R_k = \frac{k!}{(2k)!} f^{(2k)}(\xi) \qquad (E.4)$$

for a certain $\xi \in [0, \infty)$.

Table E.2

Abscissas and Weight Factors for a 6-point Laguerre Integration [E.1]

$\pm x_i$	w_i
0.222847	$4.589647 \cdot 10^{-1}$
1.188932	$4.170008 \cdot 10^{-1}$
2.992736	$1.133734 \cdot 10^{-1}$
5.775144	$1.039920 \cdot 10^{-2}$
9.837467	$2.610172 \cdot 10^{-4}$
15.982874	$8.985479 \cdot 10^{-7}$

REFERENCE

[E.1]   Handbook of Mathematical Functions, Ed. by M. Abramowitz and I. A. Stegun, New York: Dover, 1965.

Appendix F
Acronyms

AMPS	Advanced Mobile Phone System
ARTS	American Radio Telephone System
AVL	automatic vehicle location
AWGN	additive white Gaussian noise
BCMA	busy channel multiple access
BER	bit error rate
BPSK	binary phase shift keying
BFSK	binary frequency shift keying
BS	base station
CCIR	International Radio Consultative Committee
CCITT	International Telegraph and Telephone Consultative Committee
cdf	cumulative distribution function
CDMA	code division multiple access
C/I	carrier-to-interference ratio at receiver front end (RF)
C/N	carrier-to-noise ratio at receiver front end (RF)
CRA	colision resolution algorithm
CSMA	carrier sense multiple access
CW	continuous wave (opposite of burst communication)
DCA	dynamic channel assignment
DECT	Digital European Cordless Telephone
D(Q)PSK	differentially encoded (quadrature) phase shift keying
DSI	digital speech interpolation
DSP	digital signal processing
DTX	discontinuous voice transmission
EDI	electronic data interchange
TAS	time-area-spectrum (package), unit of spectrum resource
FDMA	frequency division multiple access
FEC	forward error correction coding
FEC	front end clipping (of a speech burst)

FSK	frequency shift keying
FSL	free-space loss
GSM	Global System for Mobile Radio (previously: Groupe Speciale Mobile)
GMSK	Gaussian minimum shift keying
HDLC	High-Level Data Link Control
IEE	Institution of Electrical Engineers
IEEE	Institute of Electrical and Electronics Engineers
ISI	intersymbol interference
ISMA	inhibit sense multiple access
ITU	International Telecommunication Union
IVHS	Intelligent Vehicle Highway System
LAN	local area network
MKE	multiple knife edge
MS	mobile station
MSC	mobile switch center
MSC	midspeech burst clipping
MSK	minimum shift keying
NMT	Nordic Mobile Telephone
OSI	Open Systems Interconnection (reference model)
PBX	private branch exchange
PCM	pulse-code modulation
pdf	probability density function
PM	phase modulation
POTS	plain old telephone service (wired telephone service)
pps	packets per slot (in slotted access schemes)
ppt	packets per unit of time (in slotted or unslotted access schemes)
PRMA	packet reservation multiple access
PSTN	public switched telephone network
PTT	postal, telephone, and telegraph administration
RF	radio frequency (as opposite to baseband)
SDL	Specification and Description Language
SKE	single knife edge
SNR	signal-to-noise ratio at detector output (baseband)
TACS	Total Access Communication System
TASI	time-assigned speech interpolation
TDMA	time division multiple access
VOX	voice-excited transmission
WOS	wireless office (communication) system

Appendix G
Major Symbols

Symbol	Description [unit]
a	duration of a cycle in CSMA or ISMA [*]
A	average duration of a cycle in CSMA or ISMA [*]
A,B	index indicating base station
A_{Bu}	diffraction loss according to model by Bullington [dB]
A_d	theoretical diffraction loss [dB]
A_{Dg}	diffraction loss according to model by Deygout [dB]
A_{fs}	free-space loss [dB]
a_j	distance between terminal j and receiver A [*]
A_j,B_j	event of successful reception of a packet from terminal j at receiver A
A_K	propagation attenuation in dB as predicted by model K ($K = 1, 2, \ldots$) [dB]
A_{loc}	shadow attenuation [dB]
A_m	measured propagation attenuation [dB]
A_{mh}	theoretical diffraction loss over main hill [dB]
A_R	ground-reflection loss [dB]
B	average duration of the busy period [*]
B	backlog, number of terminals in retransmission mode
B_c	coherence bandwidth [Hz]
b_j	see a_j [*]
B_T	occupied bandwidth [Hz]
c	speed of light [m/s]
C	cluster size in cellular frequency reuse
C_i	probability that one out of i packets captures the receiver
c_j	event that signal j captures receiver A
C_j	amplitude of the dominant (specular) component of jth Rician fading signal [*]
c_z	vulnerability parameter

c_1, c_2	empirical parameters in propagation model, c_1 in [dB]
d	distance between transmitter and receiver [m]
D_a	access delay [*]
d_m	drift of the backlog of packets in a multiple access net [*]
D_q	queuing delay in mobile terminal [*]
d_r, d_t	receiver, transmitter distance to obstacle [m]
d_0	event that signal 0 experiences a dropout
d_1, d_2	inhibit signaling delays [*]
e	$2.71828\ldots$ ($\ln(e) = 1$)
e	event of error
$E[\cdot]$	expectation value
$\mathrm{erf}(x)$	error function
$\mathrm{erfc}(x)$	complementary error function ($1 - \mathrm{erf}(x)$)
f_c	carrier frequency [Hz]
f_m	Doppler shift [Hz]
$f_x(x)$	probability density function of x
	N.B.: The mathematically more correct notation $f_X(x)$ is not adopted here for practical reasons.
$F_x(x)$	cumulative distribution function
$G(r)$	offered packet traffic per unit of time per unit of area at a distance r [*]
G_t	total offered (attempted) packet traffic per unit of time [*]
h	duration of the inhibited period [*]
H	average duration of the inhibited period [*]
H_B	event that test packet arrives when busy signal is received
H_d	event that test packet is bound to arrive in the vulnerable period of an initiating packet
H_I	event that test packet is bound to arrive in at idle receiver
h_m	(effective) height of an obstacle [m]
h_r	height of the receiving antenna [m]
h_t	height of the transmitting antenna [m]
i	interfering signal index
I	integral
I	average duration of idle period [*]
$i(r)$	duration of the idle period [*]
$I(r)$	average duration of the idle period [*]
i_{ON}	event that transmitter i is active
$I_0(x)$	Modified Bessel function of the first kind and zero order
j	imaginary number $\sqrt{-1}$
j	test signal index
J	number of subintervals
$J_0(x)$	Bessel function of first kind and zero order

K	index indicating propagation model, used only in Sections
K	Rician factor: ratio of the power of the dominant (specular) component and the power of the scattered component
K_i, K_n	constants describing the effect of cochannel interference in expression for BER
k_{ON}	event that transmitter k is active
L	number of bits in a packet
$\log(\cdot)$	logarithm with base 10
$\ln(\cdot)$	natural logarithm
m	backlog of a multiple access net
m	Nakagami m-parameter
m_K	mean power expressed [dB]
n	total number of interfering terminals
$n(t)$	noise signal [*]
$N(z)$	threshold crossing rate [$\sec^{-1}$]
N	number of terminals in the population
N_A	noise floor of receiver A [*]
n_I	noise sample in decision variable [*]
N_0	one-sided spectral power density of additive white Gaussian noise [*, Hz^{-1}]
O	outage probability
p	persistency of terminal in ISMA and CSMA protocols
p_{A_s}	instantaneous power received from terminal s at base station A [*]
$\bar{p}_{A_s}$	local-mean power received from terminal s at base station A [*]
$\bar{\bar{p}}_{A_s}$	area-mean power received from terminal s at base station A [*]
$P_b(e)$	bit error probability
P_c	packet error probability, block error rate
p_j	power received from terminal j [*]
$\bar{p}_j$	local-mean received power [*]
$\bar{\bar{p}}_j$	area-mean received power [*]
$P_n(n)$	probability of n arrivals
P_{NA}	probability that the wanted signal is below the noise floor at receiver A ($p_{A_j} < zN_A$)
$\Pr(\cdot)$	probability
P_r	probability of packet retransmission
p_t	joint interference power [*]
P_0	probability of packet arrival
p_0	power of wanted signal (with index 0) [*]

$Q(a,b)$	Marcum Q-function
$Q(r)$	probability of successful transmission of a data packet given the distance of the mobile terminal to the receiver
$\bar{q}_{dj}$	received power in dominant component [*]
q_n	capture probability conditional on n interfering signals
$q_n(r)$	capture probability for a packet from distance r conditional on n interfering signals
$\bar{q}_{sj}$	local-mean scattered power [*]
Q_t	probability of successful transmission of a data packet
r,r_j	distance between (jth) mobile terminal and the base station [*]
R	normalized distance, cell radius [*]
R_c	reflection coefficient
$R[\cdot,\cdot]$	linear regression
r_b	bit rate [Hz]
r_g	turnover distance free-space to groundwave propagation
R_k,R_m	remainder in numerical computation technique
r_u	normalized frequency reuse distance (distance between the centers of two cochannel cells) [*]
s	variable of Laplace transform in image domain [*]
s_i,s_s	standard deviation of shadowing of individual signal
s_j	event of success of the jth transmitter
s_K	standard deviation of received power [dB]
S_m	throughput of a random-access net with backlog m [*]
S_t	channel throughput [*]
s_t	(logarithmic) standard deviation of joint interference signal
t	time [sec]
t	time [*]
T	observation interval [sec]
t_B	instant when initiating packet arrives at base station [*]
T_b	bit duration [sec]
t_I	instant when base station starts transmitting idle signal[*]
t_p	round-trip delay [*]
T_s	duration of a data packet or block [sec]
T_0	criterion (minimum duration) of a signal dropout [sec]
$T_1 \ldots T_{10}$	time scale [sec]
u	normalized speed of light [*]
$U(t)$	unit step function
v,v_i	terminal velocity [m/s]
v	(Sections 2.2 and 2.3) diffraction parameter
$v(r)$	duration of the vulnerable period as a function of terminal distance [*]

$V(r)$	average duration of the vulnerable period $v(r)$ [*]
$v(t)$	received signal [*]
$v_t(t)$	joint received signal [*]
$W(\cdot,\cdot)$	weight function
X	theoretical propagation loss according to a certain model [dB]
y	amplitude-ratio between wanted and joint interference signals
$\dot{y}$	(time-) derivative of amplitude-ratio between wanted and joint interference signals [sec^{-1}]
z	receiver threshold (absolute value, not in dB)
α_i	bit synchronization offset of the ith interfering signal
β	attenuation exponent
γ	0.57721... (Euler's constant)
$\gamma(\cdot,\cdot)$	incomplete gamma function
$\Gamma(\cdot)$	gamma function
$\delta(\cdot)$	Dirac function
Δ	phase difference between direct and ground-reflected wave
Δ_t	delay spread [sec]
ϵ	displacement of the antenna [m]
ζ_j	inphase component of the signal amplitude [*]
η	fade margin (ratio between rms amplitude of wanted signal and rms amplitude of interference and noise)
θ	carrier phase
θ_r	phase of local oscillator in coherent receiver
κ_i	value of bit in ith signal in balanced notation ($\kappa = \pm 1$)
$\kappa_{i\leftarrow},(\kappa_{i\rightarrow})$	value of overlapping earlier (later) bit in the ith interfering signal
λ	wavelength [*]
$\lambda,\lambda(\cdot)$	Poisson arrival rate [*]
ν	receiver decision variable
ξ_j	quadrature phase component [*]
π	3.141592...
π_m	probability of a backlog m
Π	product
ρ_j	amplitude of signal j ($\rho^2 = \zeta^2 + \xi^2$) [*]
$\dot{\rho},\dot{\rho}_i$	(time-) derivative of amplitude of signal i [*, sec^{-1}]
ρ_t	amplitude of joint interference signal [*]
σ	standard deviation
$\sigma_{\rho j}$	standard deviation of (time-) derivative of the ith Rayleigh-fading signal [*, sec^{-1}]
σ_i	standard deviation of shadowing of the ith signal

σ_t	logarithmic standard deviation of the power of the joint interference signal
$\sigma_{\rho t}$	standard deviation of $\dot{\rho}_t$ [*, sec^{-1}]
σ_s, σ_i	logarithmic standard deviation of shadowing
σ_p	logarithmic standard deviation of model error
Σ	sum
τ	interarrival time [*]
$\tau_F(z)$	fade duration [sec]
$\tau_{NF}(z)$	non-fade duration [sec]
$\phi(A,i)$	Laplace image of a Suzuki pdf of power of the ith interferer at receiver A, evaluated in the point $s = z\bar{p}_{A_0}$
$\phi_\sigma(s)$	Laplace transform of a Suzuki pdf
ϕ_t	Laplace image of power of the joint interference signal
ϕ_i	Laplace image of power of interference from terminal
χ	factor in expression for average nonfade duration
$\psi_i(t)$	phase modulation of the ith
$\psi(\cdot)$	Laplace image function
$\psi(A,B,i)$	function
Ψ	empirical vegetation parameter
ω_c	angular carrier frequency ($2\pi f_c$) [rad sec^{-1}]
ω_m	angular Doppler frequency ($2\pi f_m$) [rad sec^{-1}]

I, II System *I* or **II**

$\mathcal{L}\{f,s\}$	Laplace transform of function f
$\otimes$	convolution
$x \uparrow a$	x approaches a from the lower side ($x \to a$, $x \le a$)
$x \downarrow a$	x approaches a from the upper side ($x \to a$, $x \ge a$)
Π	product
$\underline{\Delta}$	is defined as
$\leftrightarrow$	(Laplace) transform pair

[*] Note on Normalization, Units, and Dimensions

All variables and parameters concerning *physical* transmission aspects are expressed in SI-MKSA dimensions; that is, based on the *second*, in contrast to most parameters concerning *logical* protocol issues, where, for convenience, *packet duration* is taken as the normalized unit of time.

Analogously, distances relevant to the *architecture* of frequency reuse patterns are normalized to the *cell radius* and denoted by the letter r, whereas distances relevant to physical transmission are expressed in SI-MKSA dimensions (i.e., based on the meter) and denoted by the letter d.

About the Author

Jean-Paul M.G. Linnartz was born in Heerlen, in the province of Limburg (The Netherlands), on September 10, 1961. He attended Gymnasium β at the Scholengemeenschap St. Michiel in Geleen. He received his Ir. (M.Sc.E.E.) degree in electrical engineering cum laude from Eindhoven University of Technology, in December 1986. His thesis addressed the spatial distribution of packet traffic in a mobile ALOHA network. From January 1987 to April 1988, he was with The Netherlands Organization for Applied Scientific Research, Physics and Electronics Laboratory (F.E.L-T.N.O.), The Hague, where he was involved in research on UHF propagation and frequency assignment for mobile and transportable radio networks.

He joined Delft University of Technology in May 1988 as universitair docent (assistant professor) in the Telecommunications and Traffic-Control Systems Group. He received his Ph.D. in December 1991 cum laude. In June 1992, he received the Dutch Veder Prize for his research in traffic aspects in mobile radio networks. Since January 1992, he has been an assistant professor in the Department of Electrical Engineering and Computer Sciences (E.E.C.S.) at the University of California, Berkeley.

Index

The Artech House Telecommunications Library

Vinton G. Cerf, Series Editor

Expert Systems Applications in Integrated Network Management, E.C. Ericson, L.T. Ericson, and D. Minoli, eds.

FAX: Digital Facsimile Technology and Applications, Second Edition, Dennis Bodson, Kenneth McConnell, and Richard Schaphorst

Fiber Network Service Survivability, Tsong-Ho Wu

Fiber Optics and CATV Business Strategy, Robert K. Yates et al.

A Guide to Fractional T1, J.E. Trulove

Handbook of Satellite Telecommunications and Broadcasting, L.Ya. Kantor, ed.

Implementing X.400 and X.500: The PP and QUIPU Systems, Steve Kille

Inbound Call Centers: Design, Implementation, and Management, Robert A. Gable

Information Superhighways: The Economics of Advanced Public Communication Networks, Bruce Egan

Integrated Broadband Networks, Amit Bhargava

Integrated Services Digital Networks, Anthony M. Rutkowski

International Telecommunications Management, Bruce R. Elbert

International Telecommunication Standards Organizations, Andrew Macpherson

Internetworking LANs: Operation, Design, and Management, Robert Davidson and Nathan Muller

Introduction to Satellite Communication, Burce R. Elbert

Introduction to T1/T3 Networking, Regis J. (Bud) Bates

Introduction to Telecommunication Electronics, A. Michael Noll

Introduction to Telephones and Telephone Systems, Second Edition, A. Michael Noll

Introduction to X.400, Cemil Betanov

The ITU in a Changing World, George A. Codding, Jr. and Anthony M. Rutkowski

Jitter in Digital Transmission Systems, Patrick R. Trischitta and Eve L. Varma

LAN/WAN Optimization Techniques, Harrell Van Norman

LANs to WANs: Network Management in the 1990s, Nathan J. Muller and Robert P. Davidson

The Law and Regulation of International Space Communication, Harold M. White, Jr. and Rita Lauria White

Long Distance Services: A Buyer's Guide, Daniel D. Briere

Mathematical Methods of Information Transmission, K. Arbenz and J.C. Martin

Measurement of Optical Fibers and Devices, G. Cancellieri and U. Ravaioli

Meteor Burst Communication, Jacob Z. Schanker

Minimum Risk Strategy for Acquiring Communications Equipment and Services, Nathan J. Muller

Mobile Information Systems, John Walker

Narrowband Land-Mobile Radio Networks, Jean-Paul Linnartz

Networking Strategies for Information Technology, Bruce Elbert

Numerical Analysis of Linear Networks and Systems, Hermann Kremer et al.

Optimization of Digital Transmission Systems, K. Trondle and Gunter Soder

The PP and QUIPU Implementation of X.400 and X.500, Stephen Kille

Packet Switching Evolution from Narrowband to Broadband ISDN, M. Smouts

Principles of Secure Communication Systems, Second Edition, Don J. Torrieri

Principles of Signals and Systems: Deterministic Signals, B. Picinbono

Private Telecommunication Networks, Bruce Elbert

Radiodetermination Satellite Services and Standards, Martin Rothblatt

Residential Fiber Optic Networks: An Engineering and Economic Analysis, David Reed

Setting Global Telecommunication Standards: The Stakes, The Players, and The Process, Gerd Wallenstein

Signal Processing with Lapped Transforms, Henrique S. Malvar

The Telecommunications Deregulation Sourcebook, Stuart N. Brotman, ed.

Television Technology: Fundamentals and Future Prospects, A. Michael Noll

Telecommunications Technology Handbook, Daniel Minoli

Telephone Company and Cable Television Competition, Stuart N. Brotman

Terrestrial Digital Microwave Communciations, Ferdo Ivanek, ed.

Transmission Networking: SONET and the SDH, Mike Sexton and Andy Reid

Transmission Performance of Evolving Telecommunications Networks, John Gruber and Godfrey Williams

Troposcatter Radio Links, G. Roda

Virtual Networks: A Buyer's Guide, Daniel D. Briere

Voice Processing, Second Edition, Walt Tetschner

Voice Teletraffic System Engineering, James R. Boucher

Wireless Access and the Local Telephone Network, George Calhoun

For further information on these and other Artech House titles, contact:

Artech House
685 Canton Street
Norwood, MA 01602
(617) 769-9750
Fax:(617) 762-9230
Telex: 951-659

Artech House
6 Buckingham Gate
London SW1E6JP England
+44(0)71 630-0166
+44(0)71 630-0166
Telex-951-659